Fundamentals of Physics

PART 4

THE WILEY BICENTENNIAL—KNOWLEDGE FOR GENERATIONS

*E*ach generation has its unique needs and aspirations. When Charles Wiley first opened his small printing shop in lower Manhattan in 1807, it was a generation of boundless potential searching for an identity. And we were there, helping to define a new American literary tradition. Over half a century later, in the midst of the Second Industrial Revolution, it was a generation focused on building the future. Once again, we were there, supplying the critical scientific, technical, and engineering knowledge that helped frame the world. Throughout the 20th Century, and into the new millennium, nations began to reach out beyond their own borders and a new international community was born. Wiley was there, expanding its operations around the world to enable a global exchange of ideas, opinions, and know-how.

For 200 years, Wiley has been an integral part of each generation's journey, enabling the flow of information and understanding necessary to meet their needs and fulfill their aspirations. Today, bold new technologies are changing the way we live and learn. Wiley will be there, providing you the must-have knowledge you need to imagine new worlds, new possibilities, and new opportunities.

Generations come and go, but you can always count on Wiley to provide you the knowledge you need, when and where you need it!

WILLIAM J. PESCE
PRESIDENT AND CHIEF EXECUTIVE OFFICER

PETER BOOTH WILEY
CHAIRMAN OF THE BOARD

HALLIDAY / RESNICK

Fundamentals of Physics

8E

PART 4

Jearl Walker

Cleveland State University

John Wiley & Sons, Inc.

BICENTENNIAL
1807
WILEY
2007
BICENTENNIAL

ACQUISITIONS EDITOR Stuart Johnson
PROJECT EDITOR Geraldine Osnato
EDITORIAL ASSISTANT Aly Rentrop
SENIOR MARKETING MANAGER Amanda Wygal
SENIOR PRODUCTION EDITOR Elizabeth Swain
TEXT DESIGNER Madelyn Lesure
COVER DESIGNER Norm Christiansen
DUMMY DESIGNER Lee Goldstein
PHOTO EDITOR Hilary Newman
ILLUSTRATION EDITOR Anna Melhorn
ILLUSTRATION STUDIO Radiant Illustrations Inc.
SENIOR MEDIA EDITOR Thomas Kulesa
MEDIA PROJECT MANAGER Bridget O'Lavin
COVER IMAGE ©Eric Heller/Photo Researchers
BICENTENNIAL LOGO DESIGN Richard J. Pacifico

This book was set in 10/12 Times Ten by Progressive Information Technologies and printed and bound by Von Hoffmann Press. The cover was printed by Von Hoffmann Press.

This book is printed on acid free paper.

To order books or for customer service please call 1-800-CALL WILEY (225-5945).

Library of Congress Cataloging-in-Publication Data
Halliday, David
    Fundamentals of physics.—8th ed., Part 4/David Halliday, Robert Resnick, Jearl Walker.
        p. cm.
    Includes index.
    Part 4: ISBN 13: 978-0-470-04478-0 (acid-free paper)
    Also catalogued as
    Extended version: ISBN 13: 978-0-471-75801-3 (acid-free paper)
    1. Physics—Textbooks. I. Resnick, Robert II. Walker, Jearl III. Title.
    QC21.3.H35 2008
    530—dc22
                                                                                2006041375

Printed in the United States of America

10 9 8 7 6 5 4 3

Brief Contents

v

Contents

Appendices

Fun with a big challenge. That is how I have regarded physics since the day when Sharon, one of the students in a class I taught as a graduate student, suddenly demanded of me, "What has any of this got to do with my life?" Of course I immediately responded, "Sharon, this has everything to do with your life—this is physics."

She asked me for an example. I thought and thought but could not come up with a single one. That night I created *The Flying Circus of Physics* for Sharon but also for me because I realized her complaint was mine. I had spent six years slugging my way through many dozens of physics textbooks that were carefully written with the best of pedagogical plans, but there was something missing. Physics is the most interesting subject in the world because it is about how the world works, and yet the textbooks had been thoroughly wrung of any connection with the real world. The fun was missing.

I have packed a lot of real-world physics into this HRW book, connecting it with the new edition of *The Flying Circus of Physics*. Much of the material comes from the HRW classes I teach, where I can judge from the faces and blunt comments what material and presentations work and what do not. The notes I make on my successes and failures there help form the basis of this book. My message here is the same as I had with every student I've met since Sharon so long ago: "Yes, you *can* reason from basic physics concepts all the way to valid conclusions about the real world, and that understanding of the real world is where the fun is."

I have many goals in writing this book but the overriding one is to provide instructors with a tool by which they can teach students how to effectively read scientific material, identify fundamental concepts, reason through scientific ques-tions, and solve quantitative problems. This process is not easy for either students or instructors. Indeed, the course associated with this book may be one of the most challenging of all the courses taken by a student. However, it can also be one of the most rewarding because it reveals the world's fundamental clockwork from which all scientific and engineering applications spring.

Many users of the seventh edition (both instructors and students) sent in comments and suggestions to improve the book. These improvements are now incorporated into the narrative and problems throughout the book. The publisher John Wiley & Sons and I regard the book as an ongoing project and encourage more input from users. You can send suggestions, corrections, and positive or negative comments to John Wiley & Sons (http:www.wiley.com/college/halliday) or Jearl Walker (mail address: Physics Department, Cleveland State University, Cleveland, OH 44115 USA; fax number: (USA) 216 687 2424; or email address: physics@wiley.com; or the blog site at www.flyingcircus-ofphysics.com). We may not be able to respond to all suggestions, but we keep and study each of them.

Major Content Changes

• Flying Circus material has been incorporated into the text in several ways: chapter opening Puzzlers, Sample Problems, text examples, and end-of-chapter Problems. The purpose of this is two-fold: (1) make the subject more interesting and engaging, (2) show the student that the world around them can be examined and understood using the fundamental principles of physics.

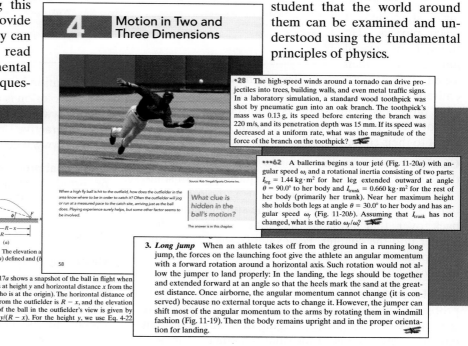

4 Motion in Two and Three Dimensions

When a high fly ball is hit to the outfield, how does the outfielder in the area know where to be in order to catch it? Often the outfielder will jog or run at a measured pace to the catch site, arriving just as the ball does. Playing experience surely helps, but some other factor seems to be involved.

What clue is hidden in the ball's motion?

The answer is in this chapter.

Source: Rob Tringali/Sports Chrome Inc.

Sample Problem 4-8

Suppose a baseball batter B hits a high fly ball to the outfield, directly toward an outfielder F and with a launch speed of $v_0 = 40$ m/s and a launch angle of $\theta_0 = 35°$. During the flight, a line from the outfielder to the ball makes an angle ϕ with the ground. Plot elevation angle ϕ versus time t, assuming that the outfielder is already positioned to catch the ball, is 6.0 m too close to the batter, and is 6.0 m too far away.

KEY IDEAS (1) If we neglect air drag, the ball is a projectile for which the vertical motion and the horizontal motion can be analyzed separately. (2) Assuming the ball is caught at approximately the height it is hit, the horizontal distance traveled by the ball is the range R, given by Eq. 4-26 ($R = (v_0^2/g) \sin 2\theta_0$).

Calculations: The ball can be caught if the outfielder's distance from the batter is equal to the range R of the ball. Using Eq. 4-26, we find

$$R = \frac{v_0^2}{g} \sin 2\theta_0 = \frac{(40 \text{ m/s})^2}{9.8 \text{ m/s}^2} \sin (70°) = 153.42 \text{ m}.$$

FIG. 4-17 The elevation a fielder is (a) defined and (b

Figure 4-17a shows a snapshot of the ball in flight when the ball is at height y and horizontal distance x from the batter (who is at the origin). The horizontal distance of the ball from the outfielder is $R - x$, and the elevation angle ϕ of the ball in the outfielder's view is given by $\tan \phi = y/(R - x)$. For the height y, we use Eq. 4-22

58

•28 The high-speed winds around a tornado can drive projectiles into trees, building walls, and even metal traffic signs. In a laboratory simulation, a standard wood toothpick was shot by pneumatic gun into an oak branch. The toothpick's mass was 0.13 g, its speed before entering the branch was 220 m/s, and its penetration depth was 15 mm. If its speed was decreased at a uniform rate, what was the magnitude of the force of the branch on the toothpick?

•••62 A ballerina begins a tour jeté (Fig. 11-20a) with angular speed ω_i and a rotational inertia consisting of two parts: $I_{leg} = 1.44$ kg·m² for her leg extended outward at angle $\theta = 90.0°$ to her body and $I_{trunk} = 0.660$ kg·m² for the rest of her body (primarily her trunk). Near her maximum height she holds both legs at angle $\theta = 30.0°$ to her body and has angular speed ω_f (Fig. 11-20b). Assuming that I_{trunk} has not changed, what is the ratio ω_f/ω_i?

3. *Long jump* When an athlete takes off from the ground in a running long jump, the forces on the launching foot give the athlete an angular momentum with a forward rotation around a horizontal axis. Such rotation would not allow the jumper to land properly: In the landing, the legs should be together and extended forward at an angle so that the heels mark the sand at the greatest distance. Once airborne, the angular momentum cannot change (it is conserved) because no external torque acts to change it. However, the jumper can shift most of the angular momentum to the arms by rotating them in windmill fashion (Fig. 11-19). Then the body remains upright and in the proper orientation for landing.

• Links to *The Flying Circus of Physics* are shown throughout the text material and end-of-chapter problems with a biplane icon. In the electronic version of this book, clicking on the icon takes you to the corresponding item in *Flying Circus*. The bibliography of *Flying Circus* (over 10 000 references to scientific and engineering journals) is located at www.flyingcircusofphysics.com.

• The Newtonian gravitational law, the Coulomb law, and the Biot-Savart law are now introduced in unit-vector notation.

• Most of the chapter-opening puzzlers (the examples of applied physics designed to entice a reader into each chapter) are new and come straight from research journals in many different fields.

• Several thousand of the end-of-chapter problems have been rewritten to streamline both the presentation and the answer. Many new problems of the moderate and difficult categories have been included.

Chapter Features

Opening puzzlers. A curious puzzling situation opens each chapter and is explained somewhere within the chapter, to entice a student to read the chapter. These features, which are a hallmark of *Fundamentals of Physics*, are based on current research as reported in scientific, engineering, medical, and legal journals.

What is physics? The narrative of every chapter now begins with this question, and with an answer that pertains to the subject of the chapter. (A plumber once asked me, "What do you do for a living?" I replied, "I teach physics." He thought for several minutes and then asked, "What is physics?" The plumber's career was entirely based on physics, yet he did not even know what physics is. Many students in introductory physics do not know what physics is but assume that it is irrelevant to their chosen career.)

Checkpoints are stopping points that effectively ask the student, "Can you answer this question with some reasoning based on the narrative or sample problem that you just read?" If not, then the student should go back over that previous material before traveling deeper into the chapter. For example, see Checkpoint 1 on page 62 and Checkpoint 2 on page 280. *Answers to all checkpoints are in the back of the book.*

Sample problems are chosen to demonstrate how problems can be solved with reasoned solutions rather than quick and simplistic plugging of numbers into an equation with no regard for what the equation means.

The sample problems with the label "Build your skill" are typically longer, with more guidance.

Key Ideas in the sample problems focus a student on the basic concepts at the root of the solution to a problem. In effect, these key ideas say, "We start our solution by using this basic concept, a procedure that prepares us for solving many other problems. We don't start by grabbing an equation for a quick plug-and-chug, a procedure that prepares us for nothing."

Problem-solving tactics contain helpful instructions to guide the beginning physics student as to how to solve problems and avoid common errors.

Review & Summary is a brief outline of the chapter contents that contains the essential concepts but which is not a substitute for reading the chapter.

Questions are like the checkpoints and require reasoning and understanding rather than calculations. *Answers to the odd-numbered questions are in the back of the book.*

Problems are grouped under section titles and are labeled according to difficulty. *Answers to the odd-numbered problems are in the back of the book.*

Icons for additional help. When worked-out solutions are provided either in print or electronically for certain of the odd-numbered problems, the statements for those problems include a trailing icon to alert both student and instructor as to where the solutions are located. An icon guide is provided here and at the beginning of each set of problems:

GO	Tutoring problem available (at instructor's discretion) in *WileyPLUS* and WebAssign
SSM	Worked-out solution available in Student Solutions Manual
• – •••	Number of dots indicates level of problem difficulty
	Additional information available in *The Flying Circus of Physics* and at flyingcircusofphysics.com
WWW	Worked-out solution is at — http://www.wiley.com/college/halliday
ILW	Interactive solution is at —

SSM Solution is in the Student Solutions Manual.
WWW Solution is at
   http://www.wiley.com/college/halliday
ILW Interactive LearningWare solution is at
   http://www.wiley.com/college/halliday

Additional problems. These problems are not ordered or sorted in any way so that a student must determine which parts of the chapter apply to any given problem.

Additional Features

Reasoning versus plug-and-chug. A primary goal of this book is to teach students to reason through challenging situations, from basic principles to a solution. Although some plug-and-chug homework problems remain in the book (on purpose), most homework problems emphasize reasoning.

Chapters of reasonable length. To avoid producing a book thick enough to stop a bullet (and thus also a student), I have made the chapters of reasonable length. I explain enough to get a student going but not so much that a student no longer must analyze and fuse ideas. After all, a student will need the skill of analyzing and fusing ideas long after this book is read and the course is completed.

Use of vector-capable calculators. When vector calculations in a sample problem can be performed directly on-screen with a vector-capable calculator, the solution of the sample problem indicates that fact but still carries through the traditional component analysis. When vector calculations cannot be performed directly on-screen, the solution explains why.

Graphs as puzzles. These are problems that give a graph and ask for a result that requires much more than just reading off a data point from the graph. Rather, the solution requires an understanding of the physical arrangement in a problem and the principles behind the associated equations. These problems are more like Sherlock Holmes puzzles because a student must decide what data are important. For examples, see problem 50 on page 80, problem 12 on page 108, and problem 22 on page 231.

Problems with applied physics, based on published research, appear in many places, either as the opening puzzler of a chapter, a sample problem, or a homework problem. For example, see the opening puzzler for Chapter 4 on page 58, Sample Problem 4-8 on pages 69–70, and homework problem 62 on page 302. For an example of homework problems that build on a continuing story, see problems 2, 39, and 61 on pages 131, 134, and 136.

Problems with novel situations. Here is one of several hundred such problems: Problem 69 on page 113 relates a true story of how Air Canada flight 143 ran out of fuel at an altitude of 7.9 km because the crew and airport personnel did not consider the units for the fuel (an important lesson for students who tend to "blow off" units).

Versions of the Text

To accommodate the individual needs of instructors and students, the eighth edition of *Fundamentals of Physics* is available in a number of different versions.

The **Regular Edition** consists of Chapters 1 through 37 (ISBN 978-0-470-04472-8).

The **Extended Edition** contains six additional chapters on quantum physics and cosmology, Chapters 1–44 (ISBN 978-0-471-75801-3).

Both editions are available as single, hard-cover books, or in the following alternative versions:

- **Volume 1** - Chapters 1–20 (Mechanics and Thermodynamics), hardcover, ISBN 978-0-47004473-5
- **Volume 2** - Chapters 21–44 (E&M, Optics, and Quantum Physics), hardcover, ISBN 978-0-470-04474-2
- **Part 1** - Chapters 1–11, paperback, ISBN 978-0-470-04475-9
- **Part 2** - Chapters 12–20, paperback, ISBN 978-0-470-04476-6
- **Part 3** - Chapters 21–32, paperback, ISBN 978-0-470-04477-3
- **Part 4** - Chapters 33–37, paperback, ISBN 978-0-470-04478-0
- **Part 5** - Chapters 38–44, paperback, ISBN 978-0-470-04479-7

WileyPLUS

There have been several significant additions to the Wiley*PLUS* course that accompanies *Fundamentals of Physics:*

- All of the end-of-chapter problems have been coded and are now available for assignment.
- Every problem has an associated Hint that can made available to the students at the instructor's discretion.
- There are approximately 400 additional Sample Problems available to the student **at the instructor's discretion.** The Sample Problems are written in the same style and format as those in the text, i.e., they are intended to give the student transferable problem-solving skills rather than specific recipes.
- For every chapter, approximately 6 problems are available in a tutorial format that provides step-by-step, interactive problem-solving guidance. Most are marked here in this book with the icon 🔵 ("Guided Online") but more are being added.
- For every chapter, approximately 6 problems are available in a version that requires the student to enter an algebraic answer.
- There are vector drawing and vector diagram problems that use "drag 'n drop" functionality to assess the students' ability to draw vectors and vector diagrams.
- There are simulation problems that require the student to work with a java applet.

The overall purpose of this new material is to move the on-line homework experience beyond simple Right/Wrong grading and provide meaningful problem solving guidance and support.

Instructor's Supplements

Instructor's Solutions Manual by Sen-Ben Liao, Lawrence Livermore National Laboratory. This manual provides worked-out solutions for all problems found at the end of each chapter.

Instructor Companion Site

http://www.wiley.com/college/halliday

• **Instructor's Manual** by J. Richard Christman, U.S. Coast Guard Academy. This resource contains lecture notes outlining the most important topics of each chapter; demonstration experiments; laboratory and computer projects; film and video sources; answers to all Questions, Exercises, Problems, and Checkpoints; and a correlation guide to the Questions, Exercises, and Problems in the previous edition. It also contains a complete list of all problems for which solutions are available to students (SSM, WWW, and ILW).

• **Lecture PowerPoint Slides** by Athos Petrou and John Cerne of the University of Buffalo. These PowerPoints cover the entire book and are heavily illustrated with figures from the text.

• **Classroom Response Systems ("Clicker") Questions** by David Marx, Illinois State University. There are two sets of questions available: Reading Quiz and Interactive Lecture. The Reading Quiz questions are intended to be relatively straightforward for any student who read the assigned material. The Interactive Lecture questions are intended to for use in an interactive lecture setting.

• **Wiley Physics Simulations** by Andrew Duffy, Boston University. 50 interactive simulations (Java applets) that can be used for classroom demonstrations.

• **Wiley Physics Demonstrations** by David Maiullo, Rutgers University. This is a collection of digital videos of 80 standard physics demonstrations. They can be shown in class or accessed from the student companion site. There is an accompanying Instructor's Guide that includes "clicker" questions.

• **Test Bank** by J. Richard Christman, U.S. Coast Guard Academy. The Test Bank includes more than 2200 multiple-choice questions. These items are also available in the Computerized Test Bank which provides full editing features to help you customize tests (available in both IBM and Macintosh versions).

• All of the *Instructor's Solutions Manual* in MSWord, and pdf files

• All text illustrations, suitable for both classroom projection and printing.

On-line homework and quizzing. In addition to Wiley*PLUS, Fundamentals of Physics*, eighth edition also supports WebAssignPLUS and CAPA, which are other programs that give instructors the ability to deliver and grade homework and quizzes on-line.

WebCT and Blackboard. A variety of materials have been prepared for easy incorporation in either WebCT or Blackboard. WebCT and Blackboard are powerful and easy-to-use web-based course-management systems that allow instructors to set up complete on-line courses with chat rooms, bulletin boards, quizzing, student tracking, etc.

Student's Supplements

Student Companion site. This web site

http://www.wiley.com/college/halliday

was developed specifically for *Fundamentals of Physics,* eighth edition, and is designed to further assist students in the study of physics. The site includes solutions to selected end-of-chapter problems (which are identified with a www icon in the text); self-quizzes; simulation exercises; tips on how to make best use of a programmable calculator; and the Interactive LearningWare tutorials that are described below.

Student Study Guide. The student study guide consists of an overview of the chapter's important concepts, hints for solving end-of-chapter questions/problems, and practice quizzes.

Student's Solutions Manual by J. Richard Christman, U.S. Coast Guard Academy and Edward Derringh, Wentworth Institute. This manual provides student with complete worked-out solutions to 15 percent of the problems found at the end of each chapter within the text. These problems are indicated with an ssm icon.

Interactive LearningWare. This software guides students through solutions to 200 of the end-of-chapter problems. These problems are indicated with an ilw icon. The solutions process is developed interactively, with appropriate feedback and access to error-specific help for the most common mistakes.

Wiley Desktop Edition. An electronic version of *Fundamentals of Physics,* eighth edition containing the complete, extended version of the text is available for download at:

www.wiley.com/college/desktop

Wiley Desktop Editions are a cost effective alternative to the printed text.

Physics as a Second Language: *Mastering Problem Solving* by Thomas Barrett, of Ohio State University. This brief paperback teaches the student how to approach problems more efficiently and effectively. The student will learn how to recognize common patterns in physics problems, break problems down into manageable steps, and apply appropriate techniques. The book takes the student step-by-step through the solutions to numerous examples.

Acknowledgments

A great many people have contributed to this book. J. Richard Christman, of the U.S. Coast Guard Academy, has once again created many fine supplements; his recommendations to this book have been invaluable. Sen-Ben Liao of Lawrence Livermore National Laboratory, James Whitenton of Southern Polytechnic State University, and Jerry Shi, of Pasadena City College, performed the Herculean task of working out solutions for every one of the homework problems in the book. At John Wiley publishers, the book received support from Stuart Johnson, the editor who oversaw the entire project, Tom Kulesa, who coordinated the state-of-the-art media package, and Geraldine Osnato, who managed a super team to create an impressive supplements package. We thank Elizabeth Swain, the production editor, for pulling all the pieces together during the complex production process. We also thank Maddy Lesure, for her design of both the text and the book cover; Lee Goldstein for her page make-up; Helen Walden for her copyediting; Anna Melhorn for managing the illustration program; and Lilian Brady for her proofreading. Hilary Newman was inspired in the search for unusual and interesting photographs. Both the publisher John Wiley & Sons, Inc. and Jearl Walker would like to thank the following for comments and ideas about the 7th edition: Richard Woodard, University of Florida; David Wick, Clarkson University; Patrick Rapp, University of Puerto Rico at Mayagüez; Nora Thornber, Raritan Valley Community College; Laurence I. Gould, University of Hartford; Greg Childers, California State University at Fullerton; Asha Khakpour of California State University at Fullerton; Joe F. McCullough, Cabrillo College. Finally, our external reviewers have been outstanding and we acknowledge here our debt to each member of that team.

Maris A. Abolins
Michigan State University

Edward Adelson
Ohio State University

Nural Akchurin
Texas Tech

Barbara Andereck
Ohio Wesleyan University

Mark Arnett
Kirkwood Community College

Arun Bansil
Northeastern University

Richard Barber
Santa Clara University

Neil Basecu
Westchester Community College

Anand Batra
Howard University

Richard Bone
Florida International University

Michael E. Browne
University of Idaho

Timothy J. Burns
Leeward Community College

Joseph Buschi
Manhattan College

Philip A. Casabella
Rensselaer Polytechnic Institute

Randall Caton
Christopher Newport College

Roger Clapp
University of South Florida

W. R. Conkie
Queen's University

Renate Crawford
University of Massachusetts-Dartmouth

Mike Crivello
San Diego State University

Robert N. Davie, Jr.
St. Petersburg Junior College

Cheryl K. Dellai
Glendale Community College

Eric R. Dietz
California State University at Chico

N. John DiNardo
Drexel University

Eugene Dunnam
University of Florida

Robert Endorf
University of Cincinnati

F. Paul Esposito
University of Cincinnati

Jerry Finkelstein
San Jose State University

Robert H. Good
California State University-Hayward

John B. Gruber
San Jose State University

Ann Hanks
American River College

Randy Harris
University of California-Davis

Samuel Harris
Purdue University

Harold B. Hart
Western Illinois University

Rebecca Hartzler
Seattle Central Community College

John Hubisz
North Carolina State University

Joey Huston
Michigan State University

David Ingram
Ohio University

Shawn Jackson
University of Tulsa

Hector Jimenez
University of Puerto Rico

Sudhakar B. Joshi
York University

Leonard M. Kahn
University of Rhode Island

Leonard Kleinman
University of Texas at Austin

Craig Kletzing
University of Iowa

Arthur Z. Kovacs
Rochester Institute of Technology

Kenneth Krane
Oregon State University

Priscilla Laws
Dickinson College

Edbertho Leal
Polytechnic University of Puerto Rico

Vern Lindberg
Rochester Institute of Technology

Peter Loly
University of Manitoba

Andreas Mandelis
University of Toronto

Robert R. Marchini
Memphis State University

Paul Marquard
Caspar College

David Marx
Illinois State University

James H. McGuire
Tulane University

David M. McKinstry
Eastern Washington University

Eugene Mosca
United States Naval Academy

James Napolitano
Rensselaer Polytechnic Institute

Michael O'Shea
Kansas State University

Patrick Papin
San Diego State University

Kiumars Parvin
San Jose State University

Robert Pelcovits
Brown University

Oren P. Quist
South Dakota State University

Joe Redish
University of Maryland

Timothy M. Ritter
University of North Carolina at Pembroke

Gerardo A. Rodriguez
Skidmore College

John Rosendahl
University of California at Irvine

Todd Ruskell
Colorado School of Mines

Michael Schatz
Georgia Institute of Technology

Darrell Seeley
Milwaukee School of Engineering

Bruce Arne Sherwood
North Carolina State University

Ross L. Spencer
Brigham Young University

Paul Stanley
Beloit College

Harold Stokes
Brigham Young University

Michael G. Strauss
University of Oklahoma

Jay D. Strieb
Villanova University

Dan Styer
Oberlin College

Michael Tammaro
University of Rhode Island

Marshall Thomsen
Eastern Michigan University

David Toot
Alfred University

Tung Tsang
Howard University

J. S. Turner
University of Texas at Austin

T. S. Venkataraman
Drexel University

Gianfranco Vidali
Syracuse University

Fred Wang
Prairie View A & M

Robert C. Webb
Texas A & M University

William M. Whelan
Ryerson Polytechnic University

George Williams
University of Utah

David Wolfe
University of New Mexico

Fundamentals of Physics

PART 4

Electromagnetic Waves

Norbert Rosing/National Geographic/Getty Images, Inc.

On bright days the sky can display a variety of bright spots and arcs (halos). Probably the most common display is a bright spot called a sun dog that appears to the left or right of the Sun and sometimes on both sides. Because sun dogs often show colors, they can be mistaken for rainbows. However, sun dogs are not produced by water drops as rainbows are.

What produces sun dogs?

The answer is in this chapter.

33-1 | WHAT IS PHYSICS?

The information age in which we live is based almost entirely on the physics of electromagnetic waves. Like it or not, we are now globally connected by television, telephones, and the Web. And like it or not, we are constantly immersed in those signals because of television, radio, and telephone transmitters.

Much of this global interconnection of information processors was not imagined by even the most visionary engineers of 20 years ago. The challenge for today's engineers is trying to envision what the global interconnection will be like 20 years from now. The starting point in meeting that challenge is understanding the basic physics of electromagnetic waves, which come in so many different types that they are poetically said to form *Maxwell's rainbow*.

33-2 | Maxwell's Rainbow

The crowning achievement of James Clerk Maxwell (see Chapter 32) was to show that a beam of light is a traveling wave of electric and magnetic fields—an **electromagnetic wave**—and thus that optics, the study of visible light, is a branch of electromagnetism. In this chapter we move from one to the other: we conclude our discussion of strictly electrical and magnetic phenomena, and we build a foundation for optics.

In Maxwell's time (the mid 1800s), the visible, infrared, and ultraviolet forms of light were the only electromagnetic waves known. Spurred on by Maxwell's work, however, Heinrich Hertz discovered what we now call radio waves and verified that they move through the laboratory at the same speed as visible light.

As Fig. 33-1 shows, we now know a wide *spectrum* (or range) of electromagnetic waves: Maxwell's rainbow. Consider the extent to which we are immersed in electromagnetic waves throughout this spectrum. The Sun, whose radiations define the environment in which we as a species have evolved and adapted, is the dominant source. We are also crisscrossed by radio and television signals. Microwaves from radar systems and from telephone relay systems may reach us. There are electromagnetic waves from lightbulbs, from the heated engine blocks of automobiles, from x-ray machines, from lightning flashes, and from buried

FIG. 33-1 The electromagnetic spectrum.

radioactive materials. Beyond this, radiation reaches us from stars and other objects in our galaxy and from other galaxies. Electromagnetic waves also travel in the other direction. Television signals, transmitted from Earth since about 1950, have now taken news about us (along with episodes of *I Love Lucy,* albeit *very* faintly) to whatever technically sophisticated inhabitants there may be on whatever planets may encircle the nearest 400 or so stars.

In the wavelength scale in Fig. 33-1 (and similarly the corresponding frequency scale), each scale marker represents a change in wavelength (and correspondingly in frequency) by a factor of 10. The scale is open-ended; the wavelengths of electromagnetic waves have no inherent upper or lower bound.

Certain regions of the electromagnetic spectrum in Fig. 33-1 are identified by familiar labels, such as *x rays* and *radio waves.* These labels denote roughly defined wavelength ranges within which certain kinds of sources and detectors of electromagnetic waves are in common use. Other regions of Fig. 33-1, such as those labeled TV channels and AM radio, represent specific wavelength bands assigned by law for certain commercial or other purposes. There are no gaps in the electromagnetic spectrum—and all electromagnetic waves, no matter where they lie in the spectrum, travel through *free space* (vacuum) with the same speed c.

The visible region of the spectrum is of course of particular interest to us. Figure 33-2 shows the relative sensitivity of the human eye to light of various wavelengths. The center of the visible region is about 555 nm, which produces the sensation that we call yellow-green.

The limits of this visible spectrum are not well defined because the eye-sensitivity curve approaches the zero-sensitivity line asymptotically at both long and short wavelengths. If we take the limits, arbitrarily, as the wavelengths at which eye sensitivity has dropped to 1% of its maximum value, these limits are about 430 and 690 nm; however, the eye can detect electromagnetic waves some-what beyond these limits if they are intense enough.

FIG. 33-2 The relative sensitivity of the average human eye to electro-magnetic waves at different wave-lengths. This portion of the electro-magnetic spectrum to which the eye is sensitive is called *visible light.*

33-3 | The Traveling Electromagnetic Wave, Qualitatively

Some electromagnetic waves, including x rays, gamma rays, and visible light, are *radiated* (emitted) from sources that are of atomic or nuclear size, where quantum physics rules. Here we discuss how other electromagnetic waves are generated. To simplify matters, we restrict ourselves to that region of the spectrum (wavelength $\lambda \approx 1$ m) in which the source of the *radiation* (the emitted waves) is both macroscopic and of manageable dimensions.

Figure 33-3 shows, in broad outline, the generation of such waves. At its heart is an *LC oscillator,* which establishes an angular frequency $\omega (= 1/\sqrt{LC})$. Charges and currents in this circuit vary sinusoidally at this frequency, as depicted in Fig. 31-1. An external source—possibly an ac generator—must be included to supply energy to compensate both for thermal losses in the circuit and for energy carried away by the radiated electromagnetic wave.

The LC oscillator of Fig. 33-3 is coupled by a transformer and a transmission line to an *antenna,* which consists essentially of two thin, solid, conducting rods.

FIG. 33-3 An arrangement for generating a trav-eling electromagnetic wave in the shortwave radio region of the spectrum: an *LC* oscillator produces a sinusoidal current in the antenna, which gener-ates the wave. *P* is a distant point at which a detec-tor can monitor the wave traveling past it.

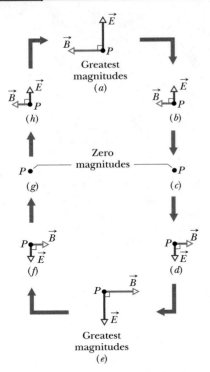

FIG. 33-4 (a)–(h) The variation in the electric field $\vec{E}$ and the magnetic field $\vec{B}$ at the distant point P of Fig. 33-3 as one wavelength of the electromagnetic wave travels past it. In this perspective, the wave is traveling directly out of the page. The two fields vary sinusoidally in magnitude and direction. Note that they are always perpendicular to each other and to the wave's direction of travel.

Through this coupling, the sinusoidally varying current in the oscillator causes charge to oscillate sinusoidally along the rods of the antenna at the angular frequency ω of the LC oscillator. The current in the rods associated with this movement of charge also varies sinusoidally, in magnitude and direction, at angular frequency ω. The antenna has the effect of an electric dipole whose electric dipole moment varies sinusoidally in magnitude and direction along the antenna.

Because the dipole moment varies in magnitude and direction, the electric field produced by the dipole varies in magnitude and direction. Also, because the current varies, the magnetic field produced by the current varies in magnitude and direction. However, the changes in the electric and magnetic fields do not happen everywhere instantaneously; rather, the changes travel outward from the antenna at the speed of light c. Together the changing fields form an electromagnetic wave that travels away from the antenna at speed c. The angular frequency of this wave is ω, the same as that of the LC oscillator.

Figure 33-4 shows how the electric field $\vec{E}$ and the magnetic field $\vec{B}$ change with time as one wavelength of the wave sweeps past the distant point P of Fig. 33-3; in each part of Fig. 33-4, the wave is traveling directly out of the page. (We choose a distant point so that the curvature of the waves suggested in Fig. 33-3 is small enough to neglect. At such points, the wave is said to be a *plane wave*, and discussion of the wave is much simplified.) Note several key features in Fig. 33-4; they are present regardless of how the wave is created:

1. The electric and magnetic fields $\vec{E}$ and $\vec{B}$ are always perpendicular to the direction in which the wave is traveling. Thus, the wave is a *transverse wave*, as discussed in Chapter 16.

2. The electric field is always perpendicular to the magnetic field.

3. The cross product $\vec{E} \times \vec{B}$ always gives the direction in which the wave travels.

4. The fields always vary sinusoidally, just like the transverse waves discussed in Chapter 16. Moreover, the fields vary with the same frequency and *in phase* (in step) with each other.

In keeping with these features, we can assume that the electromagnetic wave is traveling toward P in the positive direction of an x axis, that the electric field in Fig. 33-4 is oscillating parallel to the y axis, and that the magnetic field is then oscillating parallel to the z axis (using a right-handed coordinate system, of course). Then we can write the electric and magnetic fields as sinusoidal functions of position x (along the path of the wave) and time t:

$$E = E_m \sin(kx - \omega t), \tag{33-1}$$

$$B = B_m \sin(kx - \omega t), \tag{33-2}$$

in which E_m and B_m are the amplitudes of the fields and, as in Chapter 16, ω and k are the angular frequency and angular wave number of the wave, respectively. From these equations, we note that not only do the two fields form the electromagnetic wave but each also forms its own wave. Equation 33-1 gives the *electric wave component* of the electromagnetic wave, and Eq. 33-2 gives the *magnetic wave component*. As we shall discuss below, these two wave components cannot exist independently.

From Eq. 16-13, we know that the speed of the wave is ω/k. However, because this is an electromagnetic wave, its speed (in vacuum) is given the symbol c rather than v. In the next section you will see that c has the value

$$c = \frac{1}{\sqrt{\mu_0 \varepsilon_0}} \quad \text{(wave speed),} \tag{33-3}$$

which is about 3.0×10^8 m/s. In other words,

> All electromagnetic waves, including visible light, have the same speed c in vacuum.

You will also see that the wave speed c and the amplitudes of the electric and magnetic fields are related by

$$\frac{E_m}{B_m} = c \qquad \text{(amplitude ratio).} \qquad (33\text{-}4)$$

If we divide Eq. 33-1 by Eq. 33-2 and then substitute with Eq. 33-4, we find that the magnitudes of the fields at every instant and at any point are related by

$$\frac{E}{B} = c \qquad \text{(magnitude ratio).} \qquad (33\text{-}5)$$

We can represent the electromagnetic wave as in Fig. 33-5a, with a *ray* (a directed line showing the wave's direction of travel) or with *wavefronts* (imaginary surfaces over which the wave has the same magnitude of electric field), or both. The two wavefronts shown in Fig. 33-5a are separated by one wavelength λ ($= 2\pi/k$) of the wave. (Waves traveling in approximately the same direction form a *beam,* such as a laser beam, which can also be represented with a ray.)

We can also represent the wave as in Fig. 33-5b, which shows the electric and magnetic field vectors in a "snapshot" of the wave at a certain instant. The curves through the tips of the vectors represent the sinusoidal oscillations given by Eqs. 33-1 and 33-2; the wave components $\vec{E}$ and $\vec{B}$ are in phase, perpendicular to each other, and perpendicular to the wave's direction of travel.

Interpretation of Fig. 33-5b requires some care. The similar drawings for a transverse wave on a taut string that we discussed in Chapter 16 represented the up and down displacement of sections of the string as the wave passed (*something actually moved*). Figure 33-5b is more abstract. At the instant shown, the electric and magnetic fields each have a certain magnitude and direction (but always perpendicular to the x axis) at each point along the x axis. We choose to represent these vector quantities with a pair of arrows for each point, and so we must draw arrows of different lengths for different points, all directed away from the x axis, like thorns on a rose stem. However, the arrows represent field values only at points that are on the x axis. Neither the arrows nor the sinusoidal curves represent a sideways motion of anything, nor do the arrows connect points on the x axis with points off the axis.

Drawings like Fig. 33-5 help us visualize what is actually a very complicated situation. First consider the magnetic field. Because it varies sinusoidally, it induces (via Faraday's law of induction) a perpendicular electric field that also varies sinusoidally. However, because that electric field is varying sinusoidally, it induces (via Maxwell's law of induction) a perpendicular magnetic field that also

FIG. 33-5   (*a*) An electromagnetic wave represented with a ray and two wavefronts; the wavefronts are separated by one wavelength λ. (*b*) The same wave represented in a "snapshot" of its electric field $\vec{E}$ and magnetic field $\vec{B}$ at points on the x axis, along which the wave travels at speed c. As it travels past point P, the fields vary as shown in Fig. 33-4. The electric component of the wave consists of only the electric fields; the magnetic component consists of only the magnetic fields. The dashed rectangle at P is used in Fig. 33-6.

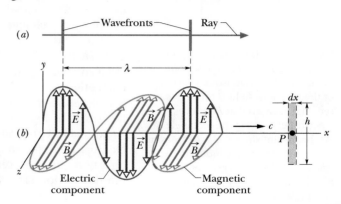

varies sinusoidally. And so on. The two fields continuously create each other via induction, and the resulting sinusoidal variations in the fields travel as a wave—the electromagnetic wave. Without this amazing result, we could not see; indeed, because we need electromagnetic waves from the Sun to maintain Earth's temperature, without this result we could not even exist.

A Most Curious Wave

The waves we discussed in Chapters 16 and 17 require a *medium* (some material) through which or along which to travel. We had waves traveling along a string, through Earth, and through the air. However, an electromagnetic wave (let's use the term *light wave* or *light*) is curiously different in that it requires no medium for its travel. It can, indeed, travel through a medium such as air or glass, but it can also travel through the vacuum of space between a star and us.

Once the special theory of relativity became accepted, long after Einstein published it in 1905, the speed of light waves was realized to be special. One reason is that light has the same speed regardless of the frame of reference from which it is measured. If you send a beam of light along an axis and ask several observers to measure its speed while they move at different speeds along that axis, either in the direction of the light or opposite it, they will all measure the *same speed* for the light. This result is an amazing one and quite different from what would have been found if those observers had measured the speed of any other type of wave; for other waves, the speed of the observers relative to the wave would have affected their measurements.

The meter has now been defined so that the speed of light (any electromagnetic wave) in vacuum has the exact value

$$c = 299\ 792\ 458 \text{ m/s},$$

which can be used as a standard. In fact, if you now measure the travel time of a pulse of light from one point to another, you are not really measuring the speed of the light but rather the distance between those two points.

33-4 | The Traveling Electromagnetic Wave, Quantitatively

We shall now derive Eqs. 33-3 and 33-4 and, even more important, explore the dual induction of electric and magnetic fields that gives us light.

Equation 33-4 and the Induced Electric Field

The dashed rectangle of dimensions dx and h in Fig. 33-6 is fixed at point P on the x axis and in the xy plane (it is shown on the right in Fig. 33-5b). As the electromagnetic wave moves rightward past the rectangle, the magnetic flux Φ_B through the rectangle changes and—according to Faraday's law of induction—induced electric fields appear throughout the region of the rectangle. We take $\vec{E}$ and $\vec{E} + d\vec{E}$ to be the induced fields along the two long sides of the rectangle. These induced electric fields are, in fact, the electrical component of the electromagnetic wave.

Note the small red portion of the magnetic field component curve far from the y axis in Fig. 33-5b. Let's consider the induced electric fields at the instant when this red portion of the magnetic component is passing through the rectangle. Just then, the magnetic field through the rectangle points in the positive z direction and is decreasing in magnitude (the magnitude was greater just before the red section arrived). Because the magnetic field is decreasing, the magnetic flux Φ_B through the rectangle is also decreasing. According to Faraday's law, this

FIG. 33-6 As the electromagnetic wave travels rightward past point P in Fig. 33-5b, the sinusoidal variation of the magnetic field $\vec{B}$ through a rectangle centered at P induces electric fields along the rectangle. At the instant shown, $\vec{B}$ is decreasing in magnitude and the induced electric field is therefore greater in magnitude on the right side of the rectangle than on the left.

change in flux is opposed by induced electric fields, which produce a magnetic field $\vec{B}$ in the positive z direction.

According to Lenz's law, this in turn means that if we imagine the boundary of the rectangle to be a conducting loop, a counterclockwise induced current would have to appear in it. There is, of course, no conducting loop; but this analysis shows that the induced electric field vectors $\vec{E}$ and $\vec{E} + d\vec{E}$ are indeed oriented as shown in Fig. 33-6, with the magnitude of $\vec{E} + d\vec{E}$ greater than that of $\vec{E}$. Otherwise, the net induced electric field would not act counterclockwise around the rectangle.

Let us now apply Faraday's law of induction,

$$\oint \vec{E} \cdot d\vec{s} = -\frac{d\Phi_B}{dt}, \tag{33-6}$$

counterclockwise around the rectangle of Fig. 33-6. There is no contribution to the integral from the top or bottom of the rectangle because $\vec{E}$ and $d\vec{s}$ are perpendicular to each other there. The integral then has the value

$$\oint \vec{E} \cdot d\vec{s} = (E + dE)h - Eh = h\,dE. \tag{33-7}$$

The flux Φ_B through this rectangle is

$$\Phi_B = (B)(h\,dx), \tag{33-8}$$

where B is the average magnitude of $\vec{B}$ within the rectangle and $h\,dx$ is the area of the rectangle. Differentiating Eq. 33-8 with respect to t gives

$$\frac{d\Phi_B}{dt} = h\,dx\,\frac{dB}{dt}. \tag{33-9}$$

If we substitute Eqs. 33-7 and 33-9 into Eq. 33-6, we find

$$h\,dE = -h\,dx\,\frac{dB}{dt}$$

or

$$\frac{dE}{dx} = -\frac{dB}{dt}. \tag{33-10}$$

Actually, both B and E are functions of *two* variables, x and t, as Eqs. 33-1 and 33-2 show. However, in evaluating dE/dx, we must assume that t is constant because Fig. 33-6 is an "instantaneous snapshot." Also, in evaluating dB/dt we must assume that x is constant because we are dealing with the time rate of change of B at a particular place, the point P in Fig. 33-5b. The derivatives under these circumstances are *partial derivatives*, and Eq. 33-10 must be written

$$\frac{\partial E}{\partial x} = -\frac{\partial B}{\partial t}. \tag{33-11}$$

The minus sign in this equation is appropriate and necessary because, although E is increasing with x at the site of the rectangle in Fig. 33-6, B is decreasing with t.

From Eq. 33-1 we have

$$\frac{\partial E}{\partial x} = kE_m \cos(kx - \omega t)$$

and from Eq. 33-2

$$\frac{\partial B}{\partial t} = -\omega B_m \cos(kx - \omega t).$$

Then Eq. 33-11 reduces to

$$kE_m \cos(kx - \omega t) = \omega B_m \cos(kx - \omega t). \tag{33-12}$$

The ratio ω/k for a traveling wave is its speed, which we are calling c. Equation

FIG. 33-7 The sinusoidal variation of the electric field through this rectangle, located (but not shown) at point P in Fig. 33-5b, induces magnetic fields along the rectangle. The instant shown is that of Fig. 33-6: $\vec{E}$ is decreasing in magnitude, and the magnitude of the induced magnetic field is greater on the right side of the rectangle than on the left.

33-12 then becomes

$$\frac{E_m}{B_m} = c \qquad \text{(amplitude ratio)}, \qquad (33\text{-}13)$$

which is just Eq. 33-4.

Equation 33-3 and the Induced Magnetic Field

Figure 33-7 shows another dashed rectangle at point P of Fig. 33-5b; this one is in the xz plane. As the electromagnetic wave moves rightward past this new rectangle, the electric flux Φ_E through the rectangle changes and—according to Maxwell's law of induction—induced magnetic fields appear throughout the region of the rectangle. These induced magnetic fields are, in fact, the magnetic component of the electromagnetic wave.

We see from Fig. 33-5b that at the instant chosen for the magnetic field represented in Fig. 33-6, marked in red on the magnetic component curve, the electric field through the rectangle of Fig. 33-7 is directed as shown. Recall that at the chosen instant, the magnetic field in Fig. 33-6 is decreasing. Because the two fields are in phase, the electric field in Fig. 33-7 must also be decreasing, and so must the electric flux Φ_E through the rectangle. By applying the same reasoning we applied to Fig. 33-6, we see that the changing flux Φ_E will induce a magnetic field with vectors $\vec{B}$ and $\vec{B} + d\vec{B}$ oriented as shown in Fig. 33-7, where field $\vec{B} + d\vec{B}$ is greater than field $\vec{B}$.

Let us apply Maxwell's law of induction,

$$\oint \vec{B} \cdot d\vec{s} = \mu_0 \varepsilon_0 \frac{d\Phi_E}{dt}, \qquad (33\text{-}14)$$

by proceeding counterclockwise around the dashed rectangle of Fig. 33-7. Only the long sides of the rectangle contribute to the integral, whose value is

$$\oint \vec{B} \cdot d\vec{s} = -(B + dB)h + Bh = -h\,dB. \qquad (33\text{-}15)$$

The flux Φ_E through the rectangle is

$$\Phi_E = (E)(h\,dx), \qquad (33\text{-}16)$$

where E is the average magnitude of $\vec{E}$ within the rectangle. Differentiating Eq. 33-16 with respect to t gives

$$\frac{d\Phi_E}{dt} = h\,dx\,\frac{dE}{dt}.$$

If we substitute this and Eq. 33-15 into Eq. 33-14, we find

$$-h\,dB = \mu_0\varepsilon_0\left(h\,dx\,\frac{dE}{dt}\right)$$

or, changing to partial-derivative notation as we did for Eq. 33-11,

$$-\frac{\partial B}{\partial x} = \mu_0\varepsilon_0\frac{\partial E}{\partial t}. \qquad (33\text{-}17)$$

Again, the minus sign in this equation is necessary because, although B is increasing with x at point P in the rectangle in Fig. 33-7, E is decreasing with t.

Evaluating Eq. 33-17 by using Eqs. 33-1 and 33-2 leads to

$$-kB_m\cos(kx - \omega t) = -\mu_0\varepsilon_0\omega E_m\cos(kx - \omega t),$$

which we can write as

$$\frac{E_m}{B_m} = \frac{1}{\mu_0\varepsilon_0(\omega/k)} = \frac{1}{\mu_0\varepsilon_0 c}.$$

Combining this with Eq. 33-13 leads at once to

$$c = \frac{1}{\sqrt{\mu_0 \varepsilon_0}} \quad \text{(wave speed),}$$

(33-18)

which is exactly Eq. 33-3.

✓ CHECKPOINT 1
The magnetic field $\vec{B}$ through the rectangle of Fig. 33-6 is shown at a different instant in part 1 of the figure here; $\vec{B}$ is directed in the xz plane, parallel to the z axis, and its magnitude is increasing. (a) Complete part 1 by drawing the induced electric fields, indicating both directions and relative magnitudes (as in Fig. 33-6). (b) For the same instant, complete part 2 of the figure by drawing the electric field of the electromagnetic wave. Also draw the induced magnetic fields, indicating both directions and relative magnitudes (as in Fig. 33-7).

(1)   (2)

33-5 | Energy Transport and the Poynting Vector

All sunbathers know that an electromagnetic wave can transport energy and deliver it to a body on which the wave falls. The rate of energy transport per unit area in such a wave is described by a vector $\vec{S}$, called the **Poynting vector** after physicist John Henry Poynting (1852–1914), who first discussed its properties. This vector is defined as

$$\vec{S} = \frac{1}{\mu_0} \vec{E} \times \vec{B} \quad \text{(Poynting vector).}$$

(33-19)

Its magnitude S is related to the rate at which energy is transported by a wave across a unit area at any instant (inst):

$$S = \left(\frac{\text{energy/time}}{\text{area}} \right)_{\text{inst}} = \left(\frac{\text{power}}{\text{area}} \right)_{\text{inst}}.$$

(33-20)

From this we can see that the SI unit for $\vec{S}$ is the watt per square meter (W/m²).

☛ The direction of the Poynting vector $\vec{S}$ of an electromagnetic wave at any point gives the wave's direction of travel and the direction of energy transport at that point.

Because $\vec{E}$ and $\vec{B}$ are perpendicular to each other in an electromagnetic wave, the magnitude of $\vec{E} \times \vec{B}$ is EB. Then the magnitude of $\vec{S}$ is

$$S = \frac{1}{\mu_0} EB,$$

(33-21)

in which S, E, and B are instantaneous values. The magnitudes E and B are so closely coupled to each other that we need to deal with only one of them; we choose E, largely because most instruments for detecting electromagnetic waves deal with the electric component of the wave rather than the magnetic component. Using $B = E/c$ from Eq. 33-5, we can rewrite Eq. 33-21 as

$$S = \frac{1}{c\mu_0} E^2 \quad \text{(instantaneous energy flow rate).}$$

(33-22)

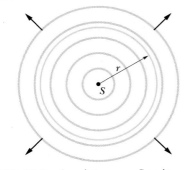

FIG. 33-8 A point source S emits electromagnetic waves uniformly in all directions. The spherical wavefronts pass through an imaginary sphere of radius r that is centered on S.

By substituting $E = E_m \sin(kx - \omega t)$ into Eq. 33-22, we could obtain an equation for the energy transport rate as a function of time. More useful in practice, however, is the average energy transported over time; for that, we need to find the time-averaged value of S, written S_{avg} and also called the **intensity** I of the wave. Thus from Eq. 33-20, the intensity I is

$$I = S_{avg} = \left(\frac{\text{energy/time}}{\text{area}}\right)_{avg} = \left(\frac{\text{power}}{\text{area}}\right)_{avg}. \quad (33\text{-}23)$$

From Eq. 33-22, we find

$$I = S_{avg} = \frac{1}{c\mu_0}[E^2]_{avg} = \frac{1}{c\mu_0}[E_m^2 \sin^2(kx - \omega t)]_{avg}. \quad (33\text{-}24)$$

Over a full cycle, the average value of $\sin^2\theta$, for any angular variable θ, is $\frac{1}{2}$ (see Fig. 31-17). In addition, we define a new quantity E_{rms}, the *root-mean-square* value of the electric field, as

$$E_{rms} = \frac{E_m}{\sqrt{2}}. \quad (33\text{-}25)$$

We can then rewrite Eq. 33-24 as

$$I = \frac{1}{c\mu_0}E_{rms}^2. \quad (33\text{-}26)$$

Because $E = cB$ and c is such a very large number, you might conclude that the energy associated with the electric field is much greater than that associated with the magnetic field. That conclusion is incorrect; the two energies are exactly equal. To show this, we start with Eq. 25-25, which gives the energy density $u\ (=\frac{1}{2}\varepsilon_0 E^2)$ within an electric field, and substitute cB for E; then we can write

$$u_E = \tfrac{1}{2}\varepsilon_0 E^2 = \tfrac{1}{2}\varepsilon_0(cB)^2.$$

If we now substitute for c with Eq. 33-3, we get

$$u_E = \tfrac{1}{2}\varepsilon_0 \frac{1}{\mu_0\varepsilon_0}B^2 = \frac{B^2}{2\mu_0}.$$

However, Eq. 30-54 tells us that $B^2/2\mu_0$ is the energy density u_B of a magnetic field $\vec{B}$; so we see that $u_E = u_B$ everywhere along an electromagnetic wave.

Variation of Intensity with Distance

How intensity varies with distance from a real source of electromagnetic radiation is often complex—especially when the source (like a searchlight at a movie premier) beams the radiation in a particular direction. However, in some situations we can assume that the source is a *point source* that emits the light *isotropically*—that is, with equal intensity in all directions. The spherical wavefronts spreading from such an isotropic point source S at a particular instant are shown in cross section in Fig. 33-8.

Let us assume that the energy of the waves is conserved as they spread from this source. Let us also center an imaginary sphere of radius r on the source, as shown in Fig. 33-8. All the energy emitted by the source must pass through the sphere. Thus, the rate at which energy passes through the sphere via the radiation must equal the rate at which energy is emitted by the source—that is, the source power P_s. The intensity I at the sphere must then be, from Eq. 33-23,

$$I = \frac{\text{power}}{\text{area}} = \frac{P_s}{4\pi r^2}, \quad (33\text{-}27)$$

where $4\pi r^2$ is the area of the sphere. Equation 33-27 tells us that the intensity of the electromagnetic radiation from an isotropic point source decreases with the square of the distance r from the source.

> ✓**CHECKPOINT 2**   The figure here gives the electric field of an electromagnetic wave at a certain point and a certain instant. The wave is transporting energy in the negative z direction. What is the direction of the magnetic field of the wave at that point and instant?

Sample Problem   33-1

When you look at the North Star (Polaris), you intercept light from a star at a distance of 431 ly and emitting energy at a rate of 2.2×10^3 times that of our Sun ($P_{sun} = 3.90 \times 10^{26}$ W). Neglecting any atmospheric absorption, find the rms values of the electric and magnetic fields when the starlight reaches you.

KEY IDEAS

1. The rms value E_{rms} of the electric field in light is related to the intensity I of the light via Eq. 33-26 ($I = E_{rms}^2/c\mu_0$).

2. Because the source is so far away and emits light with equal intensity in all directions, the intensity I at any distance r from the source is related to the source's power P_s via Eq. 33-27 ($I = P_s/4\pi r^2$).

3. The magnitudes of the electric field and magnetic field of an electromagnetic wave at any instant and at any point in the wave are related by the speed of light c according to Eq. 33-5 ($E/B = c$). Thus, the rms values of those fields are also related by Eq. 33-5.

Electric field: Putting the first two ideas together gives us

$$I = \frac{P_s}{4\pi r^2} = \frac{E_{rms}^2}{c\mu_0}$$

and

$$E_{rms} = \sqrt{\frac{P_s c\mu_0}{4\pi r^2}}.$$

Substituting $P_s = (2.2 \times 10^3)(3.90 \times 10^{26}$ W$)$, $r = 431$ ly $= 4.08 \times 10^{18}$ m, and values for the constants, we find

$$E_{rms} = 1.24 \times 10^{-3} \text{ V/m} \approx 1.2 \text{ mV/m}. \quad \text{(Answer)}$$

Magnetic field: From Eq. 33-5, we write

$$B_{rms} = \frac{E_{rms}}{c} = \frac{1.24 \times 10^{-3} \text{ V/m}}{3.00 \times 10^8 \text{ m/s}}$$

$$= 4.1 \times 10^{-12} \text{ T} = 4.1 \text{ pT}.$$

Cannot compare the fields: Note that E_{rms} ($= 1.2$ mV/m) is small as judged by ordinary laboratory standards, but B_{rms} ($= 4.1$ pT) is quite small. This difference helps to explain why most instruments used for the detection and measurement of electromagnetic waves are designed to respond to the electric component of the wave. It is wrong, however, to say that the electric component of an electromagnetic wave is "stronger" than the magnetic component. You cannot compare quantities that are measured in different units. As we have seen, the electric and magnetic components are on an equal basis as far as the propagation of the wave is concerned because their average energies, which *can* be compared, are equal.

33-6 | Radiation Pressure

Electromagnetic waves have linear momentum as well as energy. This means that we can exert a pressure—a **radiation pressure**—on an object by shining light on it. However, the pressure must be very small because, for example, you do not feel a camera flash when it is used to take your photograph (which is good because otherwise every photographic flash could be like a punch).

To find an expression for the pressure, let us shine a beam of electromagnetic radiation—light, for example—on an object for a time interval Δt. Further, let us assume that the object is free to move and that the radiation is entirely **absorbed** (taken up) by the object. This means that during the interval Δt, the object gains an energy ΔU from the radiation. Maxwell showed that the object also gains linear momentum. The magnitude Δp of the momentum change of the object is related to the energy change ΔU by

$$\Delta p = \frac{\Delta U}{c} \quad \text{(total absorption)}, \quad (33\text{-}28)$$

where c is the speed of light. The direction of the momentum change of the object is the direction of the *incident* (incoming) beam that the object absorbs.

Instead of being absorbed, the radiation can be **reflected** by the object; that is, the radiation can be sent off in a new direction as if it bounced off the object. If the radiation is entirely reflected back along its original path, the magnitude of the momentum change of the object is twice that given above, or

$$\Delta p = \frac{2 \Delta U}{c} \quad \text{(total reflection back along path).} \tag{33-29}$$

In the same way, an object undergoes twice as much momentum change when a perfectly elastic tennis ball is bounced from it as when it is struck by a perfectly inelastic ball (a lump of wet putty, say) of the same mass and velocity. If the incident radiation is partly absorbed and partly reflected, the momentum change of the object is between $\Delta U/c$ and $2 \Delta U/c$.

From Newton's second law in its linear momentum form (Section 9-4), we know that a change in momentum is related to a force by

$$F = \frac{\Delta p}{\Delta t}. \tag{33-30}$$

To find expressions for the force exerted by radiation in terms of the intensity I of the radiation, we first note that intensity is

$$I = \frac{\text{power}}{\text{area}} = \frac{\text{energy/time}}{\text{area}}.$$

Next, suppose that a flat surface of area A, perpendicular to the path of the radiation, intercepts the radiation. In time interval Δt, the energy intercepted by area A is

$$\Delta U = IA \, \Delta t. \tag{33-31}$$

If the energy is completely absorbed, then Eq. 33-28 tells us that $\Delta p = IA \, \Delta t/c$, and, from Eq. 33-30, the magnitude of the force on the area A is

$$F = \frac{IA}{c} \quad \text{(total absorption).} \tag{33-32}$$

Similarly, if the radiation is totally reflected back along its original path, Eq. 33-29 tells us that $\Delta p = 2IA \, \Delta t/c$ and, from Eq. 33-30,

$$F = \frac{2IA}{c} \quad \text{(total reflection back along path).} \tag{33-33}$$

If the radiation is partly absorbed and partly reflected, the magnitude of the force on area A is between the values of IA/c and $2IA/c$.

The force per unit area on an object due to radiation is the radiation pressure p_r. We can find it for the situations of Eqs. 33-32 and 33-33 by dividing both sides of each equation by A. We obtain

$$p_r = \frac{I}{c} \quad \text{(total absorption)} \tag{33-34}$$

and

$$p_r = \frac{2I}{c} \quad \text{(total reflection back along path).} \tag{33-35}$$

Be careful not to confuse the symbol p_r for radiation pressure with the symbol p for momentum. Just as with fluid pressure in Chapter 14, the SI unit of radiation pressure is the newton per square meter (N/m^2), which is called the pascal (Pa).

The development of laser technology has permitted researchers to achieve radiation pressures much greater than, say, that due to a camera flashlamp. This comes about because a beam of laser light—unlike a beam of light from a small lamp filament—can be focused to a tiny spot. This permits the delivery of great amounts of energy to small objects placed at that spot.

✓ **CHECKPOINT 3**   Light of uniform intensity shines perpendicularly on a totally absorbing surface, fully illuminating the surface. If the area of the surface is decreased, do (a) the radiation pressure and (b) the radiation force on the surface increase, decrease, or stay the same?

33-7 | Polarization

VHF (very high frequency) television antennas in England are oriented vertically, but those in North America are horizontal. The difference is due to the direction of oscillation of the electromagnetic waves carrying the TV signal. In England, the transmitting equipment is designed to produce waves that are **polarized** vertically; that is, their electric field oscillates vertically. Thus, for the electric field of the incident television waves to drive a current along an antenna (and provide a signal to a television set), the antenna must be vertical. In North America, the waves are polarized horizontally.

Figure 33-9a shows an electromagnetic wave with its electric field oscillating parallel to the vertical y axis. The plane containing the $\vec{E}$ vectors is called the **plane of oscillation** of the wave (hence, the wave is said to be *plane-polarized* in the y direction). We can represent the wave's *polarization* (state of being polarized) by showing the directions of the electric field oscillations in a head-on view of the plane of oscillation, as in Fig. 33-9b. The vertical double arrow in that figure indicates that as the wave travels past us, its electric field oscillates vertically—it continuously changes between being directed up and down the y axis.

Polarized Light

The electromagnetic waves emitted by a television station all have the same polarization, but the electromagnetic waves emitted by any common source of light (such as the Sun or a bulb) are **polarized randomly,** or **unpolarized** (the two terms mean the same thing). That is, the electric field at any given point is always perpendicular to the direction of travel of the waves but changes directions randomly. Thus, if we try to represent a head-on view of the oscillations over some time period, we do not have a simple drawing with a single double arrow like that of Fig. 33-9b; instead we have a mess of double arrows like that in Fig. 33-10a.

In principle, we can simplify the mess by resolving each electric field of Fig. 33-10a into y and z components. Then as the wave travels past us, the net y component oscillates parallel to the y axis and the net z component oscillates parallel to the z axis. We can then represent the unpolarized light with a pair of

FIG. 33-9   (a) The plane of oscillation of a polarized electromagnetic wave. (b) To represent the polarization, we view the plane of oscillation head-on and indicate the directions of the oscillating electric field with a double arrow.

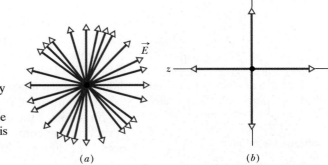

FIG. 33-10   (a) Unpolarized light consists of waves with randomly directed electric fields. Here the waves are all traveling along the same axis, directly out of the page, and all have the same amplitude E. (b) A second way of representing unpolarized light—the light is the superposition of two polarized waves whose planes of oscillation are perpendicular to each other.

(a)   (b)

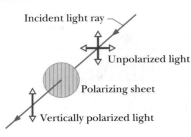

FIG. 33-11 Unpolarized light becomes polarized when it is sent through a polarizing sheet. Its direction of polarization is then parallel to the polarizing direction of the sheet, which is represented here by the vertical lines drawn in the sheet.

double arrows as shown in Fig. 33-10b. The double arrow along the y axis represents the oscillations of the net y component of the electric field. The double arrow along the z axis represents the oscillations of the net z component of the electric field. In doing all this, we effectively change unpolarized light into the superposition of two polarized waves whose planes of oscillation are perpendicular to each other—one plane contains the y axis and the other contains the z axis. One reason to make this change is that drawing Fig. 33-10b is a lot easier than drawing Fig. 33-10a.

We can draw similar figures to represent light that is **partially polarized** (its field oscillations are not completely random as in Fig. 33-10a, nor are they parallel to a single axis as in Fig. 33-9b). For this situation, we draw one of the double arrows in a perpendicular pair of double arrows longer than the other one.

We can transform unpolarized visible light into polarized light by sending it through a *polarizing sheet,* as is shown in Fig. 33-11. Such sheets, commercially known as Polaroids or Polaroid filters, were invented in 1932 by Edwin Land while he was an undergraduate student. A polarizing sheet consists of certain long molecules embedded in plastic. When the sheet is manufactured, it is stretched to align the molecules in parallel rows, like rows in a plowed field. When light is then sent through the sheet, electric field components along one direction pass through the sheet, while components perpendicular to that direction are absorbed by the molecules and disappear.

We shall not dwell on the molecules but, instead, shall assign to the sheet a *polarizing direction,* along which electric field components are passed:

> An electric field component parallel to the polarizing direction is passed (*transmitted*) by a polarizing sheet; a component perpendicular to it is absorbed.

Thus, the electric field of the light emerging from the sheet consists of only the components that are parallel to the polarizing direction of the sheet; hence the light is polarized in that direction. In Fig. 33-11, the vertical electric field components are transmitted by the sheet; the horizontal components are absorbed. The transmitted waves are then vertically polarized.

Intensity of Transmitted Polarized Light

We now consider the intensity of light transmitted by a polarizing sheet. We start with unpolarized light, whose electric field oscillations we can resolve into y and z components as represented in Fig. 33-10b. Further, we can arrange for the y axis to be parallel to the polarizing direction of the sheet. Then only the y components of the light's electric field are passed by the sheet; the z components are absorbed. As suggested by Fig. 33-10b, if the original waves are randomly oriented, the sum of the y components and the sum of the z components are equal. When the z components are absorbed, half the intensity I_0 of the original light is lost. The intensity I of the emerging polarized light is then

$$I = \tfrac{1}{2}I_0. \tag{33-36}$$

Let us call this the *one-half rule;* we can use it *only* when the light reaching a polarizing sheet is unpolarized.

Suppose now that the light reaching a polarizing sheet is already polarized. Figure 33-12 shows a polarizing sheet in the plane of the page and the electric field $\vec{E}$ of such a polarized light wave traveling toward the sheet (and thus prior to any absorption). We can resolve $\vec{E}$ into two components relative to the polarizing direction of the sheet: parallel component E_y is transmitted by the sheet, and perpendicular component E_z is absorbed. Since θ is the angle between $\vec{E}$ and the polarizing direction of the sheet, the transmitted parallel component is

$$E_y = E \cos\theta. \tag{33-37}$$

FIG. 33-12 Polarized light approaching a polarizing sheet. The electric field $\vec{E}$ of the light can be resolved into components E_y (parallel to the polarizing direction of the sheet) and E_z (perpendicular to that direction). Component E_y will be transmitted by the sheet; component E_z will be absorbed.

Recall that the intensity of an electromagnetic wave (such as our light wave) is proportional to the square of the electric field's magnitude (Eq. 33-26, $I = E_{rms}^2/c\mu_0$). In our present case then, the intensity I of the emerging wave is proportional to E_y^2 and the intensity I_0 of the original wave is proportional to E^2. Hence, from Eq. 33-37 we can write $I/I_0 = \cos^2 \theta$, or

$$I = I_0 \cos^2 \theta. \tag{33-38}$$

FIG. 33-13 The light transmitted by polarizing sheet P_1 is vertically polarized, as represented by the vertical double arrow. The amount of that light that is then transmitted by polarizing sheet P_2 depends on the angle between the polarization direction of that light and the polarizing direction of P_2 (indicated by the lines drawn in the sheet and by the dashed line).

Let us call this the *cosine-squared rule;* we can use it *only* when the light reaching a polarizing sheet is already polarized. Then the transmitted intensity I is a maximum and is equal to the original intensity I_0 when the original wave is polarized parallel to the polarizing direction of the sheet (when θ in Eq. 33-38 is 0° or 180°). The transmitted intensity is zero when the original wave is polarized perpendicular to the polarizing direction of the sheet (when θ is 90°).

Figure 33-13 shows an arrangement in which initially unpolarized light is sent through two polarizing sheets P_1 and P_2. (Often, the first sheet is called the *polarizer,* and the second the *analyzer.*) Because the polarizing direction of P_1 is vertical, the light transmitted by P_1 to P_2 is polarized vertically. If the polarizing direction of P_2 is also vertical, then all the light transmitted by P_1 is transmitted by P_2. If the polarizing direction of P_2 is horizontal, none of the light transmitted by P_1 is transmitted by P_2. We reach the same conclusions by considering only the *relative* orientations of the two sheets: If their polarizing directions are parallel, all the light passed by the first sheet is passed by the second sheet (Fig. 33-14a). If those directions are perpendicular (the sheets are said to be *crossed*), no light is passed by the second sheet (Fig. 33-14b). Finally, if the two polarizing directions of Fig. 33-13 make an angle between 0° and 90°, some of the light transmitted by P_1 will be transmitted by P_2, as set by Eq. 33-38.

Light can be polarized by means other than polarizing sheets, such as by reflection (discussed in Section 33-10) and by scattering from atoms or molecules. In *scattering,* light that is intercepted by an object, such as a molecule, is sent off in many, perhaps random, directions. An example is the scattering of sunlight by molecules in the atmosphere, which gives the sky its general glow.

Although direct sunlight is unpolarized, light from much of the sky is at least partially polarized by such scattering. Bees use the polarization of sky light in navigating to and from their hives. Similarly, the Vikings used it to navigate across the North Sea when the daytime Sun was below the horizon (because of the high latitude of the North Sea). These early seafarers had discovered certain crystals (now called cordierite) that changed color when rotated in polarized light. By looking at the sky through such a crystal while rotating it about their line of sight, they could locate the hidden Sun and thus determine which way was south.

FIG. 33-14 (*a*) Overlapping polarizing sheets transmit light fairly well when their polarizing directions have the same orientation, but (*b*) they block most of the light when they are crossed. (*Richard Megna/Fundamental Photographs.*)

(*a*)   (*b*)

✓**CHECKPOINT 4** The figure shows four pairs of polarizing sheets, seen face-on. Each pair is mounted in the path of initially unpolarized light. The polarizing direction of each sheet (indicated by the dashed line) is referenced to either a horizontal x axis or a vertical y axis. Rank the pairs according to the fraction of the initial intensity that they pass, greatest first.

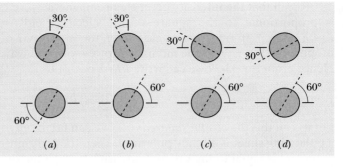

(a) (b) (c) (d)

Sample Problem 33-2 | Build your skill

Figure 33-15a, drawn in perspective, shows a system of three polarizing sheets in the path of initially unpolarized light. The polarizing direction of the first sheet is parallel to the y axis, that of the second sheet is at an angle of 60° counterclockwise from the y axis, and that of the third sheet is parallel to the x axis. What fraction of the initial intensity I_0 of the light emerges from the three-sheet system, and in which direction is that emerging light polarized?

KEY IDEAS

1. We work through the system sheet by sheet, from the first one encountered by the light to the last one.

2. To find the intensity transmitted by any sheet, we apply either the one-half rule or the cosine-squared rule, depending on whether the light reaching the sheet is unpolarized or already polarized.

3. The light that is transmitted by a polarizing sheet is always polarized parallel to the polarizing direction of the sheet.

First sheet: The original light wave is represented in Fig. 33-15b, using the head-on, double-arrow representation of Fig. 33-10b. Because the light is initially unpolarized, the intensity I_1 of the light transmitted by the first sheet is given by the one-half rule (Eq. 33-36):

$$I_1 = \tfrac{1}{2}I_0.$$

Because the polarizing direction of the first sheet is parallel to the y axis, the polarization of the light transmitted by it is also, as shown in the head-on view of Fig. 33-15c.

Second sheet: Because the light reaching the second sheet is polarized, the intensity I_2 of the light transmitted by that sheet is given by the cosine-squared rule (Eq. 33-38). The angle θ in the rule is the angle between the polarization direction of the entering light (parallel to the y axis) and the polarizing direction of the second sheet (60° counterclockwise from the y axis), and so θ is 60°. Then

$$I_2 = I_1 \cos^2 60°.$$

The polarization of this transmitted light is parallel to the polarizing direction of the sheet transmitting it—

that is, 60° counterclockwise from the y axis, as shown in the head-on view of Fig. 33-15d.

Third sheet: Because the light reaching the third sheet is polarized, the intensity I_3 of the light transmitted by that sheet is given by the cosine-squared rule. The angle θ is now the angle between the polarization direction of the entering light (Fig. 33-15d) and the polarizing direction of the third sheet (parallel to the x axis), and so $\theta = 30°$. Thus,

$$I_3 = I_2 \cos^2 30°.$$

This final transmitted light is polarized parallel to the x axis (Fig. 33-15e). We find its intensity by substituting first for I_2 and then for I_1 in the equation above:

$$I_3 = I_2 \cos^2 30° = (I_1 \cos^2 60°) \cos^2 30°$$
$$= (\tfrac{1}{2}I_0) \cos^2 60° \cos^2 30° = 0.094 I_0.$$

Thus,
$$\frac{I_3}{I_0} = 0.094. \qquad \text{(Answer)}$$

That is to say, 9.4% of the initial intensity emerges from the three-sheet system. (If we now remove the second sheet, what fraction of the initial intensity emerges from the system?)

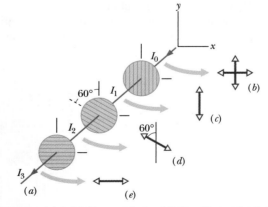

FIG. 33-15 (a) Initially unpolarized light of intensity I_0 is sent into a system of three polarizing sheets. The intensities I_1, I_2, and I_3 of the light transmitted by the sheets are labeled. Shown also are the polarizations, from head-on views, of (b) the initial light and the light transmitted by (c) the first sheet, (d) the second sheet, and (e) the third sheet.

33-8 | Reflection and Refraction

Although a light wave spreads as it moves away from its source, we can often approximate its travel as being in a straight line; we did so for the light wave in Fig. 33-5a. The study of the properties of light waves under that approximation is called *geometrical optics.* For the rest of this chapter and all of Chapter 34, we shall discuss the geometrical optics of visible light.

The photograph in Fig. 33-16a shows an example of light waves traveling in approximately straight lines. A narrow beam of light (the *incident* beam), angled downward from the left and traveling through air, encounters a *plane* (flat) water surface. Part of the light is **reflected** by the surface, forming a beam directed upward toward the right, traveling as if the original beam had bounced from the surface. The rest of the light travels through the surface and into the water, forming a beam directed downward to the right. Because light can travel through it, the water is said to be *transparent;* that is, we can see through it. (In this chapter we shall consider only transparent materials and not opaque materials, through which light cannot travel.)

The travel of light through a surface (or *interface*) that separates two media is called **refraction,** and the light is said to be *refracted.* Unless an incident beam of light is perpendicular to the surface, refraction changes the light's direction of travel. For this reason, the beam is said to be "bent" by the refraction. Note in Fig. 33-16a that the bending occurs only at the surface; within the water, the light travels in a straight line.

In Figure 33-16b, the beams of light in the photograph are represented with an *incident ray,* a *reflected ray,* and a *refracted ray* (and wavefronts). Each ray is oriented with respect to a line, called the *normal,* that is perpendicular to the surface at the point of reflection and refraction. In Fig. 33-16b, the **angle of incidence** is θ_1, the **angle of reflection** is θ_1', and the **angle of refraction** is θ_2, all measured *relative to the normal.* The plane containing the incident ray and the normal is the *plane of incidence,* which is in the plane of the page in Fig. 33-16b.

Experiment shows that reflection and refraction are governed by two laws:

Law of reflection: A reflected ray lies in the plane of incidence and has an angle of reflection equal to the angle of incidence. In Fig. 33-16b, this means that

$$\theta_1' = \theta_1 \quad \text{(reflection).} \tag{33-39}$$

(We shall now usually drop the prime on the angle of reflection.)

Law of refraction: A refracted ray lies in the plane of incidence and has an angle of refraction θ_2 that is related to the angle of incidence θ_1 by

$$n_2 \sin \theta_2 = n_1 \sin \theta_1 \quad \text{(refraction).} \tag{33-40}$$

Here each of the symbols n_1 and n_2 is a dimensionless constant, called the **index of refraction,** that is associated with a medium involved in the refraction. We derive this equation, called **Snell's law,** in Chapter 35. As we shall discuss there, the index of refraction of a medium is equal to c/v, where v is the speed of light in that medium and c is its speed in vacuum.

Table 33-1 gives the indexes of refraction of vacuum and some common substances. For vacuum, n is defined to be exactly 1; for air, n is very close to 1.0 (an approximation we shall often make). Nothing has an index of refraction below 1.

We can rearrange Eq. 33-40 as

$$\sin \theta_2 = \frac{n_1}{n_2} \sin \theta_1 \tag{33-41}$$

to compare the angle of refraction θ_2 with the angle of incidence θ_1. We can

(a)

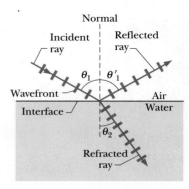

(b)

FIG. 33-16 (a) A photograph showing an incident beam of light reflected and refracted by a horizontal water surface. (b) A ray representation of (a). The angles of incidence (θ_1), reflection (θ_1'), and refraction (θ_2) are marked. (©1974 FP/Fundamentals Photography)

TABLE 33-1

Some Indexes of Refraction[a]

Medium	Index	Medium	Index
Vacuum	Exactly 1	Typical crown glass	1.52
Air (STP)[b]	1.00029	Sodium chloride	1.54
Water (20°C)	1.33	Polystyrene	1.55
Acetone	1.36	Carbon disulfide	1.63
Ethyl alcohol	1.36	Heavy flint glass	1.65
Sugar solution (30%)	1.38	Sapphire	1.77
Fused quartz	1.46	Heaviest flint glass	1.89
Sugar solution (80%)	1.49	Diamond	2.42

[a]For a wavelength of 589 nm (yellow sodium light).

[b]STP means "standard temperature (0°C) and pressure (1 atm)."

FIG. 33-17 Refraction of light traveling from a medium with an index of refraction n_1 into a medium with an index of refraction n_2. (a) The beam does not bend when $n_2 = n_1$; the refracted light then travels in the *undeflected direction* (the dotted line), which is the same as the direction of the incident beam. The beam bends (b) toward the normal when $n_2 > n_1$ and (c) away from the normal when $n_2 < n_1$.

then see that the relative value of θ_2 depends on the relative values of n_2 and n_1. In fact, we can have three basic results:

1. If n_2 is equal to n_1, then θ_2 is equal to θ_1 and refraction does not bend the light beam, which continues in the *undeflected direction,* as in Fig. 33-17a.

2. If n_2 is greater than n_1, then θ_2 is less than θ_1. In this case, refraction bends the light beam away from the undeflected direction and toward the normal, as in Fig. 33-17b.

3. If n_2 is less than n_1, then θ_2 is greater than θ_1. In this case, refraction bends the light beam away from the undeflected direction and away from the normal, as in Fig. 33-17c.

Refraction *cannot* bend a beam so much that the refracted ray is on the same side of the normal as the incident ray.

Chromatic Dispersion

The index of refraction n encountered by light in any medium except vacuum depends on the wavelength of the light. The dependence of n on wavelength implies that when a light beam consists of rays of different wavelengths, the rays will be refracted at different angles by a surface; that is, the light will be spread out by the refraction. This spreading of light is called **chromatic dispersion,** in which "chromatic" refers to the colors associated with the individual wavelengths and "dispersion" refers to the spreading of the light according to its wavelengths or colors. The refractions of Figs. 33-16 and 33-17 do not show chromatic dispersion because the beams are *monochromatic* (of a single wavelength or color).

Generally, the index of refraction of a given medium is *greater* for a shorter wavelength (corresponding to, say, blue light) than for a longer wavelength (say, red light). As an example, Fig. 33-18 shows how the index of refraction of fused quartz depends on the wavelength of light. Such dependence means that when a beam made up of waves of both blue and red light is refracted through a surface, such as from air into quartz or vice versa, the blue *component* (the ray corresponding to the wave of blue light) bends more than the red component.

A beam of *white light* consists of components of all (or nearly all) the colors in the visible spectrum with approximately uniform intensities. When you see such a beam, you perceive white rather than the individual colors. In Fig. 33-19a, a beam of white light in air is incident on a glass surface. (Because the pages of this book are white, a beam of white light is represented with a gray ray here.

FIG. 33-18 The index of refraction as a function of wavelength for fused quartz. The graph indicates that a beam of short-wavelength light, for which the index of refraction is higher, is bent more upon entering or leaving quartz than a beam of long-wavelength light.

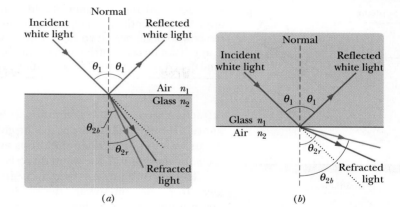

FIG. 33-19 Chromatic dispersion of white light. The blue component is bent more than the red component. (*a*) Passing from air to glass, the blue component ends up with the smaller angle of refraction. (*b*) Passing from glass to air, the blue component ends up with the greater angle of refraction. Each dotted line represents the direction in which the light would continue to travel if it were not bent by the refraction.

Also, a beam of monochromatic light is generally represented with a red ray.) Of the refracted light in Fig. 33-19*a*, only the red and blue components are shown. Because the blue component is bent more than the red component, the angle of refraction θ_{2b} for the blue component is *smaller* than the angle of refraction θ_{2r} for the red component. (Remember, angles are measured relative to the normal.) In Fig. 33-19*b*, a ray of white light in glass is incident on a glass–air interface. Again, the blue component is bent more than the red component, but now θ_{2b} is greater than θ_{2r}.

To increase the color separation, we can use a solid glass prism with a triangular cross section, as in Fig. 33-20*a*. The dispersion at the first surface (on the left in Figs. 33-20*a*, *b*) is then enhanced by the dispersion at the second surface.

Rainbows

The most charming example of chromatic dispersion is a rainbow. When sunlight (which consists of all visible colors) is intercepted by a falling raindrop, some of the light refracts into the drop, reflects once from the drop's inner surface, and then refracts out of the drop. Figure 33-21*a* shows the situation when the Sun is on the horizon at the left (and thus when the rays of sunlight are horizontal). The first refraction separates the sunlight into its component colors, and the second

(*a*)

(*b*)

FIG. 33-20 (*a*) A triangular prism separating white light into its component colors. (*b*) Chromatic dispersion occurs at the first surface and is increased at the second surface. (*Courtesy Bausch & Lomb*)

(a)

(b)

(c)

(d)

FIG. 33-21 (*a*) The separation of colors when sunlight refracts into and out of falling rain-drops leads to a primary rainbow. The antisolar point *A* is on the horizon at the right. The rainbow colors appear at an angle of 42° from the direction of *A*. (*b*) Drops at 42° from *A* in any direction can contribute to the rainbow. (*c*) The rainbow arc when the Sun is higher (and thus *A* is lower). (*d*) The separation of colors leading to a secondary rainbow.

refraction increases the separation. (Only the red and blue rays are shown in the figure.) If many falling drops are brightly illuminated, you can see the separated colors they produce when the drops are at an angle of 42° from the direction of the *antisolar point A*, the point directly opposite the Sun in your view.

To locate the drops, face away from the Sun and point both arms directly away from the Sun, toward the shadow of your head. Then move your right arm directly up, directly rightward, or in any intermediate direction until the angle between your arms is 42°. If illuminated drops happen to be in the direction of your right arm, you see color in that direction.

Because any drop at an angle of 42° in any direction from *A* can contribute to the rainbow, the rainbow is always a 42° circular arc around *A* (Fig. 33-21*b*) and the top of a rainbow is never more than 42° above the horizon. When the Sun is above the horizon, the direction of *A* is below the horizon, and only a shorter, lower rainbow arc is possible (Fig. 33-21*c*).

Because rainbows formed in this way involve one reflection of light inside each drop, they are often called *primary rainbows*. A *secondary rainbow* involves two reflections inside a drop, as shown in Fig. 33-21*d*. Colors appear in the secondary rainbow at an angle of 52° from the direction of *A*. A secondary rainbow is wider and dimmer than a primary rainbow and thus is more difficult to see. Also, the order of colors in a secondary rainbow is reversed from the order in a primary rainbow, as you can see by comparing parts *a* and *d* of Fig. 33-21.

Rainbows involving three or four reflections occur in the direction of the Sun and cannot be seen against the glare of sunshine in that part of the sky. Rainbows involving even more reflections inside the drops can occur in other parts of the sky but are always too dim to see.

✔**CHECKPOINT 5** Which of the three drawings here (if any) show physically possible refraction?

(a)   (b)   (c)

Sample Problem **33-3**

(a) In Fig. 33-22*a*, a beam of monochromatic light reflects and refracts at point *A* on the interface between material 1 with index of refraction $n_1 = 1.33$ and material 2 with index of refraction $n_2 = 1.77$. The incident beam makes an angle of 50° with the interface. What is the angle of reflection at point *A*? What is the angle of refraction there?

KEY IDEAS (1) The angle of reflection is equal to the angle of incidence, and both angles are measured relative to the normal to the surface at the point of reflection. (2) When light reaches the interface between two materials with different indexes of refraction (call them n_1 and n_2), part of the light can be refracted by the interface according to Snell's law, Eq. 33-40:

$$n_2 \sin \theta_2 = n_1 \sin \theta_1, \qquad (33\text{-}42)$$

where both angles are measured relative to the normal at the point of refraction.

Calculations: In Fig. 33-22a, the normal at point A is drawn as a dashed line through the point. Note that the angle of incidence θ_1 is not the given 50° but rather is $90° - 50° = 40°$. Thus, the angle of reflection is

$$\theta_1' = \theta_1 = 40°. \qquad \text{(Answer)}$$

The light that passes from material 1 into material 2 undergoes refraction at point A on the interface between the two materials. Again we measure angles between light rays and a normal, here at the point of refraction. Thus, in Fig. 33-22a, the angle of refraction is the angle marked θ_2. Solving Eq. 33-42 for θ_2 gives us

$$\theta_2 = \sin^{-1}\left(\frac{n_1}{n_2} \sin \theta_1\right) = \sin^{-1}\left(\frac{1.33}{1.77} \sin 40°\right)$$
$$= 28.88° \approx 29°. \qquad \text{(Answer)}$$

This result means that the beam swings toward the normal (it was at 40° to the normal and is now at 29°). The reason is that when the light travels across the interface, it moves into a material with a greater index of refraction.

(b) The light that enters material 2 at point A then reaches point B on the interface between material 2 and material 3, which is air, as shown in Fig. 33-22b. The interface through B is parallel to that through A. At B, some of the light reflects and the rest enters the air. What is the angle of reflection? What is the angle of refraction into the air?

Calculations: We first need to relate one of the angles at point B with a known angle at point A. Because the interface through point B is parallel to that through point A, the incident angle at B must be equal to the

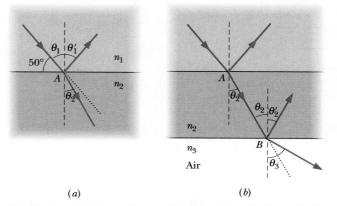

FIG. 33-22 (a) Light reflects and refracts at point A on the interface between materials 1 and 2. (b) The light that passes through material 2 reflects and refracts at point B on the interface between materials 2 and 3 (air). Each dashed line is a normal. Each dotted line gives the incident direction of travel.

angle of refraction θ_2, as shown in Fig. 33-22b. Then for reflection, we again use the law of reflection. Thus, the angle of reflection at B is

$$\theta_2' = \theta_2 = 28.88° \approx 29°. \qquad \text{(Answer)}$$

Next, the light that passes from material 2 into the air undergoes refraction at point B, with refraction angle θ_3. Thus, we again apply Snell's law of refraction, but this time we write Eq. 33-40 as

$$n_3 \sin \theta_3 = n_2 \sin \theta_2.$$

Solving for θ_3 then leads to

$$\theta_3 = \sin^{-1}\left(\frac{n_2}{n_3} \sin \theta_2\right) = \sin^{-1}\left(\frac{1.77}{1.00} \sin 28.88°\right)$$
$$= 58.75° \approx 59°. \qquad \text{(Answer)}$$

This result means that the beam swings away from the normal (it was at 29° to the normal and is now at 59°). The reason is that when the light travels across the interface, it moves into a material (air) with a lower index of refraction.

Sample Problem   33-4

Sun dogs occur when sunlight passes through atmospheric ice crystals (Fig. 33-23a) that have their hexagonal cross sections horizontal. Such a crystal can turn a ray (Fig. 33-23b) from its original direction of travel by an angle that depends on the crystal's orientation. However, there is a minimum turning angle θ_{dog} at which rays bunch up to form an especially bright spot in the sky—a sun dog. The situation is similar to how white wine in a wine glass can bunch up candlelight to form bright lines on a tablecloth. A sunray undergoing such minimum turning follows a path through the crystal that is parallel to one of the hexagonal sides, and its

entrance and exit paths are symmetric. The minimum turning angle θ_{dog} is the angle to the left or right of the Sun at which you can see a sun dog. The index of refraction of ice is 1.31. Find θ_{dog}.

KEY IDEA When light reaches the interface between two materials with different indexes of refraction (call them n_1 and n_2), part of the light can be refracted by the interface according to Snell's law, Eq. 33-40:

$$n_2 \sin \theta_2 = n_1 \sin \theta_1. \qquad (33\text{-}43)$$

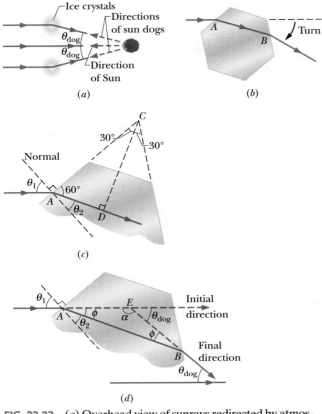

(a)

(b)

(c)

(d)

FIG. 33-23   (a) Overhead view of sunrays redirected by atmospheric ice crystals. (b) Overhead view of hexagonal cross section of a crystal. Path of ray (c) into and (d) out of the cross section.

Calculations: To get started, we mentally complete the corner of one side of the hexagon as in Fig. 33-23c and label the corner as having an angle of 60°. We are interested in the right triangle ADC because the light path forms one leg of the triangle. We can see that one of the acute angles in the triangle is 30°, so the other one must be 60°. In Fig. 33-23c, we then see that the angle of refraction θ_2 is the complement of 60°, so $\theta_2 = 30°$. We now use Snell's law to get the incident angle θ_1 for the ray that ends up in the sun dog. Solving Eq. 33-43 for θ_1, we obtain

$$\theta_1 = \sin^{-1}\left(\frac{n_2}{n_1}\sin\theta_2\right) = \sin^{-1}\left(\frac{1.31}{1.00}\sin 30°\right)$$
$$= 40.92°.$$

Next, note in Fig. 33-23d that θ_1 on the left side of point A is equal to the sum of θ_2 and ϕ on the right side:

$$\theta_1 = \theta_2 + \phi,$$

so

$$\phi = 40.92° - 30° = 10.92°.$$

Next, note that angle θ_{dog} at point E is an *exterior* angle of triangle ABE and thus (from Appendix E) we can write

$$\theta_{dog} = \phi + \phi = 2(10.92°)$$
$$= 21.8°.$$

This means that the sunray reaching you from the sun dog is turned by 21.8° from its initial direction of travel, and thus you see the sun dog at 21.8° from the direction of the Sun.

33-9 | Total Internal Reflection

Figure 33-24a shows rays of monochromatic light from a point source S in glass incident on the interface between the glass and air. For ray a, which is perpendicular to the interface, part of the light reflects at the interface and the rest travels through it with no change in direction.

For rays b through e, which have progressively larger angles of incidence at the interface, there are also both reflection and refraction at the interface. As the

(a)   (b)

FIG. 33-24   (a) Total internal reflection of light from a point source S in glass occurs for all angles of incidence greater than the critical angle θ_c. At the critical angle, the refracted ray points along the air–glass interface. (b) A source in a tank of water. *(Ken Kay/Fundamental Photographs)* angle of incidence increases, the angle of refraction increases; for ray e it is 90°, which means that the refracted ray points directly along the interface. The angle of incidence giving this situation is called the **critical angle** θ_c. For angles of incidence larger than θ_c, such as for rays f and g, there is no refracted ray and *all* the light is reflected; this effect is called **total internal reflection.**

To find θ_c, we use Eq. 33-40; we arbitrarily associate subscript 1 with the glass and subscript 2 with the air, and then we substitute θ_c for θ_1 and 90° for θ_2, finding

$$n_1 \sin \theta_c = n_2 \sin 90°,$$

which gives us

$$\theta_c = \sin^{-1} \frac{n_2}{n_1} \qquad \text{(critical angle)}. \qquad (33\text{-}44)$$

Because the sine of an angle cannot exceed unity, n_2 cannot exceed n_1 in this equation. This restriction tells us that total internal reflection cannot occur when the incident light is in the medium of lower index of refraction. If source S were in the air in Fig. 33-24a, all its rays that are incident on the air–glass interface (including f and g) would be both reflected *and* refracted at the interface.

Total internal reflection has found many applications in medical technology. For example, a physician can view the interior of an artery of a patient by running two thin bundles of *optical fibers* through the chest wall and into an artery (Fig. 33-25). Light introduced at the outer end of one bundle undergoes repeated total internal reflection within the fibers so that, even though the bundle provides a curved path, most of the light ends up exiting the other end and illuminating the interior of the artery. Some of the light reflected from the interior then comes back up the second bundle in a similar way, to be detected and converted to an image on a monitor's screen for the physician to view.

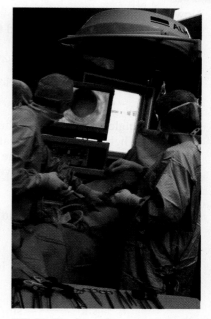

FIG. 33-25 An endoscope used to inspect an artery. *(©Laurent/Phototake)*

Sample Problem   33-5

The purpose of a diamond on a ring is, of course, to sparkle. Part of the art of cutting a diamond is to ensure that all the light entering through the top face or side facets leaves through those surfaces, to participate in the sparkle. Figure 33-26 shows part of a cross-sectional slice through a brilliant-cut diamond, with a ray entering at point A on the top face. In this type of cut, the top and bottom surfaces have normal lines that intersect at the indicated 48.84°. At point B, at least part of the light reflects and leaves the diamond properly, but part could refract and thus leak out of the diamond. Consider a light ray incident at angle $\theta_1 = 40°$ at A. Does light leak at B if air ($n_4 = 1.00$) lies next to the bottom surface? Does light leak if greasy grime ($n_4 = 1.63$) coats the surface? The index of refraction of diamond is $n_{\text{dia}} = 2.419$.

FIG. 33-26 Light entering the top of a brilliant-cut diamond. Can the light leak from the bottom surface at point B?

KEY IDEAS When light reaches the interface between two materials with different indexes of refraction (call them n_1 and n_2), part or all of the light is reflected. The reflection is total if (1) the incident light is in the material with the *higher* index of refraction ($n_1 > n_2$) and (2) the incident angle exceeds a critical value given by Eq. 33-44:

$$\theta_c = \sin^{-1} \frac{n_2}{n_1}. \qquad (33\text{-}45)$$

If those two conditions are not met, part of the light is refracted across the interface according to Snell's law, Eq. 33-40:

$$n_2 \sin \theta_2 = n_1 \sin \theta_1. \qquad (33\text{-}46)$$

Clean diamond: We need to follow the light from point A to point B, to see if it can leak at point B. The light incident at point A is in the material (air, $n_1 = 1.00$) with a *lower* index of refraction than the material on the other side of the interface (diamond, $n_{\text{dia}} = 2.419$). Thus, some of the light is refracted across the interface (the reflected portion is not shown in Fig. 33-26), and we find the angle of refraction θ_2 from Snell's law:

$$\theta_2 = \sin^{-1}\left(\frac{n_1}{n_2} \sin \theta_1\right) = \sin^{-1}\left(\frac{1.00}{2.419} \sin 40°\right)$$
$$= 15.41°.$$

Now note that the given angle of 48.84° at point C is an *exterior angle* of triangle ABC and thus (from Appendix E) we can write

$$\theta_2 + \theta_3 = 48.84°,$$

or

$$\theta_3 = 48.84° - \theta_2 = 48.84° - 15.41°$$
$$= 33.43°.$$

This is the angle of incidence of the ray at point B. Now the incident light is in the material (diamond) with the *greater* index of refraction than the material (air) on the other side of the interface. However, if we apply Snell's law at point B, we find

$$\theta_4 = \sin^{-1}\left(\frac{2.419}{1.00}\sin 33.43°\right) \qquad (33\text{-}47)$$

$$= \text{no answer.}$$

The reason is that the light is incident at an angle greater than the critical value of

$$\theta_c = \sin^{-1}\frac{n_4}{n_{\text{dia}}} = \sin^{-1}\frac{1.00}{2.419}$$

$$= 24.4°.$$

Thus, all the light reaching B reflects and none leaks out in the air.

Grimy diamond: We again follow the light from A to B, with the only difference lying in the last two calculations. Now at B, the material on the other side of the interface is grime with an index $n_4 = 1.63$. The incident light is again in the material with the greater index, but now the critical angle is

$$\theta_c = \sin^{-1}\frac{n_4}{n_{\text{dia}}} = \sin^{-1}\frac{1.63}{2.419}$$

$$= 42.4°.$$

So, the incident angle $\theta_3 = 33.43°$ is *less* than the critical angle and light leaks through the bottom of the diamond. From Eq. 33-47, the angle of refraction is

$$\theta_4 = \sin^{-1}\left(\frac{2.419}{1.63}\sin 33.43°\right)$$

$$= 54.8°.$$

Thus, to keep a diamond sparkling, clean both the top *and* bottom surfaces.

33-10 | Polarization by Reflection

You can vary the glare you see in sunlight that has been reflected from, say, water by looking through a polarizing sheet (such as a polarizing sunglass lens) and then rotating the sheet's polarizing axis around your line of sight. You can do so because any light that is reflected from a surface is either fully or partially polarized by the reflection.

Figure 33-27 shows a ray of unpolarized light incident on a glass surface. Let us resolve the electric field vectors of the light into two components. The *perpendicular components* are perpendicular to the plane of incidence and thus also to the page in Fig. 33-27; these components are represented with dots (as if we see the tips of the vectors). The *parallel components* are parallel to the plane of incidence and the page; they are represented with double-headed arrows. Because the light is unpolarized, these two components are of equal magnitude.

In general, the reflected light also has both components but with unequal magnitudes. This means that the reflected light is partially polarized — the electric fields oscillating along one direction have greater amplitudes than those oscillating along other directions. However, when the light is incident at a particular incident angle, called the *Brewster angle* θ_B, the reflected light has only perpendicular components, as shown in Fig. 33-27. The reflected light is then fully polarized perpendicular to the plane of incidence. The parallel components of the incident light do not disappear but (along with perpendicular components) refract into the glass.

Glass, water, and the other dielectric materials discussed in Section 25-7 can partially and fully polarize light by reflection. When you intercept sunlight reflected from such a surface, you see a bright spot (the glare) on the surface where the reflection takes place. If the surface is horizontal as in Fig. 33-27, the reflected light is partially or fully polarized horizontally. To eliminate such glare from horizontal surfaces, the lenses in polarizing sunglasses are mounted with their polarizing direction vertical.

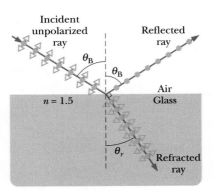

● Component perpendicular to page

⟺ Component parallel to page

FIG. 33-27 A ray of unpolarized light in air is incident on a glass surface at the Brewster angle θ_B. The electric fields along that ray have been resolved into components perpendicular to the page (the plane of incidence, reflection, and refraction) and components parallel to the page. The reflected light consists only of components perpendicular to the page and is thus polarized in that direction. The refracted light consists of the original components parallel to the page and weaker components perpendicular to the page; this light is partially polarized.

Brewster's Law

For light incident at the Brewster angle θ_B, we find experimentally that the reflected and refracted rays are perpendicular to each other. Because the reflected ray is reflected at the angle θ_B in Fig. 33-27 and the refracted ray is at an angle θ_r, we have

$$\theta_B + \theta_r = 90°. \tag{33-48}$$

These two angles can also be related with Eq. 33-40. Arbitrarily assigning subscript 1 in Eq. 33-40 to the material through which the incident and reflected rays travel, we have, from that equation,

$$n_1 \sin \theta_B = n_2 \sin \theta_r.$$

Combining these equations leads to

$$n_1 \sin \theta_B = n_2 \sin(90° - \theta_B) = n_2 \cos \theta_B,$$

which gives us

$$\theta_B = \tan^{-1} \frac{n_2}{n_1} \qquad \text{(Brewster angle).} \tag{33-49}$$

(Note carefully that the subscripts in Eq. 33-49 are *not* arbitrary because of our decision as to their meanings.) If the incident and reflected rays travel *in air*, we can approximate n_1 as unity and let n represent n_2 in order to write Eq. 33-49 as

$$\theta_B = \tan^{-1} n \qquad \text{(Brewster's law).} \tag{33-50}$$

This simplified version of Eq. 33-49 is known as **Brewster's law**. Like θ_B, it is named after Sir David Brewster, who found both experimentally in 1812.

REVIEW & SUMMARY

Electromagnetic Waves An electromagnetic wave consists of oscillating electric and magnetic fields. The various possible frequencies of electromagnetic waves form a *spectrum*, a small part of which is visible light. An electromagnetic wave traveling along an x axis has an electric field $\vec{E}$ and a magnetic field $\vec{B}$ with magnitudes that depend on x and t:

$$E = E_m \sin(kx - \omega t)$$

and

$$B = B_m \sin(kx - \omega t), \tag{33-1, 33-2}$$

where E_m and B_m are the amplitudes of $\vec{E}$ and $\vec{B}$. The electric field induces the magnetic field and vice versa. The speed of any electromagnetic wave in vacuum is c, which can be written as

$$c = \frac{E}{B} = \frac{1}{\sqrt{\mu_0 \varepsilon_0}}, \tag{33-5, 33-3}$$

where E and B are the simultaneous magnitudes of the fields.

Energy Flow The rate per unit area at which energy is transported via an electromagnetic wave is given by the Poynting vector $\vec{S}$:

$$\vec{S} = \frac{1}{\mu_0} \vec{E} \times \vec{B}. \tag{33-19}$$

The direction of $\vec{S}$ (and thus of the wave's travel and the energy transport) is perpendicular to the directions of both $\vec{E}$ and $\vec{B}$. The time-averaged rate per unit area at which energy is transported is S_{avg}, which is called the *intensity* I of the wave:

$$I = \frac{1}{c\mu_0} E_{rms}^2, \tag{33-26}$$

in which $E_{rms} = E_m/\sqrt{2}$. A *point source* of electromagnetic waves emits the waves *isotropically* — that is, with equal intensity in all directions. The intensity of the waves at distance r from a point source of power P_s is

$$I = \frac{P_s}{4\pi r^2}. \tag{33-27}$$

Radiation Pressure When a surface intercepts electromagnetic radiation, a force and a pressure are exerted on the surface. If the radiation is totally absorbed by the surface, the force is

$$F = \frac{IA}{c} \qquad \text{(total absorption),} \tag{33-32}$$

in which I is the intensity of the radiation and A is the area of the surface perpendicular to the path of the radiation. If the radiation is totally reflected back along its original path, the force is

$$F = \frac{2IA}{c} \qquad \text{(total reflection back along path).} \tag{33-33}$$

The radiation pressure p_r is the force per unit area:

$$p_r = \frac{I}{c} \qquad \text{(total absorption)} \tag{33-34}$$

and $\quad p_r = \dfrac{2I}{c} \quad$ (total reflection back along path). (33-35)

Polarization Electromagnetic waves are **polarized** if their electric field vectors are all in a single plane, called the *plane of oscillation*. Light waves from common sources are not polarized; that is, they are **unpolarized**, or **polarized randomly**.

Polarizing Sheets When a polarizing sheet is placed in the path of light, only electric field components of the light parallel to the sheet's **polarizing direction** are *transmitted* by the sheet; components perpendicular to the polarizing direction are absorbed. The light that emerges from a polarizing sheet is polarized parallel to the polarizing direction of the sheet.

If the original light is initially unpolarized, the transmitted intensity I is half the original intensity I_0:

$$I = \tfrac{1}{2}I_0. \tag{33-36}$$

If the original light is initially polarized, the transmitted intensity depends on the angle θ between the polarization direction of the original light and the polarizing direction of the sheet:

$$I = I_0 \cos^2 \theta. \tag{33-38}$$

Geometrical Optics *Geometrical optics* is an approximate treatment of light in which light waves are represented as straight-line rays.

Reflection and Refraction When a light ray encounters a boundary between two transparent media, a **reflected** ray and a **refracted** ray generally appear. Both rays remain in the plane of incidence. The **angle of reflection** is equal to the angle of incidence, and the **angle of refraction** is related to the angle of incidence by Snell's law,

$$n_2 \sin \theta_2 = n_1 \sin \theta_1 \quad \text{(refraction)}, \tag{33-40}$$

where n_1 and n_2 are the indexes of refraction of the media in which the incident and refracted rays travel.

Total Internal Reflection A wave encountering a boundary across which the index of refraction decreases will experience **total internal reflection** if the angle of incidence exceeds a **critical angle** θ_c, where

$$\theta_c = \sin^{-1} \frac{n_2}{n_1} \quad \text{(critical angle)}. \tag{33-44}$$

Polarization by Reflection A reflected wave will be fully **polarized**, with its $\vec{E}$ vectors perpendicular to the plane of incidence, if it strikes a boundary at the **Brewster angle** θ_B, where

$$\theta_B = \tan^{-1} \frac{n_2}{n_1} \quad \text{(Brewster angle)}. \tag{33-49}$$

QUESTIONS

1 If the magnetic field of a light wave oscillates parallel to a y axis and is given by $B_y = B_m \sin(kz - \omega t)$, (a) in what direction does the wave travel and (b) parallel to which axis does the associated electric field oscillate?

2 Figure 33-28 shows the electric and magnetic fields of an electromagnetic wave at a certain instant. Is the wave traveling into the page or out of it?

FIG. 33-28 Question 2.

3 In Fig. 33-15a, start with light that is initially polarized parallel to the x axis, and write the ratio of its final intensity I_3 to its initial intensity I_0 as $I_3/I_0 = A \cos^n \theta$. What are A, n, and θ if we rotate the polarizing direction of the first sheet (a) $60°$ counterclockwise and (b) $90°$ clockwise from what is shown?

4 Suppose we rotate the second sheet in Fig. 33-15a, starting with the polarization direction aligned with the y axis ($\theta = 0$) and ending with it aligned with the x axis ($\theta = 90°$). Which of the four curves in Fig. 33-29 best shows the intensity of the light through the three-sheet system during this $90°$ rotation?

FIG. 33-29 Question 4.

5 (a) Figure 33-30 shows light reaching a polarizing sheet whose polarizing direction is parallel to a y axis. We shall rotate the sheet $40°$ clockwise about the light's indicated line of travel. During this rotation, does the fraction of the initial light intensity passed by the sheet in-

FIG. 33-30
Question 5.

crease, decrease, or remain the same if the light is (a) initially unpolarized, (b) initially polarized parallel to the x axis, and (c) initially polarized parallel to the y axis?

6 In Fig. 33-31, unpolarized light is sent into a system of five polarizing sheets. Their polarizing directions, measured counterclockwise from the positive direction of the y axis, are the following: sheet 1, $35°$; sheet 2, $0°$; sheet 3, $0°$; sheet 4, $110°$; sheet 5, $45°$. Sheet 3 is then rotated $180°$ counterclockwise about the light ray. During that rotation, at what angles (measured counterclockwise from the y axis) is the transmission of light through the system eliminated?

FIG. 33-31
Question 6.

7 Figure 33-32 shows rays of monochromatic light passing through three materials a, b, and c. Rank the materials according to index of refraction, greatest first.

FIG. 33-32 Question 7.

8 Figure 33-33 shows the multiple reflections of a light ray along a glass corridor where the

FIG. 33-33 Question 8.

walls are either parallel or perpendicular to one another. If the angle of incidence at point *a* is 30°, what are the angles of reflection of the light ray at points *b*, *c*, *d*, *e*, and *f*?

9   Each part of Fig. 33-34 shows light that refracts through an interface between two materials. The incident ray (shown gray in the figure) consists of red and blue light. The approximate index of refraction for visible light is indicated for each material. Which of the three parts show physically possible refraction?

FIG. 33-34   Question 9.

10   In Fig. 33-35, light travels from material *a*, through three layers of other materials with surfaces parallel to one another, and then back into another layer of material *a*. The refractions (but not the associated reflections) at the surfaces are shown. Rank the materials according to index of refraction, greatest first.

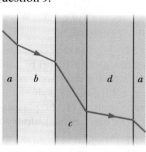

FIG. 33-35   Question 10.

11   Figure 33-36 shows four long horizontal layers *A–D* of different materials, with air above and below them. The index of refraction of each material is given. Rays of light are sent into the left end of each layer as shown. In which layer is there the possibility of totally trapping the light in that layer so that, after many reflections, all the light reaches the right end of the layer?

FIG. 33-36   Question 11.

12   The leftmost block in Fig. 33-37 depicts total internal reflection for light inside a material with an index of refraction n_1 when air is outside the material. A light ray reaching point *A* from anywhere within the shaded region at the left (such as the ray shown) fully reflects at that point and ends up in the shaded region at the right. The other blocks show similar situations for two other materials. Rank the indexes of refraction of the three materials, greatest first.

FIG. 33-37   Question 12.

PROBLEMS

sec. 33-2   Maxwell's Rainbow

•**1**   From Fig. 33-2, approximate the (a) smaller and (b) larger wavelength at which the eye of a standard observer has half the eye's maximum sensitivity. What are the (c) wavelength, (d) frequency, and (e) period of the light at which the eye is the most sensitive?

•**2**   About how far apart must you hold your hands for them to be separated by 1.0 nano-light-second (the distance light travels in 1.0 ns)?

•**3**   A certain helium–neon laser emits red light in a narrow band of wavelengths centered at 632.8 nm and with a "wavelength width" (such as on the scale of Fig. 33-1) of 0.0100 nm. What is the corresponding "frequency width" for the emission?

•**4**   Project Seafarer was an ambitious program to construct an enormous antenna, buried underground on a site about 10 000 km² in area. Its purpose was to transmit signals to submarines while they were deeply submerged. If the effective wavelength were 1.0×10^4 Earth radii, what would be the (a) frequency and (b) period of the radiations emitted? Ordinarily, electromagnetic radiations do not penetrate very far into conductors such as seawater.

sec. 33-3   The Traveling Electromagnetic Wave, Qualitatively

•**5**   What inductance must be connected to a 17 pF capacitor in an oscillator capable of generating 550 nm (i.e., visible) electromagnetic waves? Comment on your answer.   **SSM**

•**6**   What is the wavelength of the electromagnetic wave emitted by the oscillator–antenna system of Fig. 33-3 if $L = 0.253$ μH and $C = 25.0$ pF?

sec. 33-5   Energy Transport and the Poynting Vector

•**7**   Some neodymium–glass lasers can provide 100 TW of power in 1.0 ns pulses at a wavelength of 0.26 μm. How much energy is contained in a single pulse?   **ILW**

•**8**   A plane electromagnetic wave has a maximum electric field magnitude of 3.20×10^{-4} V/m. Find the magnetic field amplitude.

•**9**   A plane electromagnetic wave traveling in the positive direction of an *x* axis in vacuum has components $E_x = E_y = 0$ and $E_z = (2.0 \text{ V/m}) \cos[(\pi \times 10^{15} \text{ s}^{-1})(t - x/c)]$. (a) What is the amplitude of the magnetic field component? (b) Parallel to which axis does the magnetic field oscillate? (c) When the electric field component is in the positive direction of the *z* axis at a certain point P, what is the direction of the magnetic field component there? **ILW**

•**10** In a plane radio wave the maximum value of the electric field component is 5.00 V/m. Calculate (a) the maximum value of the magnetic field component and (b) the wave intensity.

•**11** What is the intensity of a traveling plane electromagnetic wave if B_m is 1.0×10^{-4} T?

•**12** Assume (unrealistically) that a TV station acts as a point source broadcasting isotropically at 1.0 MW. What is the intensity of the transmitted signal reaching Proxima Centauri, the star nearest our solar system, 4.3 ly away? (An alien civilization at that distance might be able to watch X Files.) A light-year (ly) is the distance light travels in one year.

••**13** The maximum electric field 10 m from an isotropic point source of light is 2.0 V/m. What are (a) the maximum value of the magnetic field and (b) the average intensity of the light there? (c) What is the power of the source?

••**14** Frank D. Drake, an investigator in the SETI (Search for Extra-Terrestrial Intelligence) program, once said that the large radio telescope in Arecibo, Puerto Rico (Fig. 33-38), "can detect a signal which lays down on the entire surface of the earth a power of only one picowatt." (a) What is the power that would be received by the Arecibo antenna for such a signal? The antenna diameter is 300 m. (b) What would be the power of an isotropic source at the center of our galaxy that could provide such a signal? The galactic center is 2.2×10^4 ly away. A light-year is the distance light travels in one year.

FIG. 33-38 Problem 14. Radio telescope at Arecibo. *(Courtesy Cornell University)*

••**15** Sunlight just outside Earth's atmosphere has an intensity of 1.40 kW/m². Calculate (a) E_m and (b) B_m for sunlight there, assuming it to be a plane wave.

••**16** An electromagnetic wave with frequency 4.00×10^{14} Hz travels through vacuum in the positive direction of an x axis. The wave has its electric field directed parallel to the y axis, with amplitude E_m. At time $t = 0$, the electric field at point P on the x axis has a value of $+E_m/4$ and is decreasing with time. What is the distance along the x axis from point P to the first point with $E = 0$ if we search in (a) the negative direction and (b) the positive direction of the x axis?

••**17** An airplane flying at a distance of 10 km from a radio transmitter receives a signal of intensity 10 μW/m². What is the amplitude of the (a) electric and (b) magnetic component of the signal at the airplane? (c) If the transmitter radiates uniformly over a hemisphere, what is the transmission power?

••**18** An isotropic point source emits light at wavelength 500 nm, at the rate of 200 W. A light detector is positioned 400 m from the source. What is the maximum rate $\partial B/\partial t$ at which the magnetic component of the light changes with time at the detector's location? **GO**

sec. 33-6 Radiation Pressure

•**19** What is the radiation pressure 1.5 m away from a 500 W lightbulb? Assume that the surface on which the pressure is exerted faces the bulb and is perfectly absorbing and that the bulb radiates uniformly in all directions. **ILW**

•**20** A black, totally absorbing piece of cardboard of area $A = 2.0$ cm² intercepts light with an intensity of 10 W/m² from a camera strobe light. What radiation pressure is produced on the cardboard by the light?

•**21** High-power lasers are used to compress a plasma (a gas of charged particles) by radiation pressure. A laser generating radiation pulses with peak power 1.5×10^3 MW is focused onto 1.0 mm² of high-electron-density plasma. Find the pressure exerted on the plasma if the plasma reflects all the light beams directly back along their paths. **SSM**

•**22** Radiation from the Sun reaching Earth (just outside the atmosphere) has an intensity of 1.4 kW/m². (a) Assuming that Earth (and its atmosphere) behaves like a flat disk perpendicular to the Sun's rays and that all the incident energy is absorbed, calculate the force on Earth due to radiation pressure. (b) For comparison, calculate the force due to the Sun's gravitational attraction.

••**23** Prove, for a plane electromagnetic wave that is normally incident on a flat surface, that the radiation pressure on the surface is equal to the energy density in the incident beam. (This relation between pressure and energy density holds no matter what fraction of the incident energy is reflected.) **SSM**

••**24** A small laser emits light at power 5.00 mW and wavelength 633 nm. The laser beam is focused (narrowed) until its diameter matches the 1266 nm diameter of a sphere placed in its path. The sphere is perfectly absorbing and has density 5.00×10^3 kg/m³. What are (a) the beam intensity at the sphere's location, (b) the radiation pressure on the sphere, (c) the magnitude of the corresponding force, and (d) the magnitude of the acceleration that force alone would give the sphere?

••**25** A plane electromagnetic wave, with wavelength 3.0 m, travels in vacuum in the positive direction of an x axis. The electric field, of amplitude 300 V/m, oscillates parallel to the y axis. What are the (a) frequency, (b) angular frequency, and (c) angular wave number of the wave? (d) What is the amplitude of the magnetic field component? (e) Parallel to which axis does the magnetic field oscillate? (f) What is the time-averaged rate of energy flow in watts per square meter associated with this wave? The wave uniformly illuminates a surface of area 2.0 m². If the surface totally absorbs the wave, what are (g) the rate at which momentum is transferred to the surface and (h) the radiation pressure on the surface? **SSM WWW**

••**26** It has been proposed that a spaceship might be propelled in the solar system by radiation pressure, using a large sail made of foil. How large must the surface area of the sail be if the radiation force is to be equal in magnitude to the Sun's gravitational attraction? Assume that the mass of the ship + sail is 1500 kg, that the sail is perfectly reflecting, and that the sail is oriented perpendicular to the Sun's rays. See Appendix C for needed data. (With a larger sail, the ship is continuously driven away from the Sun.) **GO**

••**27** A small spaceship whose mass is 1.5×10^3 kg (including an astronaut) is drifting in outer space with negligible gravitational forces acting on it. If the astronaut turns on a 10 kW laser beam, what speed will the ship attain in 1.0 day because of the momentum carried away by the beam? **SSM**

••**28** In Fig. 33-39, a laser beam of power 4.60 W and diameter $D = 2.60$ mm is directed upward at one circular face (of diameter $d < 2.60$ mm) of a perfectly reflecting cylinder. The cylinder is levitated because the upward radiation force matches the downward gravitational force. If the cylinder's density is 1.20 g/cm³, what is its height H?

FIG. 33-39
Problem 28.

••**29** Someone plans to float a small, totally absorbing sphere 0.500 m above an isotropic point source of light, so that the upward radiation force from the light matches the downward gravitational force on the sphere. The sphere's density is 19.0 g/cm³, and its radius is 2.00 mm. (a) What power would be required of the light source? (b) Even if such a source were made, why would the support of the sphere be unstable?

••**30** The intensity I of light from an isotropic point source is determined as a function of distance r from the source. Figure 33-40 gives intensity I versus the inverse square r^{-2} of that distance. The vertical axis scale is set by $I_s = 200$ W/m², and the horizontal axis scale is set by $r_s^{-2} = 8.0$ m⁻². What is the power of the source?

FIG. 33-40 Problem 30.

•••**31** As a comet swings around the Sun, ice on the comet's surface vaporizes, releasing trapped dust particles and ions. The ions, because they are

FIG. 33-41 Problem 31.

electrically charged, are forced by the electrically charged *solar wind* into a straight *ion tail* that points radially away from the Sun (Fig. 33-41). The (electrically neutral) dust particles are pushed radially outward from the Sun by the radiation force on them from sunlight. Assume that the dust particles are spherical, have density 3.5×10^3 kg/m³, and are totally absorbing. (a) What radius must a particle have in order to follow a straight path, like path 2 in the figure? (b) If its radius is larger, does its path curve away from the Sun (like path 1) or toward the Sun (like path 3)?

sec. 33-7 Polarization

•**32** In Fig. 33-42, a beam of unpolarized light, with intensity 43 W/m², is sent into a system of two polarizing sheets with

polarizing directions at angles $\theta_1 = 70°$ and $\theta_2 = 90°$ to the y axis. What is the intensity of the light transmitted by the system? **GO**

•**33** In Fig. 33-42, a beam of light, with intensity 43 W/m² and polarization parallel to a y axis, is sent into a system of two polarizing sheets with polarizing directions at angles of $\theta_1 = 70°$ and $\theta_2 = 90°$ to the y axis. What is the intensity of the light transmitted by the two-sheet system? **ILW**

FIG. 33-42
Problems 32, 33, and 36.

•**34** In Fig. 33-43, initially unpolarized light is sent into a system of three polarizing sheets whose polarizing directions make angles of $\theta_1 = \theta_2 = \theta_3 = 50°$ with the direction of the y axis. What percentage of the initial intensity is transmitted by the system? (*Hint:* Be careful with the angles.)

•**35** In Fig. 33-43, initially unpolarized light is sent into a system of three polarizing sheets whose polarizing directions make angles of $\theta_1 = 40°$, $\theta_2 = 20°$, and $\theta_3 = 40°$ with the direction of the y axis. What percentage of the light's initial intensity is transmitted by the system? (*Hint:* Be careful with the angles.) **SSM**

FIG. 33-43 Problems 34 and 35.

••**36** In Fig. 33-42, unpolarized light is sent into a system of two polarizing sheets. The angles θ_1 and θ_2 of the polarizing directions of the sheets are measured counterclockwise from the positive direction of the y axis (they are not drawn to scale in the figure). Angle θ_1 is fixed but angle θ_2 can be varied. Figure 33-44 gives the intensity of the light emerging from sheet 2 as a function of θ_2. (The scale of the intensity axis is not indicated.) What percentage of the light's initial intensity is transmitted by the two-sheet system when $\theta_2 = 90°$?

FIG. 33-44 Problem 36.

••**37** Unpolarized light of intensity 10 mW/m² is sent into a polarizing sheet as in Fig. 33-11. What are (a) the amplitude of the electric field component of the transmitted light and (b) the radiation pressure on the sheet due to its absorbing some of the light?

••**38** At a beach the light is generally partially polarized due to reflections off sand and water. At a particular beach on a particular day near sundown, the horizontal component of the electric field vector is 2.3 times the vertical component. A standing sunbather puts on polarizing sunglasses; the glasses eliminate the horizontal field component. (a) What fraction of the light intensity received before the glasses were put on now reaches the sunbather's eyes? (b) The sunbather, still wearing the glasses, lies on his side. What fraction of the light intensity received before the glasses were put on now reaches his eyes?

••**39** A beam of polarized light is sent into a system of

two polarizing sheets. Relative to the polarization direction of that incident light, the polarizing directions of the sheets are at angles θ for the first sheet and 90° for the second sheet. If 0.10 of the incident intensity is transmitted by the two sheets, what is θ?

••40 In Fig. 33-45, unpolarized light is sent into a system of three polarizing sheets. The angles θ_1, θ_2, and θ_3 of the polarizing directions are measured counterclockwise from the positive direction of the y axis (they are not drawn to scale). Angles θ_1 and θ_3 are fixed, but angle θ_2 can be varied. Figure 33-46 gives the intensity of the light emerging from sheet 3 as a function of θ_2. (The scale of the intensity axis is not indicated.) What percentage of the light's initial intensity is transmitted by the system when $\theta_2 = 30°$?

FIG. 33-45 Problems 40, 42, and 44.

••41 A beam of partially polarized light can be considered to be a mixture of polarized and unpolarized light. Suppose we send such a beam through a polarizing filter and then rotate the filter through 360° while keeping it perpendicular to the beam. If the transmitted intensity varies by a factor of 5.0 during the rotation, what fraction of the intensity of the original beam is associated with the beam's polarized light?

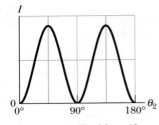

FIG. 33-46 Problem 40.

••42 In Fig. 33-45, unpolarized light is sent into a system of three polarizing sheets, which transmits 0.0500 of the initial light intensity. The polarizing directions of the first and third sheets are at angles $\theta_1 = 0°$ and $\theta_3 = 90°$. What are the (a) smaller and (b) larger possible values of angle θ_2 ($< 90°$) for the polarizing direction of sheet 2?

••43 We want to rotate the direction of polarization of a beam of polarized light through 90° by sending the beam through one or more polarizing sheets. (a) What is the minimum number of sheets required? (b) What is the minimum number of sheets required if the transmitted intensity is to be more than 60% of the original intensity? SSM WWW

••44 In Fig. 33-45, unpolarized light is sent into a system of three polarizing sheets. The angles θ_1, θ_2, and θ_3 of the polarizing directions are measured counterclockwise from the positive direction of the y axis (they are not drawn to scale). Angles θ_1 and θ_3 are fixed, but angle θ_2 can be varied. Figure 33-47 gives the intensity of the light emerging from sheet 3 as a function of θ_2. (The scale of the intensity axis is not indicated.) What percentage of the light's initial intensity is transmitted by the three-sheet system when $\theta_2 = 90°$?

FIG. 33-47 Problem 44.

sec. 33-8 Reflection and Refraction

•45 Light in vacuum is incident on the surface of a glass slab. In the vacuum the beam makes an angle of 32.0° with the normal to the surface, while in the glass it makes an angle of 21.0° with the normal. What is the index of refraction of the glass?

•46 In Fig. 33-48a, a light ray in water is incident at angle θ_1 on a boundary with an underlying material, into which some of the light refracts. There are two choices of underlying material. For each, the angle of refraction θ_2 versus the incident angle θ_1 is given in Fig. 33-48b. The vertical axis scale is set by $\theta_{2s} = 90°$. Without calculation, determine whether the index of refraction of (a) material 1 and (b) material 2 is greater or less than the index of water ($n = 1.33$). What is the index of refraction of (c) material 1 and (d) material 2?

(a)   (b)

FIG. 33-48 Problem 46.

•47 Figure 33-49 shows light reflecting from two perpendicular reflecting surfaces A and B. Find the angle between the incoming ray i and the outgoing ray r'.

FIG. 33-49 Problem 47.

•48 In Fig. 33-50a, a light ray in an underlying material is incident at angle θ_1 on a boundary with water, and some of the light refracts into the water. There are two choices of underlying material. For each, the angle of refraction θ_2 versus the incident angle θ_1 is given in Fig. 33-50b. The horizontal axis scale is set by $\theta_{1s} = 90°$. Without calculation, determine whether the index of refraction of (a) material 1 and (b) material 2 is greater or less than the index of water ($n = 1.33$). What is the index of refraction of (c) material 1 and (d) material 2?

(a)   (b)

FIG. 33-50 Problem 48.

•49 When the rectangular metal tank in Fig. 33-51 is filled to the top with an unknown liquid, observer O, with eyes level with the top of the tank, can just see corner E. A ray that refracts toward O at the top surface of the liquid is shown.

If $D = 85.0$ cm and $L = 1.10$ m, what is the index of refraction of the liquid?

••50 In Fig. 33-52a, a beam of light in material 1 is incident on a boundary at an angle of $\theta_1 = 30°$. The extent of refraction of the light into material 2 depends, in part, on the index of refraction n_2 of material 2. Figure 33-52b gives the angle of refraction θ_2 versus n_2 for a range of possible n_2 values. The vertical axis scale is set by $\theta_{2a} = 20.0°$ and $\theta_{2b} = 40.0°$. (a) What is the index of refraction of material 1? (b) If the incident angle is changed to 60° and material 2 has $n_2 = 2.4$, then what is angle θ_2?

FIG. 33-51 Problem 49.

(a)　　　(b)

FIG. 33-52 Problem 50.

••51 In Fig. 33-53, a 2.00-m-long vertical pole extends from the bottom of a swimming pool to a point 50.0 cm above the water. Sunlight is incident at angle $\theta = 55.0°$. What is the length of the shadow of the pole on the level bottom of the pool? **SSM**

FIG. 33-53 Problem 51.

••52 *Dispersion in a window pane.* In Fig. 33-54, a beam of white light is incident at angle $\theta = 50°$ on a common window pane (shown in cross section). For the pane's type of glass, the index of refraction for visible light ranges from 1.524 at the blue end of the spectrum to 1.509 at the red end. The two sides of the pane are parallel. What is the angular spread of the colors in the beam (a) when the light enters the pane and (b) when it emerges from the opposite side? (*Hint:* When you look at an object through a window pane, are the colors in the light from the object dispersed as shown in, say, Fig. 33-20?)

FIG. 33-54 Problem 52.

••53 In Fig. 33-55, light is incident at angle $\theta_1 = 40.1°$ on a boundary between two transparent materials. Some of the light travels down through the next three layers of transparent materials, while some of it reflects upward and then escapes

FIG. 33-55 Problem 53.

into the air. If $n_1 = 1.30$, $n_2 = 1.40$, $n_3 = 1.32$, and $n_4 = 1.45$, what is the value of (a) θ_5 and (b) θ_4? **GO**

••54 In Fig. 33-56a, a beam of light in material 1 is incident on a boundary at an angle $\theta_1 = 40°$. Some of the light travels through material 2, and then some of it emerges into material 3. The two boundaries between the three materials are parallel. The final direction of the beam depends, in part, on the index of refraction n_3 of the third material. Figure 33-56b gives the angle of refraction θ_3 in that material versus n_3 for a range of possible n_3 values. The vertical axis scale is set by $\theta_{3a} = 30.0°$ and $\theta_{3b} = 50.0°$. (a) What is the index of refraction of material 1, or is the index impossible to calculate without more information? (b) What is the index of refraction of material 2, or is the index impossible to calculate without more information? (c) If θ_1 is changed to 70° and the index of refraction of material 3 is 2.4, what is θ_3?

(a)　　　(b)

FIG. 33-56 Problem 54.

••55 In Fig. 33-57, a ray is incident on one face of a triangular glass prism in air. The angle of incidence θ is chosen so that the emerging ray also makes the same angle θ with the normal to the other face. Show that the index of refraction n of the glass prism is given by

$$n = \frac{\sin\frac{1}{2}(\psi + \phi)}{\sin\frac{1}{2}\phi},$$

where ϕ is the vertex angle of the prism and ψ is the *deviation angle,* the total angle through which the beam is turned in passing through the prism. (Under these conditions the deviation angle ψ has the smallest possible value, which is called the *angle of minimum deviation.*) **SSM ILW WWW**

FIG. 33-57
Problems 55 and 66.

••56 *Rainbows from square drops.* Suppose that, on some surreal world, raindrops had a square cross section and always fell with one face horizontal. Figure 33-58 shows such a falling drop, with a white beam of sunlight incident at $\theta = 70.0°$ at point P. The part of the light that enters the drop then travels to point A, where some of it refracts out into the air and the rest reflects. That reflected light then travels to point B, where again some of the light refracts out into the air and the rest reflects. What is the difference in the angles of the red light ($n = 1.331$) and the blue light ($n = 1.343$) that emerge at (a) point A and

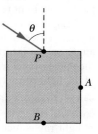

FIG. 33-58
Problem 56.

(b) point B? (This angular difference in the light emerging at, say, point A would be the angular width of the rainbow you would see were you to intercept the light emerging there.)

sec. 33-9 Total Internal Reflection

•57 A point source of light is 80.0 cm below the surface of a body of water. Find the diameter of the circle at the surface through which light emerges from the water.

•58 The index of refraction of benzene is 1.8. What is the critical angle for a light ray traveling in benzene toward a flat layer of air above the benzene?

••59 In Fig. 33-59, light initially in material 1 refracts into material 2, crosses that material, and is then incident at the critical angle on the interface between materials 2 and 3. The indexes of refraction are $n_1 =$ 1.60, $n_2 = 1.40$, and $n_3 = 1.20$. (a) What is angle θ? (b) If θ is increased, is there refraction of light into material 3? **GO**

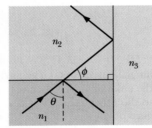

FIG. 33-59 Problem 59.

••60 A catfish is 2.00 m below the surface of a smooth lake. (a) What is the diameter of the circle on the surface through which the fish can see the world outside the water? (b) If the fish descends, does the diameter of the circle increase, decrease, or remain the same?

••61 In the ray diagram of Fig. 33-60, where the angles are not drawn to scale, the ray is incident at the critical angle on the interface between materials 2 and 3. Angle $\phi = 60.0°$, and two of the indexes of refraction are $n_1 = 1.70$ and $n_2 = 1.60$. Find (a) index of refraction n_3 and (b) angle θ. (c) If θ is decreased, does light refract into material 3?

FIG. 33-60 Problem 61.

••62 In Fig. 33-61, light from ray A refracts from material 1 ($n_1 = 1.60$) into a thin layer of material 2 ($n_2 = 1.80$), crosses that layer, and is then incident at the critical angle on the interface between materials 2 and 3 ($n_3 = 1.30$). (a) What is the value of incident angle θ_A? (b) If θ_A is decreased, does part of the light refract into material 3?

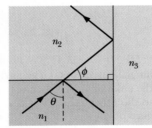

FIG. 33-61 Problem 62.

Light from ray B refracts from material 1 into the thin layer, crosses that layer, and is then incident at the critical angle on the interface between materials 2 and 3. (c) What is the value of incident angle θ_B? (d) If θ_B is decreased, does part of the light refract into material 3?

••63 Figure 33-62 depicts a simplistic optical fiber: a plastic core ($n_1 = 1.58$) is surrounded by a plastic sheath ($n_2 = 1.53$). A light ray is incident on one end of the fiber at angle θ. The ray is to undergo total internal reflection at point A, where it

encounters the core–sheath boundary. (Thus there is no loss of light through that boundary.) What is the maximum value of θ that allows total internal reflection at A? **GO**

FIG. 33-62 Problem 63.

••64 In Fig. 33-63, a light ray in air is incident at angle θ_1 on a block of transparent plastic with an index of refraction of 1.56. The dimensions indicated are $H = 2.00$ cm and $W = 3.00$ cm. The light passes through the block to one of its sides and there undergoes reflection (inside the block) and possibly refraction (out into the air). This is the point of *first reflection*. The reflected light

FIG. 33-63 Problem 64.

then passes through the block to another of its sides—a point of *second reflection*. If $\theta_1 = 40°$, on which side is the point of (a) first reflection and (b) second reflection? If there is refraction at the point of (c) first reflection and (d) second reflection, give the angle of refraction; if not, answer "none." If $\theta_1 = 70°$, on which side is the point of (e) first reflection and (f) second reflection? If there is refraction at the point of (g) first reflection and (h) second reflection, give the angle of refraction; if not, answer "none."

••65 In Fig. 33-64, a ray of light is perpendicular to the face ab of a glass prism ($n = 1.52$). Find the largest value for the angle ϕ so that the ray is totally reflected at face ac if the prism is immersed (a) in air and (b) in water. **SSM ILW**

FIG. 33-64 Problem 65.

••66 Suppose the prism of Fig. 33-57 has apex angle $\phi = 60.0°$ and index of refraction $n = 1.60$. (a) What is the smallest angle of incidence θ for which a ray can enter the left face of the prism and exit the right face? (b) What angle of incidence θ is required for the ray to exit the prism with an identical angle θ for its refraction, as it does in Fig. 33-57?

••67 In Fig. 33-65, light enters a 90° triangular prism at point P with incident angle θ, and then some of it refracts at point Q with an angle of refraction of 90°. (a) What is the index of refraction of the prism in terms of θ? (b) What, numerically, is the maximum value that the index of refraction can have? Does light emerge at Q if the incident angle at P is (c) increased slightly and (d) decreased slightly?

FIG. 33-65 Problem 67.

sec. 33-10 Polarization by Reflection

•68 (a) At what angle of incidence will the light reflected from water be completely polarized? (b) Does this angle depend on the wavelength of the light?

•69 Light traveling in water of refractive index 1.33 is incident on a plate of glass with index of refraction 1.53. At what angle of incidence is the reflected light fully polarized? **SSM**

••70 In Fig. 33-66, a light ray in air is incident on a flat layer of material 2 that has an index of refraction $n_2 = 1.5$. Beneath material 2 is material 3 with an index of refraction n_3. The ray is incident on the air–material 2 interface at the Brewster angle for that interface. The ray of

FIG. 33-66 Problem 70.

light refracted into material 3 happens to be incident on the material 2–material 3 interface at the Brewster angle for that interface. What is the value of n_3?

Additional Problems

71 *Rainbow.* Figure 33-67 shows a light ray entering and then leaving a falling, spherical raindrop after one internal reflection (see Fig. 33-21a). The final direction of travel is deviated (turned) from the initial direction of travel by angular deviation θ_{dev}. (a) Show that θ_{dev} is

$$\theta_{dev} = 180° + 2\theta_i - 4\theta_r,$$

where θ_i is the angle of incidence of the ray on the drop and θ_r is the angle of refraction of the ray within the drop. (b) Using Snell's law, substitute for θ_r in terms of θ_i and the index of refraction n of the water. Then, on a graphing calculator or with a computer graphing package, graph θ_{dev} versus θ_i for the

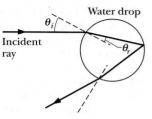

FIG. 33-67 Problem 71.

range of possible θ_i values and for $n = 1.331$ for red light and $n = 1.333$ for blue light.

The red-light curve and the blue-light curve have different minima, which means that there is a different *angle of minimum deviation* for each color. The light of any given color that leaves the drop at that color's angle of minimum deviation is especially bright because rays bunch up at that angle. Thus, the bright red light leaves the drop at one angle and the bright blue light leaves it at another angle.

Determine the angle of minimum deviation from the θ_{dev} curve for (c) red light and (d) blue light. (e) If these colors form the inner and outer edges of a rainbow (Fig. 33-21a), what is the angular width of the rainbow?

72 The *primary rainbow* described in Problem 71 is the type commonly seen in regions where rainbows appear. It is produced by light reflecting once inside the drops. Rarer is the *secondary rainbow* described in Section 33-8, produced by light reflecting twice inside the drops (Fig. 33-68a). (a) Show that the angular deviation of light entering and then leaving a

(a)   (b)

FIG. 33-68 Problem 72.

spherical water drop is

$$\theta_{dev} = (180°)k + 2\theta_i - 2(k + 1)\theta_r,$$

where k is the number of internal reflections. Using the procedure of Problem 71, find the angle of minimum deviation for (b) red light and (c) blue light in a secondary rainbow. (d) What is the angular width of that rainbow (Fig. 33-21d)?

The *tertiary rainbow* depends on three internal reflections (Fig. 33-68b). It probably occurs but, as noted in Section 33-8, cannot be seen because it is very faint and lies in the bright sky surrounding the Sun. What is the angle of minimum deviation for (e) the red light and (f) the blue light in this rainbow? (g) What is the rainbow's angular width?

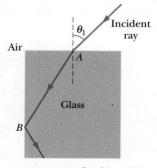

FIG. 33-69 Problem 73.

73 In Fig. 33-69, a light ray enters a glass slab at point A at incident angle $\theta_1 = 45.0°$ and then undergoes total internal reflection at point B. What minimum value for the index of refraction of the glass can be inferred from this information? **SSM**

74 In Fig. 33-70, unpolarized light with an intensity of 25 W/m² is sent into a system of four polarizing sheets with polarizing directions at angles $\theta_1 = 40°$, $\theta_2 = 20°$, $\theta_3 = 20°$, and $\theta_4 = 30°$. What is the intensity of the light that emerges from the system? **GO**

FIG. 33-70 Problem 74.

75 (a) Prove that a ray of light incident on the surface of a sheet of plate glass of thickness t emerges from the opposite face parallel to its initial direction but displaced sideways, as in Fig. 33-71. (b) Show that, for small angles of incidence θ, this displacement is given by

$$x = t\theta\,\frac{n - 1}{n},$$

where n is the index of refraction of the glass and θ is measured in radians. **SSM**

FIG. 33-71 Problem 75.

76 In Fig. 33-72, two light rays pass from air through five layers of transparent plastic and then back into air. The layers have parallel interfaces and unknown thicknesses; their indexes of refraction are $n_1 = 1.7$, $n_2 = 1.6$, $n_3 = 1.5$, $n_4 = 1.4$, and $n_5 = 1.6$. Ray b is incident at angle $\theta_b = 20°$. Relative to a normal at the last interface, at what angle do (a) ray a and (b) ray b emerge? (*Hint:* Solving the problem algebraically can save time.) If the air at the left and right sides in the figure were, instead, glass with index of refraction 1.5, at what angle would (c) ray a and (d) ray b emerge?

FIG. 33-72 Problem 76.

77 (a) How long does it take a radio signal to travel 150 km from a transmitter to a receiving antenna? (b) We see a full Moon by reflected sunlight. How much earlier did the light that enters our eye leave the Sun? The Earth–Moon and Earth–Sun distances are 3.8×10^5 km and 1.5×10^8 km, respectively. (c) What is the round-trip travel time for light between Earth and a spaceship orbiting Saturn, 1.3×10^9 km distant? (d) The Crab nebula, which is about 6500 light-years (ly) distant, is thought to be the result of a supernova explosion recorded by Chinese astronomers in A.D. 1054. In approximately what year did the explosion actually occur? **SSM**

78 In about A.D. 150, Claudius Ptolemy gave the following measured values for the angle of incidence θ_1 and the angle of refraction θ_2 for a light beam passing from air to water:

θ_1	θ_2	θ_1	θ_2
10°	8°	50°	35°
20°	15°30'	60°	40°30'
30°	22°30'	70°	45°30'
40°	29°	80°	50°

Assuming these data are consistent with the law of refraction, use them to find the index of refraction of water. These data are interesting as perhaps the oldest recorded physical measurements.

79 The electric component of a beam of polarized light is

$$E_y = (5.00 \text{ V/m}) \sin[(1.00 \times 10^6 \text{ m}^{-1})z + \omega t].$$

(a) Write an expression for the magnetic field component of the wave, including a value for ω. What are the (b) wavelength, (c) period, and (d) intensity of this light? (e) Parallel to which axis does the magnetic field oscillate? (f) In which region of the electromagnetic spectrum is this wave? **SSM**

80 The magnetic component of an electromagnetic wave in vacuum has an amplitude of 85.8 nT and an angular wave number of 4.00 m^{-1}. What are (a) the frequency of the wave, (b) the rms value of the electric component, and (c) the intensity of the light?

81 A helium–neon laser, radiating at 632.8 nm, has a power output of 3.0 mW. The beam diverges (spreads) at angle $\theta = 0.17$ mrad (Fig. 33-73). (a) What is the intensity of the beam 40 m from the laser? (b) What is the power of a point source providing that intensity at that distance?

FIG. 33-73 Problem 81.

82 The average intensity of the solar radiation that strikes normally on a surface just outside Earth's atmosphere is 1.4 kW/m^2. (a) What radiation pressure p_r is exerted on this surface, assuming complete absorption? (b) For comparison, find the ratio of p_r to Earth's sea-level atmospheric pressure, which is 1.0×10^5 Pa.

83 During a test, a NATO surveillance radar system, operating at 12 GHz at 180 kW of power, attempts to detect an incoming stealth aircraft at 90 km. Assume that the radar beam is emitted uniformly over a hemisphere. (a) What is the intensity of the beam when the beam reaches the aircraft's location? The aircraft reflects radar waves as though it has a cross-sectional area of only 0.22 m^2. (b) What is the power of the aircraft's reflection? Assume that the beam is reflected uniformly over a hemisphere. Back at the radar site, what are (c) the intensity, (d) the maximum value of the electric field vector, and (e) the rms value of the magnetic field of the reflected radar beam? **SSM**

84 An unpolarized beam of light is sent into a stack of four polarizing sheets, oriented so that the angle between the polarizing directions of adjacent sheets is 30°. What fraction of the incident intensity is transmitted by the system?

85 A beam of initially unpolarized light is sent through two polarizing sheets placed one on top of the other. What must be the angle between the polarizing directions of the sheets if the intensity of the transmitted light is to be one-third the incident intensity?

86 An electromagnetic wave is traveling in the negative direction of a y axis. At a particular position and time, the electric field is directed along the positive direction of the z axis and has a magnitude of 100 V/m. What are the (a) magnitude and (b) direction of the corresponding magnetic field?

87 Calculate the (a) upper and (b) lower limit of the Brewster angle for white light incident on fused quartz. Assume that the wavelength limits of the light are 400 and 700 nm.

88 A particle in the solar system is under the combined influence of the Sun's gravitational attraction and the radiation force due to the Sun's rays. Assume that the particle is a sphere of density 1.0×10^3 kg/m^3 and that all the incident light is absorbed. (a) Show that, if its radius is less than some critical radius R, the particle will be blown out of the solar system. (b) Calculate the critical radius.

89 The magnetic component of a polarized wave of light is

$$B_x = (4.0 \times 10^{-6} \text{ T}) \sin[(1.57 \times 10^7 \text{ m}^{-1})y + \omega t].$$

(a) Parallel to which axis is the light polarized? What are the (b) frequency and (c) intensity of the light?

90 Three polarizing sheets are stacked. The first and third are crossed; the one between has its polarizing direction at 45.0° to the polarizing directions of the other two. What fraction of the intensity of an originally unpolarized beam is transmitted by the stack?

91 A ray of white light traveling through fused quartz is incident at a quartz–air interface at angle θ_1. Assume that the index of refraction of quartz is $n = 1.456$ at the red end of the visible range and $n = 1.470$ at the blue end. If θ_1 is (a) 42.00°, (b) 43.10°, and (c) 44.00°, is the refracted light white, white dominated by the red end of the visible range, or white dominated by the blue end of the visible range, or is there no refracted light? SSM

92 In Fig. 33-74, unpolarized light is sent into the system of three polarizing sheets, where the polarizing directions of the first and third sheets are at angles $\theta_1 = 30°$ (counterclockwise) and $\theta_3 = 30°$ (clockwise). What fraction of the initial light intensity emerges from the system?

FIG. 33-74 Problem 92.

93 In a region of space where gravitational forces can be neglected, a sphere is accelerated by a uniform light beam of intensity 6.0 mW/m². The sphere is totally absorbing and has a radius of 2.0 μm and a uniform density of 5.0×10^3 kg/m³. What is the magnitude of the sphere's acceleration due to the light?

94 In Fig. 33-75, unpolarized light is sent into a system of three polarizing sheets, where the polarizing directions of the first and second sheets are at angles $\theta_1 = 20°$ and $\theta_2 = 40°$. What fraction of the initial light intensity emerges from the system?

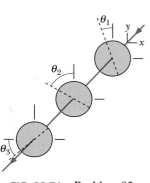

FIG. 33-75 Problem 94.

95 In Fig. 33-76, unpolarized light is sent into a system of three polarizing sheets with polarizing directions at angles $\theta_1 = 20°$, $\theta_2 = 60°$, and $\theta_3 = 40°$. What fraction of the initial light intensity emerges from the system?

96 A square, perfectly reflecting surface is oriented in space to be perpendicular to the light rays from the Sun. The surface has an edge length of 2.0 m and is located 3.0×10^{11} m from the Sun's center. What is the radiation force on the surface from the light rays?

FIG. 33-76 Problem 95.

97 The rms value of the electric field in a certain light wave is 0.200 V/m. What is the amplitude of the associated magnetic field?

98 In Fig. 33-77, an albatross glides at a constant 15 m/s horizontally above level ground, moving in a vertical plane that contains the Sun. It glides toward a wall of height $h = 2.0$ m, which it will just barely clear. At that time of day, the angle of the Sun relative to the ground is $\theta = 30°$. At what speed does the shadow of the albatross move (a) across the level ground and then (b) up the wall? Suppose that later a hawk happens to glide along the same path, also at 15 m/s. You see that when its shadow reaches the wall, the speed of the shadow noticeably increases. (c) Is the Sun now higher or lower in the sky than when the albatross flew by earlier? (d) If the speed of the hawk's shadow on the wall is 45 m/s, what is the angle θ of the Sun just then?

FIG. 33-77 Problem 98.

99 The magnetic component of a polarized wave of light is given by $B_x = (4.00 \ \mu\text{T}) \sin[ky + (2.00 \times 10^{15} \ \text{s}^{-1})t]$. (a) In which direction does the wave travel, (b) parallel to which axis is it polarized, and (c) what is its intensity? (d) Write an expression for the electric field of the wave, including a value for the angular wave number. (e) What is the wavelength? (f) In which region of the electromagnetic spectrum is this electromagnetic wave?

100 In Fig. 33-78, where $n_1 = 1.70$, $n_2 = 1.50$, and $n_3 = 1.30$, light refracts from material 1 into material 2. If it is incident at point A at the critical angle for the interface between materials 2 and 3, what are (a) the angle of refraction at point B and (b) the initial angle θ? If, instead, light is incident at B at the critical angle for the interface between materials 2 and 3, what are (c) the angle of refraction at point A and (d) the initial angle θ? If, instead of all that, light is incident at point A at Brewster's angle for the interface between materials 2 and 3, what are (e) the angle of refraction at point B and (f) the initial angle θ?

FIG. 33-78 Problem 100.

101 When red light in vacuum is incident at the Brewster angle on a certain glass slab, the angle of refraction is 32.0°. What are (a) the index of refraction of the glass and (b) the Brewster angle?

102 Start from Eqs. 33-11 and 33-17 and show that $E(x, t)$ and $B(x, t)$, the electric and magnetic field components of a plane traveling electromagnetic wave, must satisfy the "wave equations"

$$\frac{\partial^2 E}{\partial t^2} = c^2 \frac{\partial^2 E}{\partial x^2} \quad \text{and} \quad \frac{\partial^2 B}{\partial t^2} = c^2 \frac{\partial^2 B}{\partial x^2}.$$

103 (a) Show that Eqs. 33-1 and 33-2 satisfy the wave equations displayed in Problem 102. (b) Show that any expressions of the form $E = E_m f(kx \pm \omega t)$ and $B = B_m f(kx \pm \omega t)$, where $f(kx \pm \omega t)$ denotes an arbitrary function, also satisfy these wave equations. SSM

104 A point source of light emits isotropically with a power of 200 W. What is the force due to the light on a totally absorbing sphere of radius 2.0 cm at a distance of 20 m from the source?

Underwater vision is usually difficult even if you have perfect vision above water. The reason has to do with how water affects the refraction of light entering the eye. The refraction may be appropriate in the air but quite wrong in the water. However, the peculiar fish *Anableps anableps* swims with its eyes partially extending above the water surface so that it can see simultaneously above and below water.

How can its eyes function in both air and water?

The answer is in this chapter.

Paddy Ryan/Ryan Photographic

34-1 | WHAT IS PHYSICS?

One goal of physics is to discover the basic laws governing light, such as the law of refraction. A broader goal is to put those laws to use, and perhaps the most important use is the production of images. The first photographic images, made in 1824, were only novelties, but our world now thrives on images. Huge industries are based on the production of images on television, computer, and theater screens. Images from satellites guide military strategists during times of conflict and environmental strategists during times of blight. Camera surveillance can make a subway system more secure, but it can also invade the privacy of unsuspecting citizens. Physiologists and medical engineers are still puzzled by how images are produced by the human eye and the visual cortex of the brain, but they have managed to create mental images in some sightless people by electrical stimulation of the visual cortex.

Our first step in this chapter is to define and classify images. Then we examine several basic ways in which they can be produced.

34-2 | Two Types of Image

For you to see, say, a penguin, your eye must intercept some of the light rays spreading from the penguin and then redirect them onto the retina at the rear of the eye. Your visual system, starting with the retina and ending with the visual cortex at the rear of your brain, automatically and subconsciously processes the information provided by the light. That system identifies edges, orientations, textures, shapes, and colors and then rapidly brings to your consciousness an **image** (a reproduction derived from light) of the penguin; you perceive and recognize the penguin as being in the direction from which the light rays came and at the proper distance.

Your visual system goes through this processing and recognition even if the light rays do not come directly from the penguin, but instead reflect toward you from a mirror or refract through the lenses in a pair of binoculars. However, you now see the penguin in the direction from which the light rays came after they reflected or refracted, and the distance you perceive may be quite different from the penguin's true distance.

For example, if the light rays have been reflected toward you from a standard flat mirror, the penguin appears to be behind the mirror because the rays you intercept come from that direction. Of course, the penguin is not back there. This type of image, which is called a **virtual image,** truly exists only within the brain but nevertheless is *said* to exist at the perceived location.

A **real image** differs in that it can be formed on a surface, such as a card or a movie screen. You can see a real image (otherwise movie theaters would be empty), but the existence of the image does not depend on your seeing it and it is present even if you are not.

In this chapter we explore several ways in which virtual and real images are formed by reflection (as with mirrors) and refraction (as with lenses). We also distinguish between the two types of image more clearly, but here first is an example of a natural virtual image.

A Common Mirage

A common example of a virtual image is a pool of water that appears to lie on the road some distance ahead of you on a sunny day, but that you can never reach. The pool is a *mirage* (a type of illusion), formed by light rays coming from the low section of the sky in front of you (Fig. 34-1a). As the rays approach the road, they travel through progressively warmer air that has been heated by

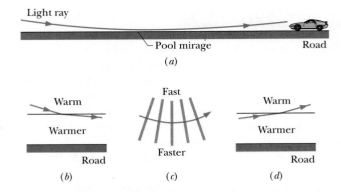

FIG. 34-1 (*a*) A ray from a low section of the sky refracts through air that is heated by a road (without reaching the road). An observer who intercepts the light perceives it to be from a pool of water on the road. (*b*) Bending (exaggerated) of a light ray descending across an imaginary boundary from warm air to warmer air. (*c*) Shifting of wavefronts and associated bending of a ray, which occur because the lower ends of wavefronts move faster in warmer air. (*d*) Bending of a ray ascending across an imaginary boundary to warm air from warmer air.

the road, which is usually relatively warm. With an increase in air temperature, the density of the air—and hence the index of refraction of the air—decreases slightly. Thus, as the rays descend, encountering progressively smaller indexes of refraction, they continuously bend toward the horizontal (Fig. 34-1*b*).

Once a ray is horizontal, somewhat above the road's surface, it still bends because the lower portion of each associated wavefront is in slightly warmer air and is moving slightly faster than the upper portion of the wavefront (Fig. 34-1*c*). This nonuniform motion of the wavefronts bends the ray upward. As the ray then ascends, it continues to bend upward through progressively greater indexes of refraction (Fig. 34-1*d*).

If you intercept some of this light, your visual system automatically infers that it originated along a backward extension of the rays you have intercepted and, to make sense of the light, assumes that it came from the road surface. If the light happens to be bluish from blue sky, the mirage appears bluish, like water. Because the air is probably turbulent due to the heating, the mirage shimmies, as if water waves were present. The bluish coloring and the shimmy enhance the illusion of a pool of water, but you are actually seeing a virtual image of a low section of the sky.

34-3 | Plane Mirrors

A **mirror** is a surface that can reflect a beam of light in one direction instead of either scattering it widely in many directions or absorbing it. A shiny metal surface acts as a mirror; a concrete wall does not. In this section we examine the images that a **plane mirror** (a flat reflecting surface) can produce.

Figure 34-2 shows a point source of light *O*, which we shall call the *object*, at a perpendicular distance *p* in front of a plane mirror. The light that is incident on the mirror is represented with rays spreading from *O*. The reflection of that light is represented with reflected rays spreading from the mirror. If we extend the reflected rays backward (behind the mirror), we find that the extensions intersect at a point that is a perpendicular distance *i* behind the mirror.

If you look into the mirror of Fig. 34-2, your eyes intercept some of the reflected light. To make sense of what you see, you perceive a point source of light located at the point of intersection of the extensions. This point source is the image *I* of object *O*. It is called a *point image* because it is a point, and it is a vir-

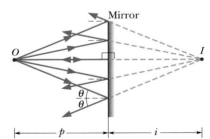

FIG. 34-2 A point source of light *O*, called the *object*, is a perpendicular distance *p* in front of a plane mirror. Light rays reaching the mirror from *O* reflect from the mirror. If your eye intercepts some of the reflected rays, you perceive a point source of light *I* to be behind the mirror, at a perpendicular distance *i*. The perceived source *I* is a virtual image of object *O*.

tual image because the rays do not actually pass through it. (As you will see, rays *do* pass through a point of intersection for a real image.)

Figure 34-3 shows two rays selected from the many rays in Fig. 34-2. One reaches the mirror at point *b*, perpendicularly. The other reaches it at an arbitrary point *a*, with an angle of incidence θ. The extensions of the two reflected rays are also shown. The right triangles *aOba* and *aIba* have a common side and three equal angles and are thus congruent (equal in size); so their horizontal sides have the same length. That is,

$$Ib = Ob, \qquad (34\text{-}1)$$

where *Ib* and *Ob* are the distances from the mirror to the image and the object, respectively. Equation 34-1 tells us that the image is as far behind the mirror as the object is in front of it. By convention (that is, to get our equations to work out), *object distances p* are taken to be positive quantities and *image distances i* for virtual images (as here) are taken to be negative quantities. Thus, Eq. 34-1 can be written as $|i| = p$ or as

$$i = -p \qquad \text{(plane mirror)}. \qquad (34\text{-}2)$$

Only rays that are fairly close together can enter the eye after reflection at a mirror. For the eye position shown in Fig. 34-4, only a small portion of the mirror near point *a* (a portion smaller than the pupil of the eye) is useful in forming the image. To find this portion, close one eye and look at the mirror image of a small object such as the tip of a pencil. Then move your fingertip over the mirror surface until you cannot see the image. Only that small portion of the mirror under your fingertip produced the image.

Extended Objects

In Fig. 34-5, an extended object *O*, represented by an upright arrow, is at perpendicular distance *p* in front of a plane mirror. Each small portion of the object that faces the mirror acts like the point source *O* of Figs. 34-2 and 34-3. If you intercept the light reflected by the mirror, you perceive a virtual image *I* that is a composite of the virtual point images of all those portions of the object. This virtual image seems to be at (negative) distance *i* behind the mirror, with *i* and *p* related by Eq. 34-2.

We can also locate the image of an extended object as we did for a point object in Fig. 34-2: we draw some of the rays that reach the mirror from the top of the object, draw the corresponding reflected rays, and then extend those reflected rays behind the mirror until they intersect to form an image of the top of the object. We then do the same for rays from the bottom of the object. As shown in Fig. 34-5, we find that virtual image *I* has the same orientation and *height* (measured parallel to the mirror) as object *O*.

FIG. 34-3 Two rays from Fig. 34-2. Ray *Oa* makes an arbitrary angle θ with the normal to the mirror surface. Ray *Ob* is perpendicular to the mirror.

FIG. 34-4 A "pencil" of rays from *O* enters the eye after reflection at the mirror. Only a small portion of the mirror near *a* is involved in this reflection. The light appears to originate at point *I* behind the mirror.

FIG. 34-5 An extended object *O* and its virtual image *I* in a plane mirror.

FIG. 34-6 A maze of mirrors. *(Courtesy Adrian Fisher. www.mazemaker.com)*

Mirror Maze

In a mirror maze (Fig. 34-6), each wall is covered, floor to ceiling, with a mirror. Walk through such a maze and what you see in most directions is a confusing montage of reflections. In some directions, however, you see a hallway that seems to offer a path through the maze. Take these hallways, though, and you soon learn, after smacking into mirror after mirror, that the hallways are largely an illusion.

Figure 34-7a is an overhead view of a simple mirror maze in which differently painted floor sections form equilateral triangles (60° angles) and walls are covered with vertical mirrors. You look into the maze while standing at point *O* at the middle of the maze entrance. In most directions, you see a confusing jumble of images. However, you see something curious in the direction of the ray shown in Fig. 34-7a. That ray leaves the middle of mirror *B* and reflects to you at the middle of mirror *A*. (The reflection obeys the law of reflection, with the angle of incidence and the angle of reflection both equal to 30°.)

To make sense of the origin of the ray reaching you, your brain automatically extends the ray backward. It appears to originate at a point lying *behind* mirror *A*. That is, you perceive a virtual image of *B* behind *A*, at a distance equal to the actual distance between *A* and *B* (Fig. 34-7b). Thus, when you face into the maze in this direction, you see *B* along an apparent straight hallway consisting of four triangular floor sections.

This story is incomplete, however, because the ray reaching you does not *originate* at mirror *B*—it only reflects there. To find the origin, we continue to apply the law of reflection as we work backwards, reflection by reflection on the mirrors. Working through the four reflections shown in Fig. 34-7c, we finally come to the origin of the ray: you! What you see when you look along the apparent hallway is a virtual image of yourself, at a distance of nine triangular floor sections from you (Fig. 34-7d). (There is a second apparent hallway extending away from point *O*. Which way must you face to look along it?)

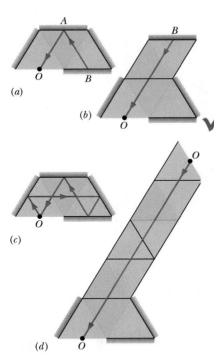

(a)
(b)
(c)
(d)

FIG. 34-7 (a) Overhead view of a mirror maze. A ray from mirror *B* reaches you at *O* by reflecting from mirror *A*. (b) Mirror *B* appears to be behind *A*. (c) The ray reaching you comes from you. (d) You see a virtual image of yourself at the end of an apparent hallway.

✓ **CHECKPOINT 1** In the figure you are in a system of two vertical parallel mirrors *A* and *B* separated by distance *d*. A grinning gargoyle is perched at point *O*, a distance $0.2d$ from mirror *A*. Each mirror produces a *first* (least deep) image of the gargoyle. Then each mirror produces a *second* image with the object being the first image in the opposite mirror. Then each mirror produces a *third* image with the object being the second image in the opposite mirror, and so on—you might see hundreds of grinning gargoyle images. How deep behind mirror *A* are the first, second, and third images in mirror *A*?

34-4 | Spherical Mirrors

We turn now from images produced by plane mirrors to images produced by mirrors with curved surfaces. In particular, we consider spherical mirrors, which are simply mirrors in the shape of a small section of the surface of a sphere. A plane mirror is in fact a spherical mirror with an infinitely large *radius of curvature* and thus an approximately flat surface.

Making a Spherical Mirror

We start with the plane mirror of Fig. 34-8a, which faces leftward toward an object *O* that is shown and an observer that is not shown. We make a **concave**

mirror by curving the mirror's surface so it is *concave* ("caved in") as in Fig. 34-8*b*. Curving the surface in this way changes several characteristics of the mirror and the image it produces of the object:

1. The *center of curvature C* (the center of the sphere of which the mirror's surface is part) was infinitely far from the plane mirror; it is now closer but still in front of the concave mirror.

2. The *field of view*—the extent of the scene that is reflected to the observer—was wide; it is now smaller.

3. The image of the object was as far behind the plane mirror as the object was in front; the image is farther behind the concave mirror; that is, |*i*| is greater.

4. The height of the image was equal to the height of the object; the height of the image is now greater. This feature is why many makeup mirrors and shaving mirrors are concave—they produce a larger image of a face.

We can make a **convex mirror** by curving a plane mirror so its surface is *convex* ("flexed out") as in Fig. 34-8*c*. Curving the surface in this way (1) moves the center of curvature *C* to *behind* the mirror and (2) *increases* the field of view. It also (3) moves the image of the object *closer* to the mirror and (4) *shrinks* it. Store surveillance mirrors are usually convex to take advantage of the increase in the field of view—more of the store can then be seen with a single mirror.

Focal Points of Spherical Mirrors

For a plane mirror, the magnitude of the image distance *i* is always equal to the object distance *p*. Before we can determine how these two distances are related for a spherical mirror, we must consider the reflection of light from an object *O* located an effectively infinite distance in front of a spherical mirror, on the mirror's *central axis*. That axis extends through the center of curvature *C* and the center *c* of the mirror. Because of the great distance between the object and the mirror, the light waves spreading from the object are plane waves when they reach the mirror along the central axis. This means that the rays representing the light waves are all parallel to the central axis when they reach the mirror.

When these parallel rays reach a concave mirror like that of Fig. 34-9*a*, those near the central axis are reflected through a common point *F*; two of these reflected rays are shown in the figure. If we placed a (small) card at *F*, a point image of the infinitely distant object *O* would appear on the card. (This would occur for any infinitely distant object.) Point *F* is called the **focal point** (or **focus**) of the mirror, and its distance from the center of the mirror *c* is the **focal length** *f* of the mirror.

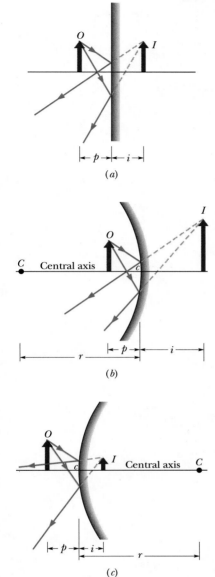

FIG. 34-8   (*a*) An object *O* forms a virtual image *I* in a plane mirror. (*b*) If the mirror is bent so that it becomes *concave*, the image moves farther away and becomes larger. (*c*) If the plane mirror is bent so that it becomes *convex*, the image moves closer and becomes smaller.

FIG. 34-9   (*a*) In a concave mirror, incident parallel light rays are brought to a real focus at *F*, on the same side of the mirror as the incident light rays. (*b*) In a convex mirror, incident parallel light rays seem to diverge from a virtual focus at *F*, on the side of the mirror opposite the light rays.

If we now substitute a convex mirror for the concave mirror, we find that the parallel rays are no longer reflected through a common point. Instead, they diverge as shown in Fig. 34-9b. However, if your eye intercepts some of the reflected light, you perceive the light as originating from a point source behind the mirror. This perceived source is located where extensions of the reflected rays pass through a common point (*F* in Fig. 34-9b). That point is the focal point (or focus) *F* of the convex mirror, and its distance from the mirror surface is the focal length *f* of the mirror. If we placed a card at this focal point, an image of object *O* would *not* appear on the card; so this focal point is not like that of a concave mirror.

To distinguish the actual focal point of a concave mirror from the perceived focal point of a convex mirror, the former is said to be a *real focal point* and the latter is said to be a *virtual focal point*. Moreover, the focal length *f* of a concave mirror is taken to be a positive quantity, and that of a convex mirror a negative quantity. For mirrors of both types, the focal length *f* is related to the radius of curvature *r* of the mirror by

$$f = \tfrac{1}{2}r \qquad \text{(spherical mirror)}, \qquad (34\text{-}3)$$

where, consistent with the signs for the focal length, *r* is a positive quantity for a concave mirror and a negative quantity for a convex mirror.

34-5 | Images from Spherical Mirrors

With the focal point of a spherical mirror defined, we can find the relation between image distance *i* and object distance *p* for concave and convex spherical mirrors. We begin by placing the object *O* *inside the focal point* of the concave mirror—that is, between the mirror and its focal point *F* (Fig. 34-10a). An observer can then see a virtual image of *O* in the mirror: The image appears to be behind the mirror, and it has the same orientation as the object.

If we now move the object away from the mirror until it is at the focal point, the image moves farther and farther back from the mirror until, when the object is at the focal point, the image is at infinity (Fig. 34-10b). The image is then ambiguous and imperceptible because neither the rays reflected by the mirror nor the ray extensions behind the mirror cross to form an image of *O*.

If we next move the object *outside the focal point*—that is, farther away from the mirror than the focal point—the rays reflected by the mirror converge to form an *inverted* image of object *O* (Fig. 34-10c) in front of the mirror. That image moves in from infinity as we move the object farther outside *F*. If you were to hold a card at the position of the image, the image would show up on the card—the image is said to be *focused* on the card by the mirror. (The verb "focus," which in this context means to produce an image, differs from the noun "focus," which is another name for the focal point.) Because this image can actually appear on a surface, it is a real image—the rays actually intersect to create the image,

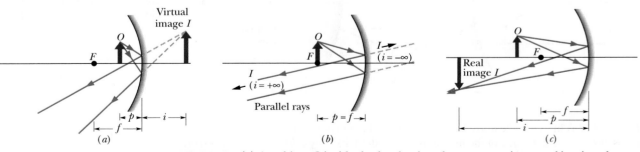

FIG. 34-10 (a) An object *O* inside the focal point of a concave mirror, and its virtual image *I*. (b) The object at the focal point *F*. (c) The object outside the focal point, and its real image *I*.

regardless of whether an observer is present. The image distance i of a real image is a positive quantity, in contrast to that for a virtual image. We also see that

> Real images form on the side of a mirror where the object is, and virtual images form on the opposite side.

As we shall prove in Section 34-9, when light rays from an object make only small angles with the central axis of a spherical mirror, a simple equation relates the object distance p, the image distance i, and the focal length f:

$$\frac{1}{p} + \frac{1}{i} = \frac{1}{f} \qquad \text{(spherical mirror).} \qquad (34\text{-}4)$$

We assume such small angles in figures such as Fig. 34-10, but for clarity the rays are drawn with exaggerated angles. With that assumption, Eq. 34-4 applies to any concave, convex, or plane mirror. For a convex or plane mirror, only a virtual image can be formed, regardless of the object's location on the central axis. As shown in the example of a convex mirror in Fig. 34-8c, the image is always on the opposite side of the mirror from the object and has the same orientation as the object.

The size of an object or image, as measured *perpendicular* to the mirror's central axis, is called the object or image *height*. Let h represent the height of the object, and h' the height of the image. Then the ratio h'/h is called the **lateral magnification** m produced by the mirror. However, by convention, the lateral magnification always includes a plus sign when the image orientation is that of the object and a minus sign when the image orientation is opposite that of the object. For this reason, we write the formula for m as

$$|m| = \frac{h'}{h} \qquad \text{(lateral magnification).} \qquad (34\text{-}5)$$

We shall soon prove that the lateral magnification can also be written as

$$m = -\frac{i}{p} \qquad \text{(lateral magnification).} \qquad (34\text{-}6)$$

For a plane mirror, for which $i = -p$, we have $m = +1$. The magnification of 1 means that the image is the same size as the object. The plus sign means that the image and the object have the same orientation. For the concave mirror of Fig. 34-10c, $m \approx -1.5$.

Equations 34-3 through 34-6 hold for all plane mirrors, concave spherical mirrors, and convex spherical mirrors. In addition to those equations, you have been asked to absorb a lot of information about these mirrors, and you should organize it for yourself by filling in Table 34-1. Under Image Location, note whether the image is on the *same* side of the mirror as the object or on the *oppo-*

TABLE 34-1

Your Organizing Table for Mirrors

Mirror Type	Object Location	Image Location	Image Type	Image Orientation	Sign of f	Sign of r	Sign of m
Plane	Anywhere						
Concave	Inside F						
	Outside F						
Convex	Anywhere						

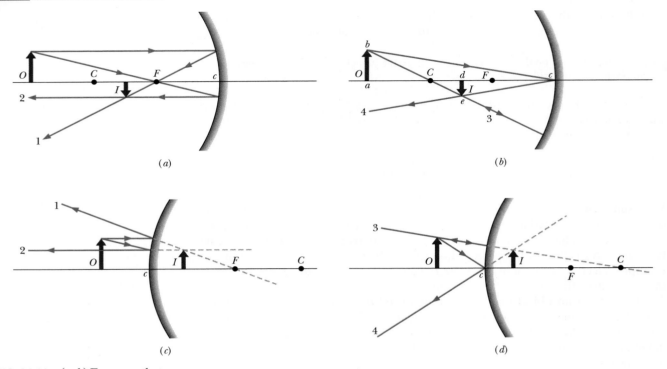

(a)

(b)

(c)

(d)

FIG. 34-11 (a, b) Four rays that may be drawn to find the image formed by a concave mirror. For the object position shown, the image is real, inverted, and smaller than the object. (c, d) Four similar rays for the case of a convex mirror. For a convex mirror, the image is always virtual, oriented like the object, and smaller than the object. [In (c), ray 2 is initially directed toward focal point F. In (d), ray 3 is initially directed toward center of curvature C.]

site side. Under Image Type, note whether the image is *real* or *virtual*. Under Image Orientation, note whether the image has the *same* orientation as the object or is *inverted*. Under Sign, give the sign of the quantity or fill in ± if the sign is ambiguous. You will need this organization to tackle homework or a test.

Locating Images by Drawing Rays

Figures 34-11a and b show an object O in front of a concave mirror. We can graphically locate the image of any off-axis point of the object by drawing a *ray diagram* with any two of four special rays through the point:

1. A ray that is initially parallel to the central axis reflects through the focal point F (ray 1 in Fig. 34-11a).
2. A ray that reflects from the mirror after passing through the focal point emerges parallel to the central axis (ray 2 in Fig. 34-11a).
3. A ray that reflects from the mirror after passing through the center of curvature C returns along itself (ray 3 in Fig. 34-11b).
4. A ray that reflects from the mirror at point c is reflected symmetrically about that axis (ray 4 in Fig. 34-11b).

The image of the point is at the intersection of the two special rays you choose. The image of the object can then be found by locating the images of two or more of its off-axis points. You need to modify the descriptions of the rays slightly to apply them to convex mirrors, as in Figs. 34-11c and d.

Proof of Equation 34-6

We are now in a position to derive Eq. 34-6 ($m = -i/p$), the equation for the lateral magnification of an object reflected in a mirror. Consider ray 4 in Fig. 34-11b. It is reflected at point c so that the incident and reflected rays make equal angles with the axis of the mirror at that point.

The two right triangles *abc* and *dec* in the figure are similar (have the same set of angles); so we can write

$$\frac{de}{ab} = \frac{cd}{ca}.$$

The quantity on the left (apart from the question of sign) is the lateral magnification *m* produced by the mirror. Because we indicate an inverted image as a *negative* magnification, we symbolize this as $-m$. However, $cd = i$ and $ca = p$; so we have

$$m = -\frac{i}{p} \quad \text{(magnification)}, \qquad (34\text{-}7)$$

which is the relation we set out to prove.

✓**CHECKPOINT 2**   A Central American vampire bat, dozing on the central axis of a spherical mirror, is magnified by $m = -4$. Is its image (a) real or virtual, (b) inverted or of the same orientation as the bat, and (c) on the same side of the mirror as the bat or on the opposite side?

Sample Problem   34-1

A tarantula of height h sits cautiously before a spherical mirror whose focal length has absolute value $|f| = 40$ cm. The image of the tarantula produced by the mirror has the same orientation as the tarantula and has height $h' = 0.20h$.

(a) Is the image real or virtual, and is it on the same side of the mirror as the tarantula or the opposite side?

Reasoning: Because the image has the same orientation as the tarantula (the object), it must be virtual and on the opposite side of the mirror. (You can easily see this result if you have filled out Table 34-1.)

(b) Is the mirror concave or convex, and what is its focal length f, sign included?

KEY IDEA   We *cannot* tell the type of mirror from the type of image because both types of mirror can produce virtual images. Similarly, we cannot tell the type of mirror from the sign of the focal length f, as obtained from Eq. 34-3 or Eq. 34-4, because we lack enough information to use either equation. However, we can make use of the magnification information.

Calculations: We know that the ratio of image height h' to object height h is 0.20. Thus, from Eq. 34-5 we have

$$|m| = \frac{h'}{h} = 0.20.$$

Because the object and image have the same orientation, we know that m must be positive: $m = +0.20$. Substituting this into Eq. 34-6 and solving for, say, i gives us

$$i = -0.20p,$$

which does not appear to be of help in finding f. However, it is helpful if we substitute it into Eq. 34-4. That equation gives us

$$\frac{1}{f} = \frac{1}{i} + \frac{1}{p} = \frac{1}{-0.20p} + \frac{1}{p} = \frac{1}{p}(-5 + 1),$$

from which we find

$$f = -p/4.$$

Now we have it: Because p is positive, f must be negative, which means that the mirror is convex with

$$f = -40 \text{ cm}. \qquad \text{(Answer)}$$

34-6 I Spherical Refracting Surfaces

We now turn from images formed by reflection to images formed by refraction through surfaces of transparent materials, such as glass. We shall consider only spherical surfaces, with radius of curvature r and center of curvature C. The light will be emitted by a point object O in a medium with index of refraction n_1; it will refract through a spherical surface into a medium of index of refraction n_2.

Our concern is whether the light rays, after refracting through the surface, form a real image (no observer necessary) or a virtual image (assuming that an observer intercepts the rays). The answer depends on the relative values of n_1 and n_2 and on the geometry of the situation.

FIG. 34-12 Six possible ways in which an image can be formed by refraction through a spherical surface of radius r and center of curvature C. The surface separates a medium with index of refraction n_1 from a medium with index of refraction n_2. The point object O is always in the medium with n_1, to the left of the surface. The material with the lesser index of refraction is unshaded (think of it as being air, and the other material as being glass). Real images are formed in (a) and (b); virtual images are formed in the other four situations.

Six possible results are shown in Fig. 34-12. In each part of the figure, the medium with the greater index of refraction is shaded, and object O is always in the medium with index of refraction n_1, to the left of the refracting surface. In each part, a representative ray is shown refracting through the surface. (That ray and a ray along the central axis suffice to determine the position of the image in each case.)

At the point of refraction of each ray, the normal to the refracting surface is a radial line through the center of curvature C. Because of the refraction, the ray bends toward the normal if it is entering a medium of greater index of refraction and away from the normal if it is entering a medium of lesser index of refraction. If the bending sends the ray toward the central axis, that ray and others (undrawn) form a real image on that axis. If the bending sends the ray away from the central axis, the ray cannot form a real image; however, backward extensions of it and other refracted rays can form a virtual image, provided (as with mirrors) some of those rays are intercepted by an observer.

Real images I are formed (at image distance i) in parts a and b of Fig. 34-12, where the refraction directs the ray *toward* the central axis. Virtual images are formed in parts c and d, where the refraction directs the ray *away* from the central axis. Note, in these four parts, that real images are formed when the object is relatively far from the refracting surface and virtual images are formed when the object is nearer the refracting surface. In the final situations (Figs. 34-12e and f), refraction always directs the ray away from the central axis and virtual images are always formed, regardless of the object distance.

Note the following major difference from reflected images:

> Real images form on the side of a refracting surface that is opposite the object, and virtual images form on the same side as the object.

In Section 34-9, we shall show that (for light rays making only small angles with the central axis)

$$\frac{n_1}{p} + \frac{n_2}{i} = \frac{n_2 - n_1}{r}. \tag{34-8}$$

Just as with mirrors, the object distance p is positive, and the image distance i is positive for a real image and negative for a virtual image. However, to keep all the signs correct in Eq. 34-8, we must use the following rule for the sign of the radius of curvature r:

> When the object faces a convex refracting surface, the radius of curvature r is positive. When it faces a concave surface, r is negative.

Be careful: This is just the reverse of the sign convention we have for mirrors.

This insect has been entombed in amber for about 25 million years. Because we view the insect through a curved refracting surface, the location of the image we see does not coincide with the location of the insect (see Fig. 34-12d). *(Dr. Paul A. Zahl/Photo Researchers)*

✓ **CHECKPOINT 3**   A bee is hovering in front of the concave spherical refracting surface of a glass sculpture. (a) Which of the general situations of Fig. 34-11 is like this situation? (b) Is the image produced by the surface real or virtual, and is it on the same side as the bee or the opposite side?

Sample Problem 34-2

A Jurassic mosquito is discovered embedded in a chunk of amber, which has index of refraction 1.6. One surface of the amber is spherically convex with radius of curvature 3.0 mm (Fig. 34-13). The mosquito's head happens to be on the central axis of that surface and, when viewed along the axis, appears to be buried 5.0 mm into the amber. How deep is it really?

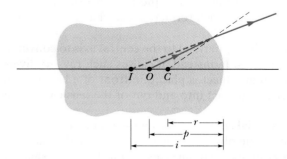

FIG. 34-13   A piece of amber with a mosquito from the Jurassic period, with the head buried at point O. The spherical refracting surface at the right end, with center of curvature C, provides an image I to an observer intercepting rays from the object at O.

KEY IDEAS   The head appears to be 5.0 mm into the amber only because the light rays that the observer intercepts are bent by refraction at the convex amber surface. The image distance i differs from the object distance p according to Eq. 34-8. To use that equation to find the object distance, we first note:

1. Because the object (the head) and its image are on the same side of the refracting surface, the image must be virtual and so $i = -5.0$ mm.

2. Because the object is always taken to be in the medium of index of refraction n_1, we must have $n_1 = 1.6$ and $n_2 = 1.0$.

3. Because the *object* faces a concave refracting surface, the radius of curvature r is negative, and so $r = -3.0$ mm.

Calculations: Making these substitutions in Eq. 34-8,

$$\frac{n_1}{p} + \frac{n_2}{i} = \frac{n_2 - n_1}{r},$$

yields

$$\frac{1.6}{p} + \frac{1.0}{-5.0 \text{ mm}} = \frac{1.0 - 1.6}{-3.0 \text{ mm}}$$

and

$$p = 4.0 \text{ mm}. \qquad \text{(Answer)}$$

Sample Problem 34-3

Figure 34-14 shows a vertical cross section through an eye of the *Anableps anableps* fish that swims with each eye half in and half out of the water, with a pigment band separating the two halves at the water surface. The front of the eye (the cornea) is a spherically convex refracting surface of radius $r = 1.95$ mm and index of refraction $n_2 = 1.335$. The refraction at the cornea is the first step in the eye's focusing of a *real* image onto the back of the eye (the retina), where visual processing begins. If the cornea faces an insect (lunch) at object distance $p = 0.20$ m, what is the image distance i of that refraction for the cornea in air ($n_1 = 1.000$) and in water ($n_1 = 1.333$)?

KEY IDEAS   (1) Because the object and its image are on opposite sides of the refracting surface with its convex side facing the object, the situation is like either Fig. 34-12a (for a real image) or Fig. 34-12c (for a virtual image). (2) Image distance and object distance are related by Eq. 34-8.

Calculations: Solving Eq. 34-8 for i gives us

$$i = \frac{n_2}{\dfrac{n_2 - n_1}{r} - \dfrac{n_1}{p}}.$$

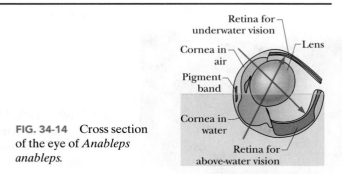

FIG. 34-14   Cross section of the eye of *Anableps anableps.*

Substituting the given values and using the index of refraction $n_1 = 1.000$ for air, we find

$$i = \frac{1.335}{\dfrac{1.335 - 1.000}{0.00195} - \dfrac{1.000}{0.20}}$$

$$= 8.00 \text{ mm}. \qquad \text{(Answer)}$$

Repeating the calculation but using the index of refraction $n_1 = 1.333$ for water, we find

$$i = \frac{1.335}{\dfrac{1.335 - 1.333}{0.00195} - \dfrac{1.333}{0.20}}$$

$$= -0.237 \text{ m}. \qquad \text{(Answer)}$$

After light is refracted by the cornea, it is further refracted by a lens to give a final real image on the retina. (The function of a lens is discussed in the next section.) Our first answer is positive (indicating a real image, as needed) and about twice the diameter of the eye. Thus the role of the lens in focusing the light onto the retina is moderate because so much refraction occurs at the cornea. Our second answer is quite different: It is negative (indicating a virtual image) and much larger. So, far more focusing is required by the lens to put a real image onto the retina. To provide moderate focusing of light from the air and much stronger focusing of light from the water, the eye lens in *Anableps anableps* is egg shaped, with much greater curvature in the bottom half than in the top.

34-7 | Thin Lenses

A **lens** is a transparent object with two refracting surfaces whose central axes coincide. The common central axis is the central axis of the lens. When a lens is surrounded by air, light refracts from the air into the lens, crosses through the lens, and then refracts back into the air. Each refraction can change the direction of travel of the light.

A lens that causes light rays initially parallel to the central axis to converge is (reasonably) called a **converging lens.** If, instead, it causes such rays to diverge, the lens is a **diverging lens.** When an object is placed in front of a lens of either type, light rays from the object that refract into and out of the lens can produce an image of the object.

We shall consider only the special case of a **thin lens** — that is, a lens in which the thickest part is thin relative to the object distance p, the image distance i, and the radii of curvature r_1 and r_2 of the two surfaces of the lens. We shall also consider only light rays that make small angles with the central axis (they are exaggerated in the figures here). In Section 34-9 we shall prove that for such rays, a thin lens has a focal length f. Moreover, i and p are related to each other by

$$\frac{1}{f} = \frac{1}{p} + \frac{1}{i} \qquad \text{(thin lens),} \qquad (34\text{-}9)$$

which is the same as we had for mirrors. We shall also prove that when a thin lens with index of refraction n is surrounded by air, this focal length f is given by

$$\frac{1}{f} = (n-1)\left(\frac{1}{r_1} - \frac{1}{r_2}\right) \qquad \text{(thin lens in air),} \qquad (34\text{-}10)$$

which is often called the *lens maker's equation.* Here r_1 is the radius of curvature of the lens surface nearer the object and r_2 is that of the other surface. The signs of these radii are found with the rules in Section 34-6 for the radii of spherical refracting surfaces. If the lens is surrounded by some medium other than air (say, corn oil) with index of refraction n_{medium}, we replace n in Eq. 34-10 with n/n_{medium}. Keep in mind the basis of Eqs. 34-9 and 34-10:

A lens can produce an image of an object only because the lens can bend light rays, but it can bend light rays only if its index of refraction differs from that of the surrounding medium.

A fire is being started by focusing sunlight onto newspaper by means of a converging lens made of clear ice. The lens was made by melting both sides of a flat piece of ice into a convex shape in the shallow vessel (which has a curved bottom). *(Courtesy Matthew G. Wheeler)*

Figure 34-15*a* shows a thin lens with convex refracting surfaces, or *sides.* When rays that are parallel to the central axis of the lens are sent through the lens, they refract twice, as is shown enlarged in Fig. 34-15*b.* This double refraction causes the rays to converge and pass through a common point F_2 at a distance f from the center of the lens. Hence, this lens is a converging lens; further, a *real* focal point (or focus) exists at F_2 (because the rays really do pass through it), and

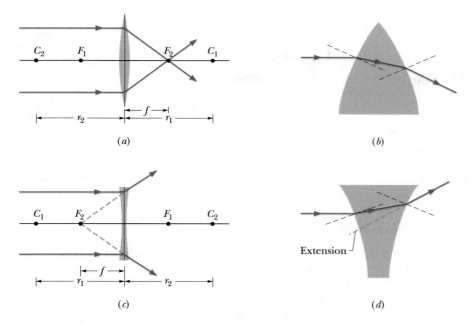

FIG. 34-15 (a) Rays initially parallel to the central axis of a converging lens are made to converge to a real focal point F_2 by the lens. The lens is thinner than drawn, with a width like that of the vertical line through it. We shall consider all the bending of rays as occurring at this central line. (b) An enlargement of the top part of the lens of (a); normals to the surfaces are shown dashed. Note that both refractions bend the ray downward, toward the central axis. (c) The same initially parallel rays are made to diverge by a diverging lens. Extensions of the diverging rays pass through a virtual focal point F_2. (d) An enlargement of the top part of the lens of (c). Note that both refractions bend the ray upward, away from the central axis.

the associated focal length is f. When rays parallel to the central axis are sent in the opposite direction through the lens, we find another real focal point at F_1 on the other side of the lens. For a thin lens, these two focal points are equidistant from the lens.

Because the focal points of a converging lens are real, we take the associated focal lengths f to be positive, just as we do with a real focus of a concave mirror. However, signs in optics can be tricky; so we had better check this in Eq. 34-10. The left side of that equation is positive if f is positive; how about the right side? We examine it term by term. Because the index of refraction n of glass or any other material is greater than 1, the term $(n-1)$ must be positive. Because the source of the light (which is the object) is at the left and faces the convex left side of the lens, the radius of curvature r_1 of that side must be positive according to the sign rule for refracting surfaces. Similarly, because the object faces a concave right side of the lens, the radius of curvature r_2 of that side must be negative according to that rule. Thus, the term $(1/r_1 - 1/r_2)$ is positive, the whole right side of Eq. 34-10 is positive, and all the signs are consistent.

Figure 34-15c shows a thin lens with concave sides. When rays that are parallel to the central axis of the lens are sent through this lens, they refract twice, as is shown enlarged in Fig. 34-15d; these rays *diverge,* never passing through any common point, and so this lens is a diverging lens. However, extensions of the rays do pass through a common point F_2 at a distance f from the center of the lens. Hence, the lens has a *virtual* focal point at F_2. (If your eye intercepts some of the diverging rays, you perceive a bright spot to be at F_2, as if it is the source of the light.) Another virtual focus exists on the opposite side of the lens at F_1, symmetrically placed if the lens is thin. Because the focal points of a diverging lens are virtual, we take the focal length f to be negative.

FIG. 34-16 (*a*) A real, inverted image *I* is formed by a converging lens when the object *O* is outside the focal point F_1. (*b*) The image *I* is virtual and has the same orientation as *O* when *O* is inside the focal point. (*c*) A diverging lens forms a virtual image *I*, with the same orientation as the object *O*, whether *O* is inside or outside the focal point of the lens.

Images from Thin Lenses

We now consider the types of image formed by converging and diverging lenses. Figure 34-16*a* shows an object *O* outside the focal point F_1 of a converging lens. The two rays drawn in the figure show that the lens forms a real, inverted image *I* of the object on the side of the lens opposite the object.

When the object is placed inside the focal point F_1, as in Fig. 34-16*b*, the lens forms a virtual image *I* on the same side of the lens as the object and with the same orientation. Hence, a converging lens can form either a real image or a virtual image, depending on whether the object is outside or inside the focal point, respectively.

Figure 34-16*c* shows an object *O* in front of a diverging lens. Regardless of the object distance (regardless of whether *O* is inside or outside the virtual focal point), this lens produces a virtual image that is on the same side of the lens as the object and has the same orientation.

As with mirrors, we take the image distance *i* to be positive when the image is real and negative when the image is virtual. However, the locations of real and virtual images from lenses are the reverse of those from mirrors:

> Real images form on the side of a lens that is opposite the object, and virtual images form on the side where the object is.

The lateral magnification *m* produced by converging and diverging lenses is given by Eqs. 34-5 and 34-6, the same as for mirrors.

You have been asked to absorb a lot of information in this section, and you should organize it for yourself by filling in Table 34-2 for thin *symmetric lenses* (both sides are convex or both sides are concave). Under Image Location note whether the image is on the *same* side of the lens as the object or on the *opposite* side. Under Image Type note whether the image is *real* or *virtual*. Under Image Orientation note whether the image has the *same* orientation as the object or is *inverted*.

TABLE 34-2

Your Organizing Table for Thin Lenses

Lens Type	Object Location	Image Location	Image Type	Image Orientation	Sign of *f*	Sign of *r*	Sign of *m*
Converging	Inside *F*						
	Outside *F*						
Diverging	Anywhere						

TACTIC 1: *Signs of Trouble with Mirrors and Lenses* Be careful: A mirror with a convex surface has a negative focal length f, just the opposite of a lens with convex surfaces. A mirror with a concave surface has a positive focal length f, just the opposite of a lens with concave surfaces. Confusing lens properties with mirror properties is a common mistake.

Locating Images of Extended Objects by Drawing Rays

Figure 34-17a shows an object O outside focal point F_1 of a converging lens. We can graphically locate the image of any off-axis point on such an object (such as the tip of the arrow in Fig. 34-17a) by drawing a ray diagram with any two of three special rays through the point. These special rays, chosen from all those that pass through the lens to form the image, are the following:

1. A ray that is initially parallel to the central axis of the lens will pass through focal point F_2 (ray 1 in Fig. 34-17a).

2. A ray that initially passes through focal point F_1 will emerge from the lens parallel to the central axis (ray 2 in Fig. 34-17a).

3. A ray that is initially directed toward the center of the lens will emerge from the lens with no change in its direction (ray 3 in Fig. 34-17a) because the ray encounters the two sides of the lens where they are almost parallel.

The image of the point is located where the rays intersect on the far side of the lens. The image of the object is found by locating the images of two or more of its points.

Figure 34-17b shows how the extensions of the three special rays can be used to locate the image of an object placed inside focal point F_1 of a converging lens. Note that the description of ray 2 requires modification (it is now a ray whose backward extension passes through F_1).

You need to modify the descriptions of rays 1 and 2 to use them to locate an image placed (anywhere) in front of a diverging lens. In Fig. 34-17c, for example, we find the point where ray 3 intersects the backward extensions of rays 1 and 2.

Two-Lens Systems

When an object O is placed in front of a system of two lenses whose central axes coincide, we can locate the final image of the system (that is, the image produced by the lens farther from the object) by working in steps. Let lens 1 be the nearer lens and lens 2 the farther lens.

Step 1 We let p_1 represent the distance of object O from lens 1. We then find the distance i_1 of the image produced by lens 1, either by use of Eq. 34-9 or by drawing rays.

Step 2 Now, ignoring the presence of lens 1, we treat the image found in step 1

(a)

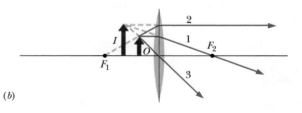

(b)

(c)

FIG. 34-17 Three special rays allow us to locate an image formed by a thin lens whether the object O is (a) outside or (b) inside the focal point of a converging lens, or (c) anywhere in front of a diverging lens.

as the object for lens 2. If this new object is located beyond lens 2, the object distance p_2 for lens 2 is taken to be negative. (Note this exception to the rule that says the object distance is positive; the exception occurs because the object here is on the side opposite the source of light.) Otherwise, p_2 is taken to be positive as usual. We then find the distance i_2 of the (final) image produced by lens 2 by use of Eq. 34-9 or by drawing rays.

A similar step-by-step solution can be used for any number of lenses or if a mirror is substituted for lens 2.

The *overall*, or *net*, lateral magnification M produced by a system of two lenses is the product of the lateral magnifications m_1 and m_2 produced by the two lenses:

$$M = m_1 m_2. \tag{34-11}$$

✓ **CHECKPOINT 4**   A thin symmetric lens provides an image of a fingerprint with a magnification of +0.2 when the fingerprint is 1.0 cm farther from the lens than the focal point of the lens. What are the type and orientation of the image, and what is the type of lens?

Sample Problem   34-4

A praying mantis preys along the central axis of a thin symmetric lens, 20 cm from the lens. The lateral magnification of the mantis provided by the lens is $m = -0.25$, and the index of refraction of the lens material is 1.65.

(a) Determine the type of image produced by the lens, the type of lens, whether the object (mantis) is inside or outside the focal point, on which side of the lens the image appears, and whether the image is inverted.

Reasoning: We can tell a lot about the lens and the image from the given value of m. From it and Eq. 34-6 ($m = -i/p$), we see that

$$i = -mp = 0.25p.$$

Even without finishing the calculation, we can answer the questions. Because p is positive, i here must be positive. That means we have a real image, which means we have a converging lens (the only lens that can produce a real image). The object must be outside the focal point (the only way a real image can be produced). Also, the image is inverted and on the side of the lens opposite the object. (That is how a converging lens makes a real image.)

(b) What are the two radii of curvature of the lens?

KEY IDEAS

1. Because the lens is symmetric, r_1 (for the surface nearer the object) and r_2 have the same magnitude r.

2. Because the lens is a converging lens, the object faces a convex surface on the nearer side and so $r_1 = +r$. Similarly, it faces a concave surface on the farther side; so $r_2 = -r$.

3. We can relate these radii of curvature to the focal length f via the lens maker's equation, Eq. 34-10 (our only equation involving the radii of curvature of a lens).

4. We can relate f to the object distance p and image distance i via Eq. 34-9.

Calculations: We know p, but we do not know i. Thus, our starting point is to finish the calculation for i in part (a); we obtain

$$i = (0.25)(20 \text{ cm}) = 5.0 \text{ cm}.$$

Now Eq. 34-9 gives us

$$\frac{1}{f} = \frac{1}{p} + \frac{1}{i} = \frac{1}{20 \text{ cm}} + \frac{1}{5.0 \text{ cm}},$$

from which we find $f = 4.0$ cm.

Equation 34-10 then gives us

$$\frac{1}{f} = (n-1)\left(\frac{1}{r_1} - \frac{1}{r_2}\right) = (n-1)\left(\frac{1}{+r} - \frac{1}{-r}\right)$$

or, with known values inserted,

$$\frac{1}{4.0 \text{ cm}} = (1.65 - 1)\frac{2}{r},$$

which yields

$$r = (0.65)(2)(4.0 \text{ cm}) = 5.2 \text{ cm}. \quad \text{(Answer)}$$

Sample Problem   34-5   Build your skill

Figure 34-18*a* shows a jalapeño seed O_1 that is placed in front of two thin symmetrical coaxial lenses 1 and 2, with focal lengths $f_1 = +24$ cm and $f_2 = +9.0$ cm, re-

spectively, and with lens separation $L = 10$ cm. The seed is 6.0 cm from lens 1. Where does the system of two lenses produce an image of the seed?

KEY IDEA We could locate the image produced by the system of lenses by tracing light rays from the seed through the two lenses. However, we can, instead, calculate the location of that image by working through the system in steps, lens by lens. We begin with the lens closer to the seed. The image we seek is the final one — that is, image I_2 produced by lens 2.

Lens 1: Ignoring lens 2, we locate the image I_1 produced by lens 1 by applying Eq. 34-9 to lens 1 alone:

$$\frac{1}{p_1} + \frac{1}{i_1} = \frac{1}{f_1}.$$

The object O_1 for lens 1 is the seed, which is 6.0 cm from the lens; thus, we substitute $p_1 = +6.0$ cm. Also substituting the given value of f_1, we then have

$$\frac{1}{+6.0 \text{ cm}} + \frac{1}{i_1} = \frac{1}{+24 \text{ cm}},$$

which yields $i_1 = -8.0$ cm.

This tells us that image I_1 is 8.0 cm from lens 1 and virtual. (We could have guessed that it is virtual by noting that the seed is inside the focal point of lens 1, that is, between the lens and its focal point.) Because I_1 is virtual, it is on the same side of the lens as object O_1 and has the same orientation as the seed, as shown in Fig. 34-18b.

Lens 2: In the second step of our solution, we treat image I_1 as an object O_2 for the second lens and now ignore lens 1. We first note that this object O_2 is outside the focal point of lens 2. So the image I_2 produced by lens 2 must be real, inverted, and on the side of the lens opposite O_2. Let us see.

The distance p_2 between this object O_2 and lens 2 is, from Fig. 34-18c,

$$p_2 = L + |i_1| = 10 \text{ cm} + 8.0 \text{ cm} = 18 \text{ cm}.$$

Then Eq. 34-9, now written for lens 2, yields

$$\frac{1}{+18 \text{ cm}} + \frac{1}{i_2} = \frac{1}{+9.0 \text{ cm}}.$$

Hence, $i_2 = +18$ cm.   (Answer)

(a)

(b)

(c)

FIG. 34-18 (a) Seed O_1 is distance p_1 from a two-lens system with lens separation L. We use the arrow to orient the seed. (b) The image I_1 produced by lens 1 alone. (c) Image I_1 acts as object O_2 for lens 2 alone, which produces the final image I_2.

The plus sign confirms our guess: Image I_2 produced by lens 2 is real, inverted, and on the side of lens 2 opposite O_2, as shown in Fig. 34-18c.

34-8 | Optical Instruments

The human eye is a remarkably effective organ, but its range can be extended in many ways by optical instruments such as eyeglasses, microscopes, and telescopes. Many such devices extend the scope of our vision beyond the visible range; satellite-borne infrared cameras and x-ray microscopes are just two examples.

The mirror and thin-lens formulas can be applied only as approximations to most sophisticated optical instruments. The lenses in typical laboratory microscopes are by no means "thin." In most optical instruments the lenses are compound lenses; that is, they are made of several components, the interfaces rarely being exactly spherical. Now we discuss three optical instruments, assuming, for simplicity, that the thin-lens formulas apply.

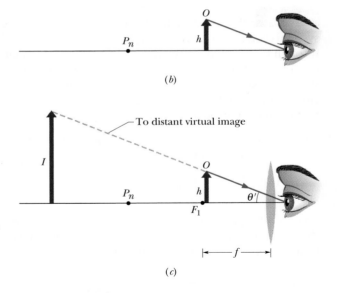

FIG. 34-19 (a) An object O of height h placed at the near point of a human eye occupies angle θ in the eye's view. (b) The object is moved closer to increase the angle, but now the observer cannot bring the object into focus. (c) A converging lens is placed between the object and the eye, with the object just inside the focal point F_1 of the lens. The image produced by the lens is then far enough away to be focused by the eye, and the image occupies a larger angle θ' than object O does in (a).

Simple Magnifying Lens

The normal human eye can focus a sharp image of an object on the retina (at the rear of the eye) if the object is located anywhere from infinity to a certain point called the *near point* P_n. If you move the object closer to the eye than the near point, the perceived retinal image becomes fuzzy. The location of the near point normally varies with age. We have all heard about people who claim not to need glasses but read their newspapers at arm's length; their near points are receding. To find your own near point, remove your glasses or contacts if you wear any, close one eye, and then bring this page closer to your open eye until it becomes indistinct. In what follows, we take the near point to be 25 cm from the eye, a bit more than the typical value for 20-year-olds.

Figure 34-19a shows an object O placed at the near point P_n of an eye. The size of the image of the object produced on the retina depends on the angle θ that the object occupies in the field of view from that eye. By moving the object closer to the eye, as in Fig. 34-19b, you can increase the angle and, hence, the possibility of distinguishing details of the object. However, because the object is then closer than the near point, it is no longer *in focus;* that is, the image is no longer clear.

You can restore the clarity by looking at O through a converging lens, placed so that O is just inside the focal point F_1 of the lens, which is at focal length f (Fig. 34-19c). What you then see is the virtual image of O produced by the lens. That image is farther away than the near point; thus, the eye can see it clearly.

Moreover, the angle θ' occupied by the virtual image is larger than the largest angle θ that the object alone can occupy and still be seen clearly. The *angular magnification* m_θ (not to be confused with lateral magnification m) of what is seen is

$$m_\theta = \theta'/\theta.$$

In words, the angular magnification of a simple magnifying lens is a comparison of the angle occupied by the image the lens produces with the angle occupied by the object when the object is moved to the near point of the viewer.

From Fig. 34-19, assuming that O is at the focal point of the lens, and approximating tan θ as θ and tan θ' as θ' for small angles, we have

$$\theta \approx h/25 \text{ cm} \quad \text{and} \quad \theta' \approx h/f.$$

We then find that

$$m_\theta \approx \frac{25 \text{ cm}}{f} \quad \text{(simple magnifier).} \tag{34-12}$$

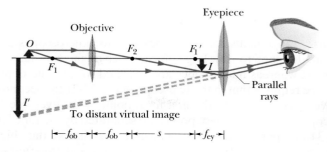

Compound Microscope

Figure 34-20 shows a thin-lens version of a compound microscope. The instrument consists of an *objective* (the front lens) of focal length f_{ob} and an *eyepiece* (the lens near the eye) of focal length f_{ey}. It is used for viewing small objects that are very close to the objective.

The object O to be viewed is placed just outside the first focal point F_1 of the objective, close enough to F_1 that we can approximate its distance p from the lens as being f_{ob}. The separation between the lenses is then adjusted so that the enlarged, inverted, real image I produced by the objective is located just inside the first focal point F_1' of the eyepiece. The *tube length s* shown in Fig. 34-20 is actually large relative to f_{ob}, and therefore we can approximate the distance i between the objective and the image I as being length s.

From Eq. 34-6, and using our approximations for p and i, we can write the lateral magnification produced by the objective as

$$m = -\frac{i}{p} = -\frac{s}{f_{ob}}. \tag{34-13}$$

Because the image I is located just inside the focal point F_1' of the eyepiece, the eyepiece acts as a simple magnifying lens, and an observer sees a final (virtual, inverted) image I' through it. The overall magnification of the instrument is the product of the lateral magnification m produced by the objective, given by Eq. 34-13, and the angular magnification m_θ produced by the eyepiece, given by Eq. 34-12; that is,

$$M = mm_\theta = -\frac{s}{f_{ob}}\frac{25\text{ cm}}{f_{ey}} \quad \text{(microscope).} \tag{34-14}$$

Refracting Telescope

Telescopes come in a variety of forms. The form we describe here is the simple refracting telescope that consists of an objective and an eyepiece; both are represented in Fig. 34-21 with simple lenses, although in practice, as is also true for most microscopes, each lens is actually a compound lens system.

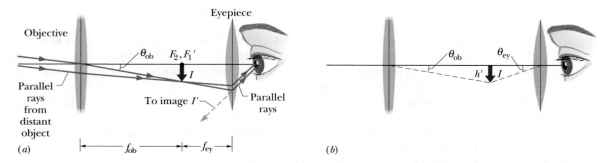

FIG. 34-21   (a) A thin-lens representation of a refracting telescope. From rays that are approximately parallel when they reach the objective, the objective produces a real image I of a distant source of light (the object). (One end of the object is assumed to lie on the central axis.) Image I, formed at the common focal points F_2 and F_1', acts as an object for the eyepiece, which produces a virtual final image I' at a great distance from the observer. The objective has focal length f_{ob}; the eyepiece has focal length f_{ey}. (b) Image I has height h' and takes up angle θ_{ob} measured from the objective and angle θ_{ey} measured from the eyepiece.

The lens arrangements for telescopes and for microscopes are similar, but telescopes are designed to view large objects, such as galaxies, stars, and planets, at large distances, whereas microscopes are designed for just the opposite purpose. This difference requires that in the telescope of Fig. 34-21 the second focal point of the objective F_2 coincide with the first focal point of the eyepiece F_1', whereas in the microscope of Fig. 34-20 these points are separated by the tube length s.

In Fig. 34-21a, parallel rays from a distant object strike the objective, making an angle θ_{ob} with the telescope axis and forming a real, inverted image I at the common focal point F_2, F_1'. This image I acts as an object for the eyepiece, through which an observer sees a distant (still inverted) virtual image I'. The rays defining the image make an angle θ_{ey} with the telescope axis.

The angular magnification m_θ of the telescope is θ_{ey}/θ_{ob}. From Fig. 34-21b, for rays close to the central axis, we can write $\theta_{ob} = h'/f_{ob}$ and $\theta_{ey} \approx h'/f_{ey}$, which gives us

$$m_\theta = -\frac{f_{ob}}{f_{ey}} \quad \text{(telescope)}, \tag{34-15}$$

where the minus sign indicates that I' is inverted. In words, the angular magnification of a telescope is a comparison of the angle occupied by the image the telescope produces with the angle occupied by the distant object as seen without the telescope.

Magnification is only one of the design factors for an astronomical telescope and is indeed easily achieved. A good telescope needs *light-gathering power,* which determines how bright the image is. This is important for viewing faint objects such as distant galaxies and is accomplished by making the objective diameter as large as possible. A telescope also needs *resolving power,* which is the ability to distinguish between two distant objects (stars, say) whose angular separation is small. *Field of view* is another important design parameter. A telescope designed to look at galaxies (which occupy a tiny field of view) is much different from one designed to track meteors (which move over a wide field of view).

The telescope designer must also take into account the difference between real lenses and the ideal thin lenses we have discussed. A real lens with spherical surfaces does not form sharp images, a flaw called *spherical aberration.* Also, because refraction by the two surfaces of a real lens depends on wavelength, a real lens does not focus light of different wavelengths to the same point, a flaw called *chromatic aberration.*

This brief discussion by no means exhausts the design parameters of astronomical telescopes—many others are involved. We could make a similar listing for any other high-performance optical instrument.

34-9 | Three Proofs

The Spherical Mirror Formula (Eq. 34-4)

Figure 34-22 shows a point object O placed on the central axis of a concave spherical mirror, outside its center of curvature C. A ray from O that makes an angle α with the axis intersects the axis at I after reflection from the mirror at a. A ray that leaves O along the axis is reflected back along itself at c and also passes through I. Thus, I is the image of O; it is a *real* image because light actually passes through it. Let us find the image distance i.

A trigonometry theorem that is useful here tells us that an exterior angle of a triangle is equal to the sum of the two opposite interior angles. Applying this to triangles OaC and OaI in Fig. 34-22 yields

$$\beta = \alpha + \theta \quad \text{and} \quad \gamma = \alpha + 2\theta.$$

If we eliminate θ between these two equations, we find

$$\alpha + \gamma = 2\beta. \tag{34-16}$$

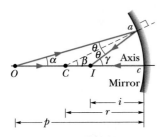

FIG. 34-22 A concave spherical mirror forms a real point image I by reflecting light rays from a point object O.

We can write angles α, β, and γ, in radian measure, as

$$\alpha \approx \frac{\widehat{ac}}{cO} = \frac{\widehat{ac}}{p}, \qquad \beta = \frac{\widehat{ac}}{cC} = \frac{\widehat{ac}}{r},$$

and

$$\gamma \approx \frac{\widehat{ac}}{cI} = \frac{\widehat{ac}}{i}, \qquad (34\text{-}17)$$

where the overhead symbol means "arc." Only the equation for β is exact, because the center of curvature of $\widehat{ac}$ is at C. However, the equations for α and γ are approximately correct if these angles are small enough (that is, for rays close to the central axis). Substituting Eqs. 34-17 into Eq. 34-16, using Eq. 34-3 to replace r with $2f$, and canceling $\widehat{ac}$ lead exactly to Eq. 34-4, the relation that we set out to prove.

FIG. 34-23   A real point image I of a point object O is formed by refraction at a spherical convex surface between two media.

The Refracting Surface Formula (Eq. 34-8)

The incident ray from point object O in Fig. 34-23 that falls on point a of a spherical refracting surface is refracted there according to Eq. 33-40,

$$n_1 \sin \theta_1 = n_2 \sin \theta_2.$$

If α is small, θ_1 and θ_2 will also be small and we can replace the sines of these angles with the angles themselves. Thus, the equation above becomes

$$n_1 \theta_1 \approx n_2 \theta_2. \qquad (34\text{-}18)$$

We again use the fact that an exterior angle of a triangle is equal to the sum of the two opposite interior angles. Applying this to triangles COa and ICa yields

$$\theta_1 = \alpha + \beta \quad \text{and} \quad \beta = \theta_2 + \gamma. \qquad (34\text{-}19)$$

If we use Eqs. 34-19 to eliminate θ_1 and θ_2 from Eq. 34-18, we find

$$n_1 \alpha + n_2 \gamma = (n_2 - n_1)\beta. \qquad (34\text{-}20)$$

In radian measure the angles α, β, and γ are

$$\alpha \approx \frac{\widehat{ac}}{p}; \qquad \beta = \frac{\widehat{ac}}{r}; \qquad \gamma \approx \frac{\widehat{ac}}{i}. \qquad (34\text{-}21)$$

Only the second of these equations is exact. The other two are approximate because I and O are not the centers of circles of which $\widehat{ac}$ is a part. However, for α small enough (for rays close to the axis), the inaccuracies in Eqs. 34-21 are small. Substituting Eqs. 34-21 into Eq. 34-20 leads directly to Eq. 34-8, as we wanted.

The Thin-Lens Formulas (Eqs. 34-9 and 34-10)

Our plan is to consider each lens surface as a separate refracting surface, and to use the image formed by the first surface as the object for the second.

We start with the thick glass "lens" of length L in Fig. 34-24a whose left and right refracting surfaces are ground to radii r' and r''. A point object O' is placed near the left surface as shown. A ray leaving O' along the central axis is not deflected on entering or leaving the lens.

A second ray leaving O' at an angle α with the central axis intersects the left surface at point a', is refracted, and intersects the second (right) surface at point a''. The ray is again refracted and crosses the axis at I'', which, being the intersection of two rays from O', is the image of point O', formed after refraction at two surfaces.

Figure 34-24b shows that the first (left) surface also forms a virtual image of O' at I'. To locate I', we use Eq. 34-8,

$$\frac{n_1}{p} + \frac{n_2}{i} = \frac{n_2 - n_1}{r}.$$

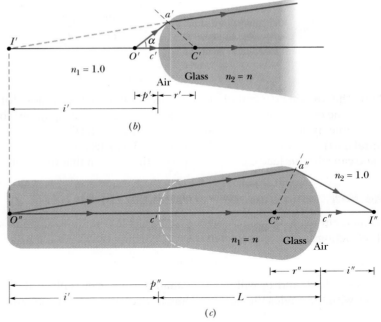

FIG. 34-24 (a) Two rays from point object O' form a real image I'' after refracting through two spherical surfaces of a lens. The object faces a convex surface at the left side of the lens and a concave surface at the right side. The ray traveling through points a' and a'' is actually close to the central axis through the lens. (b) The left side and (c) the right side of the lens in (a), shown separately.

Putting $n_1 = 1$ for air and $n_2 = n$ for lens glass and bearing in mind that the image distance is negative (that is, $i = -i'$ in Fig. 34-24b), we obtain

$$\frac{1}{p'} - \frac{n}{i'} = \frac{n-1}{r'}. \tag{34-22}$$

In this equation i' will be a positive number because we have already introduced the minus sign appropriate to a virtual image.

Figure 34-24c shows the second surface again. Unless an observer at point a'' were aware of the existence of the first surface, the observer would think that the light striking that point originated at point I' in Fig. 34-24b and that the region to the left of the surface was filled with glass as indicated. Thus, the (virtual) image I' formed by the first surface serves as a real object O'' for the second surface. The distance of this object from the second surface is

$$p'' = i' + L. \tag{34-23}$$

To apply Eq. 34-8 to the second surface, we must insert $n_1 = n$ and $n_2 = 1$ because the object now is effectively imbedded in glass. If we substitute with Eq. 34-23, then Eq. 34-8 becomes

$$\frac{n}{i' + L} + \frac{1}{i''} = \frac{1-n}{r''}. \tag{34-24}$$

Let us now assume that the thickness L of the "lens" in Fig. 34-24a is so small that we can neglect it in comparison with our other linear quantities (such as p', i', p'', i'', r', and r''). In all that follows we make this *thin-lens approximation*. Putting $L = 0$ in Eq. 34-24 and rearranging the right side lead to

$$\frac{n}{i'} + \frac{1}{i''} = -\frac{n-1}{r''}. \tag{34-25}$$

Adding Eqs. 34-22 and 34-25 leads to

$$\frac{1}{p'} + \frac{1}{i''} = (n-1)\left(\frac{1}{r'} - \frac{1}{r''}\right).$$

Finally, calling the original object distance simply p and the final image distance simply i leads to

$$\frac{1}{p} + \frac{1}{i} = (n-1)\left(\frac{1}{r'} - \frac{1}{r''}\right), \qquad (34\text{-}26)$$

which, with a small change in notation, is Eqs. 34-9 and 34-10, the relations we set out to prove.

REVIEW & SUMMARY

Real and Virtual Images An *image* is a reproduction of an object via light. If the image can form on a surface, it is a *real image* and can exist even if no observer is present. If the image requires the visual system of an observer, it is a *virtual image*.

Image Formation *Spherical mirrors, spherical refracting surfaces,* and *thin lenses* can form images of a source of light—the object—by redirecting rays emerging from the source. The image occurs where the redirected rays cross (forming a real image) or where backward extensions of those rays cross (forming a virtual image). If the rays are sufficiently close to the *central axis* through the spherical mirror, refracting surface, or thin lens, we have the following relations between the *object distance p* (which is positive) and the *image distance i* (which is positive for real images and negative for virtual images):

1. **Spherical Mirror:**

$$\frac{1}{p} + \frac{1}{i} = \frac{1}{f} = \frac{2}{r}, \qquad (34\text{-}4, 34\text{-}3)$$

where f is the mirror's focal length and r is its radius of curvature. A *plane mirror* is a special case for which $r \to \infty$, so that $p = -i$. Real images form on the side of a mirror where the object is located, and virtual images form on the opposite side.

2. **Spherical Refracting Surface:**

$$\frac{n_1}{p} + \frac{n_2}{i} = \frac{n_2 - n_1}{r} \qquad \text{(single surface),} \qquad (34\text{-}8)$$

where n_1 is the index of refraction of the material where the object is located, n_2 is the index of refraction of the material on the other side of the refracting surface, and r is the radius of curvature of the surface. When the object faces a convex refracting surface, the radius r is positive. When it faces a concave surface, r is negative. Real images form on the side of a refracting surface that is opposite the object, and virtual images form on the same side as the object.

3. **Thin Lens:**

$$\frac{1}{p} + \frac{1}{i} = \frac{1}{f} = (n-1)\left(\frac{1}{r_1} - \frac{1}{r_2}\right), \qquad (34\text{-}9, 34\text{-}10)$$

where f is the lens's focal length, n is the index of refraction of the lens material, and r_1 and r_2 are the radii of curvature of the two sides of the lens, which are spherical surfaces. A convex lens surface that faces the object has a positive radius of curvature; a concave lens surface that faces the object has a negative radius of curvature. Real images form on the side of a lens that is opposite the object, and virtual images form on the same side as the object.

Lateral Magnification The *lateral magnification m* produced by a spherical mirror or a thin lens is

$$m = -\frac{i}{p}. \qquad (34\text{-}6)$$

The magnitude of m is given by

$$|m| = \frac{h'}{h}, \qquad (34\text{-}5)$$

where h and h' are the heights (measured perpendicular to the central axis) of the object and image, respectively.

Optical Instruments Three optical instruments that extend human vision are:

1. The *simple magnifying lens,* which produces an *angular magnification* m_θ given by

$$m_\theta = \frac{25 \text{ cm}}{f}, \qquad (34\text{-}12)$$

where f is the focal length of the magnifying lens.

2. The *compound microscope,* which produces an *overall magnification M* given by

$$M = mm_\theta = -\frac{s}{f_{\text{ob}}} \frac{25 \text{ cm}}{f_{\text{ey}}}, \qquad (34\text{-}14)$$

where m is the lateral magnification produced by the objective, m_θ is the angular magnification produced by the eyepiece, s is the tube length, and f_{ob} and f_{ey} are the focal lengths of the objective and eyepiece, respectively.

3. The *refracting telescope,* which produces an *angular magnification* m_θ given by

$$m_\theta = -\frac{f_{\text{ob}}}{f_{\text{ey}}}. \qquad (34\text{-}15)$$

QUESTIONS

1 Figure 34-25 is an overhead view of a mirror maze based on floor sections that are equilateral triangles. Every wall within the maze is mirrored. If you stand at entrance x, (a) which of the maze monsters a, b, and c hiding in the maze can you see along the virtual hallways extending from entrance x; (b) how many times does each visible monster appear in a hallway; and (c) what is at the far end of a hallway?

FIG. 34-25 Question 1.

2 A penguin waddles along the central axis of a concave mirror, from the focal point to an effectively infinite distance. (a) How does its image move? (b) Does the height of its image increase continuously, decrease continuously, or change in some more complicated manner?

3 Figure 34-26 shows a fish and a fish stalker in water. (a) Does the stalker see the fish in the general region of point a or point b? (b) Does the fish see the (wild) eyes of the stalker in the general region of point c or point d?

FIG. 34-26 Question 3.

4 In Fig. 34-27, stick figure O stands in front of a spherical mirror that is mounted within the boxed region; the central axis through the mirror is shown. The four stick figures I_1 to I_4 suggest general locations and orientations for the images that might be produced by the mirror. (The figures are only sketched in; neither their heights nor their distances from the mirror are drawn to scale.) (a) Which of the stick figures could not possibly represent images? Of the possible images, (b) which would be due to a concave mirror, (c) which would be due to a convex mirror, (d) which would be virtual, and (e) which would involve negative magnification?

FIG. 34-27
Questions 4 and 8.

5 Figure 34-28 shows four thin lenses, all of the same material, with sides that either are flat or have a radius of curvature of magnitude 10 cm. Without written calculation, rank the lenses according to the magnitude of the focal length, greatest first.

(a) (b) (c) (d)

FIG. 34-28 Question 5.

6 An object is placed against the center of a concave mirror and then moved along the central axis until it is 5.0 m from the mirror. During the motion, the distance |i| between the mirror and the image it produces is measured. The procedure is then repeated with a convex mirror and a plane mirror. Figure 34-29 gives the results versus object distance p. Which curve corresponds to which mirror? (Curve 1 has two segments.)

FIG. 34-29 Questions 6 and 10.

7 When a *T. rex* pursues a jeep in the movie *Jurassic Park*, we see a reflected image of the *T. rex* via a side-view mirror, on which is printed the (then darkly humorous) warning: "Objects in mirror are closer than they appear." Is the mirror flat, convex, or concave?

8 In Fig. 34-27, stick figure O stands in front of a thin, symmetric lens that is mounted within the boxed region; the central axis through the lens is shown. The four stick figures I_1 to I_4 suggest general locations and orientations for the images that might be produced by the lens. (The figures are only sketched in; neither their height nor their distance from the lens is drawn to scale.) (a) Which of the stick figures could not possibly represent images? Of the possible images, (b) which would be due to a converging lens, (c) which would be due to a diverging lens, (d) which would be virtual, and (e) which would involve negative magnification?

9 The table details six variations of the basic arrangement of two thin lenses represented in Fig. 34-30. (The points labeled F_1 and F_2 are the focal points of lenses 1 and 2.) An object is distance p_1 to the left of lens 1, as in Fig. 34-18. (a) For which variations can we tell, *without calculation*, whether the final image (that due to lens 2) is to the left or right of lens 2 and whether it has the same orientation as the object? (b) For those "easy" variations, give the image location as "left" or "right" and the orientation as "same" or "inverted."

FIG. 34-30 Question 9.

Variation	Lens 1	Lens 2	
1	Converging	Converging	$p_1 < |f_1|$
2	Converging	Converging	$p_1 > |f_1|$
3	Diverging	Converging	$p_1 < |f_1|$
4	Diverging	Converging	$p_1 > |f_1|$
5	Diverging	Diverging	$p_1 < |f_1|$
6	Diverging	Diverging	$p_1 > |f_1|$

10 An object is placed against the center of a converging lens and then moved along the central axis until it is 5.0 m from the lens. During the motion, the distance |i| between the lens and the image it produces is measured. The procedure is then repeated with a diverging lens. Which of the curves in Fig. 34-29 best gives |i| versus the object distance p for these lenses? (Curve 1 consists of two segments. Curve 3 is straight.)

PROBLEMS

GO   Tutoring problem available (at instructor's discretion) in *WileyPLUS* and WebAssign

SSM   Worked-out solution available in Student Solutions Manual   WWW   Worked-out solution is at ───

• – •••   Number of dots indicates level of problem difficulty   ILW   Interactive solution is at ────   http://www.wiley.com/college/halliday

  Additional information available in *The Flying Circus of Physics* and at flyingcircusofphysics.com

sec. 34-3 Plane Mirrors

•1   A moth at about eye level is 10 cm in front of a plane mirror; you are behind the moth, 30 cm from the mirror. What is the distance between your eyes and the apparent position of the moth's image in the mirror?   ILW

•2   You look through a camera toward an image of a hummingbird in a plane mirror. The camera is 4.30 m in front of the mirror. The bird is at camera level, 5.00 m to your right and 3.30 m from the mirror. What is the distance between the camera and the apparent position of the bird's image in the mirror?

••3   Figure 34-31 shows an overhead view of a corridor with a plane mirror M mounted at one end. A burglar B sneaks along the corridor directly toward the center of the mirror. If $d = 3.0$ m, how far from the mirror will she be when the security guard S can first see her in the mirror?

FIG. 34-31   Problem 3.

••4   In Fig. 34-32, an isotropic point source of light S is positioned at distance d from a viewing screen A and the light intensity I_P at point P (level with S) is measured. Then a plane mirror M is placed behind S at distance d. By how much is I_P multiplied by the presence of the mirror?

FIG. 34-32 Problem 4.

•••5   Figure 34-33 shows a small lightbulb suspended at distance $d_1 = 250$ cm above the surface of the water in a swimming pool where the water depth is $d_2 = 200$ cm. The bottom of the pool is a large mirror. How far below the mirror surface is the image of the bulb? (*Hint:* Construct a diagram of two rays like that of Fig. 34-3, but take into account the bending of light rays by refraction at the water surface. Assume that the rays are close to a vertical axis through the bulb, and use the small-angle approximation in which $\sin \theta \approx \tan \theta \approx \theta$.)   SSM WWW

sec. 34-5 Images from Spherical Mirrors

•6   An object is placed against the center of a spherical mirror and then moved 70 cm from it along the central axis as the image distance i is measured. Figure 34-34 gives i versus object distance p out to $p_s = 40$ cm. What is the image distance when the object is on the central axis and 70 cm from the mirror?

•7   A concave shaving mirror has a radius of curvature of 35.0 cm. It is positioned so that the (upright) image of a man's face is 2.50 times the size of the face. How far is the mirror from the face?

•8   An object is moved along the central axis of a spherical mirror while the lateral magnification m of it is measured. Figure 34-35 gives m versus object distance p for the range $p_a = 2.0$ cm to $p_b = 8.0$ cm. What is the magnification of the object when the object is 14.0 cm from the mirror?

FIG. 34-33   Problem 5.

FIG. 34-34   Problem 6.

FIG. 34-35   Problem 8.

••9 through 16   *Spherical mirrors.* Object O stands on the central axis of a spherical mirror. For this situation, each problem in Table 34-3 gives object distance p_s (centimeters), the

TABLE 34-3

Problems 9 through 16: Spherical Mirrors. See the setup for these problems.

	p	Mirror	(a) r	(b) i	(c) m	(d) R/V	(e) I/NI	(f) Side
9	+12	Concave, 18						
10	+24	Concave, 36						
11	+18	Concave, 12						
12	+15	Concave, 10						
13	+10	Convex, 8.0						
14	+17	Convex, 14						
15	+8.0	Convex, 10						
16	+22	Convex, 35						

type of mirror, and then the distance (centimeters, without proper sign) between the focal point and the mirror. Find (a) the radius of curvature r (including sign), (b) the image distance i, and (c) the lateral magnification m. Also, determine whether the image is (d) real (R) or virtual (V), (e) inverted (I) from object O or noninverted (NI), and (f) on the *same* side of the mirror as O or on the *opposite* side. SSM 9, 11, 15 GO 10

••17 (a) A luminous point is moving at speed v_O toward a spherical mirror with radius of curvature r, along the central axis of the mirror. Show that the image of this point is moving at speed

$$v_I = -\left(\frac{r}{2p - r}\right)^2 v_O,$$

where p is the distance of the luminous point from the mirror at any given time. Now assume the mirror is concave, with $r = 15$ cm, and let $v_O = 5.0$ cm/s. Find v_I when (b) $p = 30$ cm (far outside the focal point), (c) $p = 8.0$ cm (just outside the focal point), and (d) $p = 10$ mm (very near the mirror).

••18 Figure 34-36 gives the lateral magnification m of an object versus the object distance p from a spherical mirror as the object is moved along the mirror's central axis through a range of values for p. The horizontal scale is set by $p_s = 10.0$ cm. What is the magnification of the object when the object is 21 cm from the mirror?

FIG. 34-36   Problem 18.

••19 through 31   *More mirrors.* Object O stands on the central axis of a spherical or plane mirror. For this situation, each problem in Table 34-4 refers to (a) the type of mirror, (b) the focal distance f, (c) the radius of curvature r, (d) the object distance p, (e) the image distance i, and (f) the lateral magnification m. (All distances are in centimeters.) It also refers to whether (g) the image is real (R) or virtual (V), (h)

inverted (I) or noninverted (NI) from O, and (i) on the *same* side of the mirror as object O or on the *opposite* side. Fill in the missing information. Where only a sign is missing, answer with the sign. SSM 27, 29 GO 26

sec. 34-6   Spherical Refracting Surfaces

•32   A glass sphere has radius $R = 5.0$ cm and index of refraction 1.6. A paperweight is constructed by slicing through the sphere along a plane that is 2.0 cm from the center of the sphere, leaving height $h = 3.0$ cm. The paperweight is placed on a table and viewed from directly above by an observer who is distance $d = 8.0$ cm from the table-top (Fig. 34-37). When viewed through the paperweight, how far away does the tabletop appear to be to the observer?

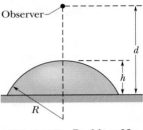

FIG. 34-37   Problem 32.

•33   In Fig. 34-38, a beam of parallel light rays from a laser is incident on a solid transparent sphere of index of refraction n. (a) If a point image is produced at the back of the sphere, what is the index of refraction of the sphere? (b) What index of refraction, if any, will produce a point image at the center of the sphere?

FIG. 34-38   Problem 33.

••34 through 40   *Spherical refracting surfaces.* An object O stands on the central axis of a spherical refracting surface. For this situation, each problem in Table 34-5 refers to the index of refraction n_1 where the object is located, (a) the index of refraction n_2 on the other side of the refracting surface, (b) the object distance p, (c) the radius of curvature r of the surface, and (d) the image distance i. (All distances are in centimeters.) Fill in the missing information, including whether the image is

---

TABLE 34-4

Problems 19 through 31: More Mirrors. See the setup for these problems.

	(a) Type	(b) f	(c) r	(d) p	(e) i	(f) m	(g) R/V	(h) I/NI	(i) Side
19	Concave	20		+10					
20				+60		−0.50			
21			−40		−10				
22				+10		+1.0			
23		+20		+30					
24				+24		0.50		I	
25		−30	·		−15				
26		20				+0.10			
27	Convex		40		4.0				
28				+40		−0.70			
29		30				+0.20			
30		20		+60					Same
31				+30		0.40		I	

TABLE 34-5

Problems 34 through 40: Spherical Refracting Surfaces. See the setup for these problems.

	(a) n_1	(b) n_2	(c) p	(d) r	(e) i	(e) R/V	(f) Side
34	1.0	1.5		+30	+600		
35	1.0	1.5	+10		−13		
36	1.0	1.5	+10	+30			
37	1.5	1.0	+70	+30			
38	1.5		+100	−30	+600		
39	1.5	1.0	+10		−6.0		
40	1.5	1.0		−30	−7.5		

(e) real (R) or virtual (V) and (f) on the *same* side of the surface as object O or on the *opposite* side. **SSM** 35,37 **GO** 39

sec. 34-7 Thin Lenses

•**41** A double-convex lens is to be made of glass with an index of refraction of 1.5. One surface is to have twice the radius of curvature of the other and the focal length is to be 60 mm. What is the (a) smaller and (b) larger radius? **SSM WWW**

•**42** An object is placed against the center of a thin lens and then moved away from it along the central axis as the image distance i is measured. Figure 34-39 gives i versus object distance p out to $p_s = 60$ cm. What is the image distance when $p = 100$ cm?

FIG. 34-39 Problem 42.

•**43** You produce an image of the Sun on a screen, using a thin lens whose focal length is 20.0 cm. What is the diameter of the image? (See Appendix C for needed data on the Sun.)

•**44** An object is placed against the center of a thin lens and then moved 70 cm from it along the central axis as the image distance i is measured. Figure 34-40 gives i versus object distance p out to $p_s = 40$ cm. What is the image distance when $p = 70$ cm?

FIG. 34-40 Problem 44.

•**45** A movie camera with a (single) lens of focal length 75 mm takes a picture of a person standing 27 m away. If the person is 180 cm tall, what is the height of the image on the film?

•**46** An object is moved along the central axis of a thin lens while the lateral magnification m is measured. Figure

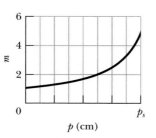

FIG. 34-41 Problem 46.

34-41 gives m versus object distance p out to $p_s = 8.0$ cm. What is the magnification of the object when the object is 14.0 cm from the lens?

•**47** An illuminated slide is held 44 cm from a screen. How far from the slide must a lens of focal length 11 cm be placed (between the slide and the screen) to form an image of the slide's picture on the screen? **SSM**

•**48** Figure 34-42 gives the lateral magnification m of an object versus the object distance p from a lens as the object is moved along the central axis of the lens through a range of values for p out to $p_s = 20.0$ cm. What is the magnification of the object when the object is 35 cm from the lens?

FIG. 34-42 Problem 48.

•**49** A lens is made of glass having an index of refraction of 1.5. One side of the lens is flat, and the other is convex with a radius of curvature of 20 cm. (a) Find the focal length of the lens. (b) If an object is placed 40 cm in front of the lens, where will the image be located?

••**50 through 57** *Thin lenses.* Object O stands on the central axis of a thin symmetric lens. For this situation, each problem in Table 34-6 gives object distance p (centimeters), the type of lens (C stands for converging and D for diverging), and then the distance (centimeters, without proper sign) between a focal point and the lens. Find (a) the image distance i and (b) the lateral magnification m of the object, including signs. Also, determine whether the image is (c) real (R) or virtual (V), (d) inverted (I) from object O or noninverted (NI), and (e) on the *same* side of the lens as object O or on the *opposite* side. **SSM** 51,55 **GO** 57

••**58** In Fig. 34-43, a real inverted image I of an object O is formed by a certain lens (not shown); the object–image separation is $d = 40.0$ cm, measured along the central axis of the lens. The image is just half the

FIG. 34-43 Problem 58.

TABLE 34-6

Problems 50 through 57: Thin Lenses. See the setup for these problems.

	p	Lens	(a) i	(b) m	(c) R/V	(d) I/NI	(e) Side
50	+10	D, 6.0					
51	+8.0	D, 12					
52	+16	C, 4.0					
53	+12	C, 16					
54	+12	D, 31					
55	+45	C, 20					
56	+25	C, 35					
57	+22	D, 14					

size of the object. (a) What kind of lens must be used to produce this image? (b) How far from the object must the lens be placed? (c) What is the focal length of the lens?

••59 through 68 *Lenses with given radii.* Object *O* stands in front of a thin lens, on the central axis. For this situation, each problem in Table 34-7 gives object distance *p*, index of refraction *n* of the lens, radius r_1 of the nearer lens surface, and radius r_2 of the farther lens surface. (All distances are in centimeters.) Find (a) the image distance *i* and (b) the lateral magnification *m* of the object, including signs. Also, determine whether the image is (c) real (R) or virtual (V), (d) inverted (I) from object *O* or noninverted (NI), and (e) on the *same* side of the lens as object *O* or on the *opposite* side. **SSM** 61 **GO** 62

••69 through 79 *More lenses.* Object *O* stands on the central axis of a thin symmetric lens. For this situation, each problem in Table 34-8 refers to (a) the lens type, converging (C) or diverging (D), (b) the focal distance *f*, (c) the object distance *p*, (d) the image distance *i*, and (e) the lateral magnification *m*. (All distances are in centimeters.) It also refers to whether (f) the image is real (R) or virtual (V), (g) inverted (I) or noninverted (NI) from *O*, and (h) on the *same* side of the lens as *O* or on the *opposite* side. Fill in the missing information, including

the value of *m* when only an inequality is given. Where only a sign is missing, answer with the sign. **SSM** 75, 79 **GO** 78

••80 through 87 *Two-lens systems.* In Fig. 34-44, stick figure *O* (the object) stands on the common central axis of two thin, symmetric lenses, which are mounted in the boxed regions. Lens 1 is mounted within the boxed region closer to *O*, which is at object distance p_1. Lens 2 is mounted within the farther boxed region, at distance *d*. Each problem in Table 34-9 refers to a different combination of lenses and different values for distances, which are given in centimeters. The type of lens is indicated by C for converging and D for diverging; the number after C or D is the distance between a lens and either of its focal points (the proper sign of the focal distance is not indicated).

FIG. 34-44 Problems 80 through 87.

Find (a) the image distance i_2 for the image produced by lens 2 (the final image produced by the system) and (b) the overall lateral magnification *M* for the system, including signs. Also, determine whether the final image is (c) real (R) or

TABLE 34-7

Problems 59 through 68: Lenses with Given Radii. See the setup for these problems.

	p	n	r_1	r_2	(a) i	(b) m	(c) R/V	(d) I/NI	(e) Side
59	+35	1.70	+42	+33					
60	+29	1.65	+35	∞					
61	+75	1.55	+30	−42					
62	+18	1.60	−27	+24					
63	+60	1.50	+35	−35					
64	+6.0	1.70	+10	−12					
65	+24	1.50	−15	−25					
66	+10	1.50	−30	−60					
67	+10	1.50	−30	+30					
68	+10	1.50	+30	−30					

TABLE 34-8

Problems 69 through 79: More Lenses. See the setup for these problems.

	(a) Type	(b) f	(c) p	(d) i	(e) m	(f) R/V	(g) I/NI	(h) Side
69			+16		+0.25			
70			+16		−0.25			
71		20	+8.0		>1.0			
72		20	+8.0		<1.0		NI	
73		+10	+5.0					
74		10	+5.0		>1.0			
75			+16		+1.25			
76	C	10	+20					
77			+10		−0.50			
78			+10		0.50		NI	
79		10	+5.0		<1.0			Same

virtual (V), (d) inverted (I) from object O or noninverted (NI), and (e) on the *same* side of lens 2 as object O or on the *opposite* side. SSM WWW 81 GO 87

sec. 34-8 Optical Instruments

•88 If the angular magnification of an astronomical telescope is 36 and the diameter of the objective is 75 mm, what is the minimum diameter of the eyepiece required to collect all the light entering the objective from a distant point source on the telescope axis?

•89 In a microscope of the type shown in Fig. 34-20, the focal length of the objective is 4.00 cm, and that of the eyepiece is 8.00 cm. The distance between the lenses is 25.0 cm. (a) What is the tube length s? (b) If image I in Fig. 34-20 is to be just inside focal point F'_1, how far from the objective should the object be? What then are (c) the lateral magnification m of the objective, (d) the angular magnification m_θ of the eyepiece, and (e) the overall magnification M of the microscope? SSM

••90 An object is 10.0 mm from the objective of a certain compound microscope. The lenses are 300 mm apart, and the intermediate image is 50.0 mm from the eyepiece. What overall magnification is produced by the instrument?

••91 Someone with a near point P_n of 25 cm views a thimble through a simple magnifying lens of focal length 10 cm by placing the lens near his eye. What is the angular magnification of the thimble if it is positioned so that its image appears at (a) P_n and (b) infinity?

••92 Figure 34-45a shows the basic structure of a camera. A lens can be moved forward or back to produce an image on film at the back of the camera. For a certain camera, with the distance i between the lens and the film set at $f = 5.0$ cm, parallel light rays from a very distant object O converge to a point image on the film, as shown. The object is now brought closer, to a distance of $p = 100$ cm, and the lens–film dis-

(a)

(b)

FIG. 34-45 Problem 92.

TABLE 34-9

Problems 80 through 87: Two-Lens Systems. See the setup for these problems.

	p_1	Lens 1	d	Lens 2	(a) i_2	(b) M	(c) R/V	(d) I/NI	(e) Side
80	+8.0	D, 6.0	12	C, 6.0					
81	+20	C, 9.0	8.0	C, 5.0					
82	+12	C, 8.0	30	D, 8.0					
83	+12	C, 8.0	32	C, 6.0					
84	+10	C, 15	10	C, 8.0					
85	+20	D, 12	10	D, 8.0					
86	+15	C, 12	67	C, 10					
87	+4.0	C, 6.0	8.0	D, 6.0					

tance is adjusted so that an inverted real image forms on the film (Fig. 34-45b). (a) What is the lens–film distance i now? (b) By how much was distance i changed?

••93 Figure 34-46a shows the basic structure of a human eye. Light refracts into the eye through the cornea and is then further redirected by a lens whose shape (and thus ability to focus the light) is controlled by muscles. We can treat the cornea and eye lens as a single effective thin lens (Fig. 34-46b). A "normal" eye can focus parallel light rays from a distant object O to a point on the retina at the back of the eye, where processing of the visual information begins. As an object is brought close to the eye, however, the muscles must change the shape of the lens so that rays form an inverted real image on the retina (Fig. 34-46c). (a) Suppose that for the parallel rays of Figs. 34-46a and b, the focal length f of the effective thin lens of the eye is 2.50 cm. For an object at distance $p =$ 40.0 cm, what focal length f' of the effective lens is required for the object to be seen clearly? (b) Must the eye muscles increase or decrease the radii of curvature of the eye lens to give focal length f'? **SSM**

FIG. 34-46 Problem 93.

Additional Problems

94 An object is placed against the center of a spherical mirror and then moved 70 cm from it along the central axis as the image distance i is measured. Figure 34-47 gives i versus object distance p out to $p_s = 40$ cm. What is the image distance when the object is 70 cm from the mirror?

FIG. 34-47 Problem 94.

95 In Fig. 34-48, a box is somewhere at the left, on the central axis of the thin converging lens. The image I_m of the box produced by the plane mirror is 4.00 cm "inside" the mirror. The lens–mirror separation is 10.0 cm, and the focal length of the lens is 2.00 cm. (a) What is the distance between the box and the lens? Light reflected by the mirror travels back through the lens, which produces a final image of the box. (b) What is the distance between the lens and that final image?

FIG. 34-48 Problem 95.

96 Two plane mirrors are placed parallel to each other and 40 cm apart. An object is placed 10 cm from one mirror.

Determine the (a) smallest, (b) second smallest, (c) third smallest (occurs twice), and (d) fourth smallest distance between the object and image of the object.

97 Figure 34-49 shows a *beam expander* made with two coaxial converging lenses of focal lengths f_1 and f_2 and separation $d = f_1 + f_2$. The device can expand a laser beam while keeping the light rays in the beam parallel to the central axis through the lenses. Suppose a uniform laser beam of width $W_i = 2.5$ mm and intensity $I_i = 9.0$ kW/m^2 enters a beam expander for which $f_1 = 12.5$ cm and $f_2 = 30.0$ cm. What are (a) W_f and (b) I_f of the beam leaving the expander? (c) What value of d is needed for the beam expander if lens 1 is replaced with a diverging lens of focal length $f_1 = -26.0$ cm?

FIG. 34-49 Problem 97.

98 In Fig. 34-50, an object is placed in front of a converging lens at a distance equal to twice the focal length f_1 of the lens. On the other side of the lens is a concave mirror of focal length f_2 separated from the lens by a distance $2(f_1 + f_2)$. Light from the object passes rightward through the lens, reflects from the mirror, passes leftward through the lens, and forms a final image of the object. What are (a) the distance between the lens and that final image and (b) the overall lateral magnification M of the object? Is the image (c) real or virtual (if it is virtual, it requires someone looking through the lens toward the mirror), (d) to the left or right of the lens, and (e) inverted

FIG. 34-50 Problem 98.

or noninverted relative to the object?

99 In Fig. 34-51, a fish watcher at point P watches a fish through a glass wall of a fish tank. The watcher is level with the fish; the index of refraction of the glass is 8/5, and that of the water is 4/3. The distances are $d_1 = 8.0$ cm, $d_2 = 3.0$ cm, and $d_3 = 6.8$ cm. (a) To the fish, how far away does the watcher appear to be? (*Hint:* The watcher is the object. Light from that object passes through the wall's outside surface, which acts as a refracting surface. Find the image produced by that surface. Then treat that image as an object whose light passes through the wall's inside surface, which acts as another refracting surface. Find the image produced by that surface, and there is the answer.) (b) To the watcher, how far away does the fish appear to be?

FIG. 34-51 Problem 99.

100 Figure 34-52a is an overhead view of two vertical plane mirrors with an object O placed between them. If you look into the mirrors, you see multiple images of O. You can find them by drawing the reflection in each mirror of the angular region between the mirrors, as is done in Fig. 34-52b for the left-hand mirror. Then draw the reflection of the reflection. Continue this on the left and on the right until the reflections meet or overlap at the rear of the mirrors. Then you can count the number of images of O. How many images of O would you see if θ is (a) $90°$, (b) $45°$, and (c) $60°$? If $\theta = 120°$, determine the (d) smallest and (e) largest number of images that can be seen, depending on your perspective and the location of O. (f) In each situation, draw the image locations and orientations as in Fig. 34-52b.

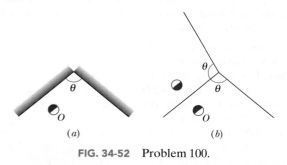

(a)   (b)

FIG. 34-52 Problem 100.

101 A point object is 10 cm away from a plane mirror, and the eye of an observer (with pupil diameter 5.0 mm) is 20 cm away. Assuming the eye and the object to be on the same line perpendicular to the mirror surface, find the area of the mirror used in observing the reflection of the point. (*Hint:* Adapt Fig. 34-4.)

102 You grind the lenses shown in Fig. 34-53 from flat glass disks ($n = 1.5$) using a machine that can grind a radius of curvature of either 40 cm or 60 cm. In a lens where either radius is appropriate, you select the 40 cm radius. Then you hold each lens in sunshine to form an image of the Sun. What are the (a) focal length f and (b) image type (real or virtual) for (*bi-convex*) lens 1, (c) f and (d) image type for (*plane-convex*) lens 2, (e) f and (f) image type for (*meniscus convex*) lens 3, (g) f and (h) image type for (*bi-concave*) lens 4, (i) f and (j) image type for (*plane-concave*) lens 5, and (k) f and (l) image type for (*meniscus concave*) lens 6?

(1)   (2)   (3)

(4)   (5)   (6)

FIG. 34-53 Problem 102.

103 The formula $1/p + 1/i = 1/f$ is called the *Gaussian* form of the thin-lens formula. Another form of this formula, the *Newtonian* form, is obtained by considering the distance x from the object to the first focal point and the distance x' from the second focal point to the image. Show that $xx' = f^2$ is the Newtonian form of the thin-lens formula. **SSM**

104 Show that the distance between an object and its real image formed by a thin converging lens is always greater than or equal to four times the focal length of the lens.

105 Two thin lenses of focal lengths f_1 and f_2 are in contact. Show that they are equivalent to a single thin lens for which the focal length is $f = f_1 f_2/(f_1 + f_2)$. **SSM**

106 A luminous object and a screen are a fixed distance D apart. (a) Show that a converging lens of focal length f, placed between object and screen, will form a real image on the screen for two lens positions that are separated by a distance $d = \sqrt{D(D - 4f)}$. (b) Show that

$$\left(\frac{D - d}{D + d}\right)^2$$

gives the ratio of the two image sizes for these two positions of the lens.

107 A fruit fly of height H sits in front of lens 1 on the central axis through the lens. The lens forms an image of the fly at a distance $d = 20$ cm from the fly; the image has the fly's orientation and height $H_I = 2.0H$. What are (a) the focal length f_1 of the lens and (b) the object distance p_1 of the fly? The fly then leaves lens 1 and sits in front of lens 2, which also forms an image at $d = 20$ cm that has the same orientation as the fly, but now $H_I = 0.50H$. What are (c) f_2 and (d) p_2? **SSM**

108 A goldfish in a spherical fish bowl of radius R is at the level of the center C of the bowl and at distance $R/2$ from the glass (Fig. 34-54). What magnification of the fish is produced by the water in the bowl for a viewer looking along a line that includes the fish and the center, with the fish on the near side of the center? The index of refraction of the water is 1.33. Neglect the glass wall of the bowl. Assume the viewer looks with one eye. (*Hint:* Equation 34-5 holds, but Eq. 34-6 does not. You need to work with a ray diagram of the situation and assume that the rays are close to the observer's line of sight — that is, they deviate from that line by only small angles.)

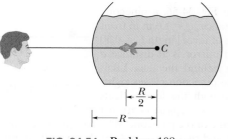

FIG. 34-54 Problem 108.

109 An eraser of height 1.0 cm is placed 10.0 cm in front of a two-lens system. Lens 1 (nearer the eraser) has focal length $f_1 = -15$ cm, lens 2 has $f_2 = 12$ cm, and the lens separation is $d = 12$ cm. For the image produced by lens 2, what are (a) the image distance i_2 (including sign), (b) the image height, (c) the image type (real or virtual), and (d) the image orientation (inverted relative to the eraser or not inverted)?

110 A peanut is placed 40 cm in front of a two-lens system: lens 1 (nearer the peanut) has focal length $f_1 = +20$ cm, lens 2 has $f_2 = -15$ cm, and the lens separation is $d = 10$ cm. For the image produced by lens 2, what are (a) the image distance i_2 (including sign), (b) the image orientation (inverted relative to the peanut or not inverted), and (c) the image type (real or virtual)? (d) What is the net lateral magnification?

111 A coin is placed 20 cm in front of a two-lens system. Lens 1 (nearer the coin) has focal length $f_1 = +10$ cm, lens 2 has $f_2 = +12.5$ cm, and the lens separation is $d = 30$ cm. For the image produced by lens 2, what are (a) the image distance

i_2 (including sign), (b) the overall lateral magnification, (c) the image type (real or virtual), and (d) the image orientation (inverted relative to the coin or not inverted)?

112 An object is 20 cm to the left of a thin diverging lens that has a 30 cm focal length. (a) What is the image distance i? (b) Draw a ray diagram showing the image position.

113 through 118 *Three-lens systems.* In Fig. 34-55, stick fig-ure O (the object) stands on the common central axis of three thin, symmetric lenses, which are mounted in the boxed regions. Lens 1 is mounted within the boxed region closest

FIG. 34-55 Problems 113 through 118.

to O, which is at object distance p_1. Lens 2 is mounted within the middle boxed region, at distance d_{12} from lens 1. Lens 3 is mounted in the farthest boxed region, at distance d_{23} from lens 2. Each problem in Table 34-10 refers to a different combination of lenses and different values for distances, which are given in centimeters. The type of lens is indicated by C for converging and D for diverging; the number after C or D is the distance between a lens and either of the focal points (the proper sign of the focal distance is not indicated).

Find (a) the image distance i_3 for the (final) image pro-duced by lens 3 (the final image produced by the system) and (b) the overall lateral magnification M for the system, includ-ing signs. Also, determine whether the final image is (c) real (R) or virtual (V), (d) inverted (I) from object O or non-inverted (NI), and (e) on the same side of lens 3 as object O or on the opposite side. [GO] 113

119 In Fig. 34-56, a pinecone is at distance $p_1 = 1.0$ m in front of a lens of focal length $f_1 = 0.50$ m; a flat mirror is at distance d = 2.0 m behind the lens. Light from the pinecone passes right-ward through the lens, reflects

FIG. 34-56 Problem 119.

from the mirror, passes leftward through the lens, and forms a final image of the pinecone. What are (a) the distance between the lens and that image and (b) the overall lateral magnifica-tion of the pinecone? Is the image (c) real or virtual (if it is vir-tual, it requires someone looking through the lens toward the mirror), (d) to the left or right of the lens, and (e) inverted rel-ative to the pinecone or not inverted?

120 One end of a long glass rod ($n = 1.5$) is a convex sur-face of radius 6.0 cm. An object is located in air along the axis of the rod, at a distance of 10 cm from the convex end. (a) How far apart are the object and the image formed by the glass rod? (b) Within what range of distances from the end of the rod must the object be located in order to produce a virtual image?

121 A short straight object of length L lies along the central axis of a spherical mirror, a distance p from the mirror. (a) Show that its image in the mirror has a length L', where

$$L' = L \left(\frac{f}{p - f} \right)^2 .$$

(*Hint:* Locate the two ends of the object.) (b) Show that the longitudinal magnification m' ($= L'/L$) is equal to m^2, where m is the lateral magnification.

122 You look down at a coin that lies at the bottom of a pool of liquid of depth d and index of refraction n (Fig. 34-57). Because you view with two eyes, which intercept different rays of light from the coin, you perceive the coin to be where extensions of the intercepted rays cross, at depth d_a instead of d. Assuming that the inter-cepted rays in Fig. 34-57 are close to a vertical axis through the coin, show that $d_a = d/n$. (*Hint:* Use the small-angle ap-proximation $\sin \theta \approx \tan \theta \approx \theta$.)

FIG. 34-57 Problem 122.

123 Prove that if a plane mirror is rotated through an angle α, the reflected beam is rotated through an angle 2α. Show that this result is reasonable for $\alpha = 45°$.

124 An object is 30.0 cm from a spherical mirror, along the mirror's central axis. The mirror produces an inverted image with a lateral magnification of absolute value 0.500. What is the focal length of the mirror?

125 A concave mirror has a radius of curvature of 24 cm. How far is an object from the mirror if the image formed is (a) virtual and 3.0 times the size of the object, (b) real and 3.0 times the size of the object, and (c) real and 1/3 the size of the object?

126 A pepper seed is placed in front of a lens. The lateral mag-nification of the seed is +0.300. The absolute value of the lens's focal length is 40.0 cm. How far from the lens is the image?

TABLE 34-10

Problems 113 through 118: Three-Lens Systems. See the setup for these problems.

	p_1	Lens 1	d_{12}	Lens 2	d_{23}	Lens 3	(a) i_3	(b) M	(c) R/V	(d) I/NI	(e) Side
113	+12	C, 8.0	28	C, 6.0	8.0	C, 6.0					
114	+4.0	C, 6.0	8.0	D, 4.0	5.7	D, 12					
115	+18	C, 6.0	15	C, 3.0	11	C, 3.0					
116	+4.0	D, 6.0	9.6	C, 6.0	14	C, 4.0					
117	+8.0	D, 8.0	8.0	D, 16	5.1	C, 8.0					
118	+2.0	C, 6.0	15	C, 6.0	19	C, 5.0					

127 The equation $1/p + 1/i = 2/r$ for spherical mirrors is an approximation that is valid if the image is formed by rays that make only small angles with the central axis. In reality, many of the angles are large, which smears the image a little. You can determine how much. Refer to Fig. 34-22 and consider a ray that leaves a point source (the object) on the central axis and that makes an angle α with that axis.

First, find the point of intersection of the ray with the mirror. If the coordinates of this intersection point are x and y and the origin is placed at the center of curvature, then $y = (x + p - r)\tan\alpha$ and $x^2 + y^2 = r^2$, where p is the object distance and r is the mirror's radius of curvature. Next, use $\tan\beta = y/x$ to find the angle β at the point of intersection, and then use $\alpha + \gamma = 2\beta$ to find the value of γ. Finally, use the relation $\tan\gamma = y/(x + i - r)$ to find the distance i of the image.

(a) Suppose $r = 12$ cm and $p = 20$ cm. For each of the following values of α, find the position of the image—that is, the position of the point where the reflected ray crosses the central axis: 0.500, 0.100, 0.0100 rad. Compare the results with those obtained with the equation $1/p + 1/i = 2/r$. (b) Repeat the calculations for $p = 4.00$ cm.

128 A small cup of green tea is positioned on the central axis of a spherical mirror. The lateral magnification of the cup is +0.250, and the distance between the mirror and its focal point is 2.00 cm. (a) What is the distance between the mirror and the image it produces? (b) Is the focal length positive or negative? (c) Is the image real or virtual?

129 A 20-mm-thick layer of water ($n = 1.33$) floats on a 40-mm-thick layer of carbon tetrachloride ($n = 1.46$) in a tank. A coin lies at the bottom of the tank. At what depth below the top water surface do you perceive the coin? (*Hint:* Use the result and assumptions of Problem 122 and work with a ray diagram of the situation.)

130 A millipede sits 1.0 m in front of the nearest part of the surface of a shiny sphere of diameter 0.70 m. (a) How far from the surface does the millipede's image appear? (b) If the millipede's height is 2.0 mm, what is the image height? (c) Is the image inverted?

131 (a) Show that if the object O in Fig. 34-19c is moved from focal point F_1 toward the observer's eye, the image moves in from infinity and the angle θ' (and thus the angular magnification m_θ) increases. (b) If you continue this process, where is the image when m_θ has its maximum usable value? (You can then still increase m_θ, but the image will no longer be clear.) (c) Show that the maximum usable value of m_θ is $1 + (25 \text{ cm})/f$. (d) Show that in this situation the angular magnification is equal to the lateral magnification.

132 Isaac Newton, having convinced himself (erroneously as it turned out) that chromatic aberration is an inherent property of refracting telescopes, invented the reflecting telescope, shown schematically in Fig. 34-58. He presented his second

FIG. 34-58
Problem 132.

model of this telescope, with a magnifying power of 38, to the Royal Society (of London), which still has it. In Fig. 34-58 incident light falls, closely parallel to the telescope axis, on the objective mirror M. After reflection from small mirror M' (the figure is not to scale), the rays form a real, inverted image in the *focal plane* (the plane perpendicular to the line of sight, at focal point F). This image is then viewed through an eyepiece. (a) Show that the angular magnification m_θ for the device is given by Eq. 34-15:

$$m_\theta = -f_{ob}/f_{ey},$$

where f_{ob} is the focal length of the objective mirror and f_{ey} is that of the eyepiece. (b) The 200 in. mirror in the reflecting telescope at Mt. Palomar in California has a focal length of 16.8 m. Estimate the size of the image formed by this mirror when the object is a meter stick 2.0 km away. Assume parallel incident rays. (c) The mirror of a different reflecting astronomical telescope has an effective radius of curvature of 10 m ("effective" because such mirrors are ground to a parabolic rather than a spherical shape, to eliminate spherical aberration defects). To give an angular magnification of 200, what must be the focal length of the eyepiece?

133 A narrow beam of parallel light rays is incident on a glass sphere from the left, directed toward the center of the sphere. (The sphere is a lens but certainly not a *thin* lens.) Approximate the angle of incidence of the rays as 0°, and assume that the index of refraction of the glass is $n < 2.0$. (a) In terms of n and the sphere radius r, what is the distance between the image produced by the sphere and the right side of the sphere? (b) Is the image to the left or right of that side? (*Hint:* Apply Eq. 34-8 to locate the image that is produced by refraction at the left side of the sphere; then use that image as the object for refraction at the right side of the sphere to locate the final image. In the second refraction, is the object distance p positive or negative?)

134 A *corner reflector*, much used in optical, microwave, and other applications, consists of three plane mirrors fastened together to form the corner of a cube. Show that after three reflections, an incident ray is returned with its direction exactly reversed.

135 A cheese enchilada is 4.00 cm in front of a converging lens. The magnification of the enchilada is −2.00. What is the focal length of the lens?

136 A grasshopper hops to a point on the central axis of a spherical mirror. The absolute magnitude of the mirror's focal length is 40.0 cm, and the lateral magnification of the image produced by the mirror is +0.200. (a) Is the mirror convex or concave? (b) How far from the mirror is the grasshopper?

137 In Fig. 34-59, a sand grain is 3.00 cm from thin lens 1, on the central axis through the two symmetric lenses. The distance between focal point and lens is 4.00 cm for both lenses; the lenses are separated by 8.00 cm. (a) What is the distance between lens 2 and the image it produces of the sand grain? Is that image (b) to the left or right of lens 2, (c) real or virtual, and (d) inverted relative to the sand grain or not inverted?

FIG. 34-59 Problem 137.

Amanda Hall/Robert Harding World Imagery/Getty Images, Inc.

Governments worldwide scurry to stay ahead of counterfeiters who are quick to use the latest technology to duplicate paper currencies. Some of the security measures now used to thwart counterfeiters are security threads and special watermarks (both of which can be seen if the currency is held up against a light) and microprinting (which consists of dots too small to be picked up by a scanner). The feature that is probably the most difficult for a counterfeiter to duplicate is the variable tint that results from color-shifting inks. For example, the "20" in the lower right of the front face of a U.S. $20 bill contains color-shifting ink. If you look directly down on the number, it is red or red-yellow. If you then tilt the bill and look at it obliquely, the color shifts to green. A copy machine can duplicate color from only one perspective and therefore cannot duplicate this shift in color you see when you change your perspective.

How do color-shifting inks shift colors?

The answer is in this chapter.

35-1   WHAT IS PHYSICS?

One of the major goals of physics is to understand the nature of light. This goal has been difficult to achieve (and has not yet fully been achieved) because light is complicated. However, this complication means that light offers many opportunities for applications, and some of the richest opportunities involve the interference of light waves—**optical interference.**

Nature has long used optical interference for coloring. For example, the wings of a *Morpho* butterfly are a dull, uninspiring brown, as can be seen on the bottom wing surface, but the brown is hidden on the top surface by an arresting blue due to the interference of light reflecting from that surface (Fig. 35-1). Moreover, the top surface is color-shifting; if you change your perspective or if the wing moves, the tint of the color changes. Similar color shifting is used in the inks on many currencies to thwart counterfeiters, whose copy machines can duplicate color from only one perspective and therefore cannot duplicate any shift in color caused by a change in perspective.

Before we get to color shifting in butterfly wings and currency inks, we first must discuss the basic physics of optical interference. That means we must largely abandon the simplicity of geometrical optics (in which we describe light as rays) and return to the wave nature of light.

FIG. 35-1 The blue of the top surface of a *Morpho* butterfly wing is due to optical interference and shifts in color as your viewing perspective changes. *(Philippe Colombi/PhotoDisc//Getty Images)*

35-2 | Light as a Wave

The first person to advance a convincing wave theory for light was Dutch physicist Christian Huygens, in 1678. Although much less comprehensive than the later electromagnetic theory of Maxwell, Huygens' theory was simpler mathematically and remains useful today. Its great advantages are that it accounts for the laws of reflection and refraction in terms of waves and gives physical meaning to the index of refraction.

Huygens' wave theory is based on a geometrical construction that allows us to tell where a given wavefront will be at any time in the future if we know its present position. This construction is based on **Huygens' principle,** which is:

> All points on a wavefront serve as point sources of spherical secondary wavelets. After a time *t*, the new position of the wavefront will be that of a surface tangent to these secondary wavelets.

Here is a simple example. At the left in Fig. 35-2, the present location of a wavefront of a plane wave traveling to the right in vacuum is represented by plane *ab*, perpendicular to the page. Where will the wavefront be at time Δt later? We let several points on plane *ab* (the dots) serve as sources of spherical secondary wavelets that are emitted at $t = 0$. At time Δt, the radius of all these spherical wavelets will have grown to $c \Delta t$, where c is the speed of light in vacuum. We draw plane *de* tangent to these wavelets at time Δt. This plane represents the wavefront of the plane wave at time Δt; it is parallel to plane *ab* and a perpendicular distance $c \Delta t$ from it.

The Law of Refraction

We now use Huygens' principle to derive the law of refraction, Eq. 33-40 (Snell's law). Figure 35-3 shows three stages in the refraction of several wavefronts at a flat interface between air (medium 1) and glass (medium 2). We arbitrarily choose the wavefronts in the incident light beam to be separated by λ_1, the wavelength in medium 1. Let the speed of light in air be v_1 and that in glass be v_2. We assume that $v_2 < v_1$, which happens to be true.

Angle θ_1 in Fig. 35-3a is the angle between the wavefront and the interface; it has the same value as the angle between the *normal* to the wavefront (that is, the incident ray) and the *normal* to the interface. Thus, θ_1 is the angle of incidence.

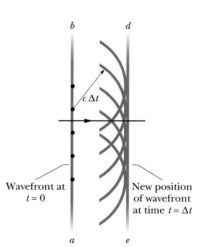

Wavefront at $t = 0$

New position of wavefront at time $t = \Delta t$

FIG. 35-2 The propagation of a plane wave in vacuum, as portrayed by Huygens' principle.

(a)

(b)

(c)

FIG. 35-3 The refraction of a plane wave at an air–glass interface, as portrayed by Huygens' principle. The wavelength in glass is smaller than that in air. For simplicity, the reflected wave is not shown. Parts (a) through (c) represent three successive stages of the refraction.

As the wave moves into the glass, a Huygens wavelet at point e in Fig. 35-3b will expand to pass through point c, at a distance of λ_1 from point e. The time interval required for this expansion is that distance divided by the speed of the wavelet, or λ_1/v_1. Now note that in this same time interval, a Huygens wavelet at point h will expand to pass through point g, at the reduced speed v_2 and with wavelength λ_2. Thus, this time interval must also be equal to λ_2/v_2. By equating these times of travel, we obtain the relation

$$\frac{\lambda_1}{\lambda_2} = \frac{v_1}{v_2}, \tag{35-1}$$

which shows that the wavelengths of light in two media are proportional to the speeds of light in those media.

By Huygens' principle, the refracted wavefront must be tangent to an arc of radius λ_2 centered on h, say at point g. The refracted wavefront must also be tangent to an arc of radius λ_1 centered on e, say at c. Then the refracted wavefront must be oriented as shown. Note that θ_2, the angle between the refracted wavefront and the interface, is actually the angle of refraction.

For the right triangles hce and hcg in Fig. 35-3b we may write

$$\sin \theta_1 = \frac{\lambda_1}{hc} \quad \text{(for triangle } hce\text{)}$$

and

$$\sin \theta_2 = \frac{\lambda_2}{hc} \quad \text{(for triangle } hcg\text{).}$$

Dividing the first of these two equations by the second and using Eq. 35-1, we find

$$\frac{\sin \theta_1}{\sin \theta_2} = \frac{\lambda_1}{\lambda_2} = \frac{v_1}{v_2}. \tag{35-2}$$

We can define the **index of refraction** n for each medium as the ratio of the speed of light in vacuum to the speed of light v in the medium. Thus,

$$n = \frac{c}{v} \quad \text{(index of refraction).} \tag{35-3}$$

In particular, for our two media, we have

$$n_1 = \frac{c}{v_1} \quad \text{and} \quad n_2 = \frac{c}{v_2}. \tag{35-4}$$

If we combine Eqs. 35-2 and 35-4, we find

$$\frac{\sin \theta_1}{\sin \theta_2} = \frac{c/n_1}{c/n_2} = \frac{n_2}{n_1} \tag{35-5}$$

or

$$n_1 \sin \theta_1 = n_2 \sin \theta_2 \quad \text{(law of refraction),} \tag{35-6}$$

as introduced in Chapter 33.

✓ **CHECKPOINT 1** The figure shows a monochromatic ray of light traveling across parallel interfaces, from an original material a, through layers of materials b and c, and then back into material a. Rank the materials according to the speed of light in them, greatest first.

Wavelength and Index of Refraction

We have now seen that the wavelength of light changes when the speed of the light changes, as happens when light crosses an interface from one medium into another. Further, the speed of light in any medium depends on the index of refraction of the medium, according to Eq. 35-3. Thus, the wavelength of light in any medium depends on the index of refraction of the medium. Let a certain monochromatic light have wavelength λ and speed c in vacuum and wavelength λ_n and speed v in a medium with an index of refraction n. Now we can rewrite Eq. 35-1 as

$$\lambda_n = \lambda \frac{v}{c}. \qquad (35\text{-}7)$$

FIG. 35-4 Two light rays travel through two media having different indexes of refraction.

Using Eq. 35-3 to substitute $1/n$ for v/c then yields

$$\lambda_n = \frac{\lambda}{n}. \qquad (35\text{-}8)$$

This equation relates the wavelength of light in any medium to its wavelength in vacuum. It tells us that the greater the index of refraction of a medium, the smaller the wavelength of light in that medium.

What about the frequency of the light? Let f_n represent the frequency of the light in a medium with index of refraction n. Then from the general relation of Eq. 16-13 ($v = \lambda f$), we can write

$$f_n = \frac{v}{\lambda_n}.$$

Substituting Eqs. 35-3 and 35-8 then gives us

$$f_n = \frac{c/n}{\lambda/n} = \frac{c}{\lambda} = f,$$

where f is the frequency of the light in vacuum. Thus, although the speed and wavelength of light in the medium are different from what they are in vacuum, *the frequency of the light in the medium is the same as it is in vacuum.*

The fact that the wavelength of light depends on the index of refraction via Eq. 35-8 is important in certain situations involving the interference of light waves. For example, in Fig. 35-4, the *waves of the rays* (that is, the waves represented by the rays) have identical wavelengths λ and are initially in phase in air ($n \approx 1$). One of the waves travels through medium 1 of index of refraction n_1 and length L. The other travels through medium 2 of index of refraction n_2 and the same length L. When the waves leave the two media, they will have the same wavelength—their wavelength λ in air. However, because their wavelengths differed in the two media, the two waves may no longer be in phase.

> The phase difference between two light waves can change if the waves travel through different materials having different indexes of refraction.

As we shall discuss soon, this change in the phase difference can determine how the light waves will interfere if they reach some common point.

To find their new phase difference in terms of wavelengths, we first count the number N_1 of wavelengths there are in the length L of medium 1. From Eq. 35-8, the wavelength in medium 1 is $\lambda_{n1} = \lambda/n_1$; so

$$N_1 = \frac{L}{\lambda_{n1}} = \frac{Ln_1}{\lambda}. \qquad (35\text{-}9)$$

Similarly, we count the number N_2 of wavelengths there are in the length L of medium 2, where the wavelength is $\lambda_{n2} = \lambda/n_2$:

$$N_2 = \frac{L}{\lambda_{n2}} = \frac{Ln_2}{\lambda}. \qquad (35\text{-}10)$$

To find the new phase difference between the waves, we subtract the smaller of N_1 and N_2 from the larger. Assuming $n_2 > n_1$, we obtain

$$N_2 - N_1 = \frac{Ln_2}{\lambda} - \frac{Ln_1}{\lambda} = \frac{L}{\lambda}(n_2 - n_1). \qquad (35\text{-}11)$$

FIG. 35-5 A primary rainbow and the faint supernumeraries below it are due to optical interference.

Suppose Eq. 35-11 tells us that the waves now have a phase difference of 45.6 wavelengths. That is equivalent to taking the initially in-phase waves and shifting one of them by 45.6 wavelengths. However, a shift of an integer number of wavelengths (such as 45) would put the waves back in phase; so it is only the decimal fraction (here, 0.6) that is important. A phase difference of 45.6 wavelengths is equivalent to an *effective phase difference* of 0.6 wavelength.

A phase difference of 0.5 wavelength puts two waves exactly out of phase. If the waves had equal amplitudes and were to reach some common point, they would then undergo fully destructive interference, producing darkness at that point. With a phase difference of 0.0 or 1.0 wavelength, they would, instead, undergo fully constructive interference, resulting in brightness at the common point. Our phase difference of 0.6 wavelength is an intermediate situation but closer to fully destructive interference, and the waves would produce a dimly illuminated common point.

We can also express phase difference in terms of radians and degrees, as we have done already. A phase difference of one wavelength is equivalent to phase differences of 2π rad and 360°.

✓ **CHECKPOINT 2** The light waves of the rays in Fig. 35-4 have the same wavelength and amplitude and are initially in phase. (a) If 7.60 wavelengths fit within the length of the top material and 5.50 wavelengths fit within that of the bottom material, which material has the greater index of refraction? (b) If the rays are angled slightly so that they meet at the same point on a distant screen, will the interference there result in the brightest possible illumination, bright intermediate illumination, dark intermediate illumination, or darkness?

Rainbows and Optical Interference

In Section 33-8, we discussed how the colors of sunlight are separated into a rainbow when sunlight travels through falling raindrops. We dealt with a simplified situation in which a single ray of white light entered a drop. Actually, light waves pass into a drop along the entire side that faces the Sun. Here we cannot discuss the details of how these waves travel through the drop and then emerge, but we can see that different parts of an incoming wave will travel different paths within the drop. That means waves will emerge from the drop with different phases. Thus, we can see that at some angles the emerging light will be in phase and give constructive interference. The rainbow is the result of such constructive interference. For example, the red of the rainbow appears because waves of red light emerge in phase from each raindrop in the direction in which you see that part of the rainbow.

If you are lucky and look carefully below a primary rainbow, you can see dimmer colored arcs called *supernumeraries* (Fig. 35-5). Like the main arcs of the rainbow, the supernumeraries are due to waves that emerge from each drop approximately in phase with one another to give constructive interference. If you are very lucky and look very carefully above a secondary rainbow, you might see even more (but even dimmer) supernumeraries. Keep in mind that both types of rainbows and both sets of supernumeraries are naturally occurring examples of optical interference and naturally occurring evidence that light consists of waves.

Sample Problem | **35-1**

In Fig. 35-4, the two light waves that are represented by the rays have wavelength 550.0 nm before entering media 1 and 2. They also have equal amplitudes and are in phase. Medium 1 is now just air, and medium 2 is a transparent plastic layer of index of refraction 1.600 and thickness 2.600 μm.

(a) What is the phase difference of the emerging waves in wavelengths, radians, and degrees? What is their effective phase difference (in wavelengths)?

KEY IDEA The phase difference of two light waves can change if they travel through different media, with different indexes of refraction. The reason is that their wavelengths are different in the different media. We can calculate the change in phase difference by counting the number of wavelengths that fits into each medium and then subtracting those numbers.

Calculations: When the path lengths of the waves in the

two media are identical, Eq. 35-11 gives the result of the subtraction. Here we have $n_1 = 1.000$ (for the air), $n_2 = 1.600$, $L = 2.600$ μm, and $\lambda = 550.0$ nm. Thus, Eq. 35-11 yields

$$N_2 - N_1 = \frac{L}{\lambda}(n_2 - n_1)$$

$$= \frac{2.600 \times 10^{-6}\,\text{m}}{5.500 \times 10^{-7}\,\text{m}}(1.600 - 1.000)$$

$$= 2.84. \qquad \text{(Answer)}$$

Thus, the phase difference of the emerging waves is 2.84 wavelengths. Because 1.0 wavelength is equivalent to 2π rad and 360°, you can show that this phase difference is equivalent to

$$\text{phase difference} = 17.8\,\text{rad} \approx 1020°. \quad \text{(Answer)}$$

The effective phase difference is the decimal part of the actual phase difference *expressed in wavelengths.* Thus, we have

$$\text{effective phase difference} = 0.84\,\text{wavelength.} \quad \text{(Answer)}$$

You can show that this is equivalent to 5.3 rad and about 300°. *Caution:* We do *not* find the effective phase difference by taking the decimal part of the actual phase difference as expressed in radians or degrees. For example, we do *not* take 0.8 rad from the actual phase difference of 17.8 rad.

(b) If the waves reached the same point on a distant screen, what type of interference would they produce?

Reasoning: We need to compare the effective phase difference of the waves with the phase differences that give the extreme types of interference. Here the effective phase difference of 0.84 wavelength is between 0.5 wavelength (for fully destructive interference, or the darkest possible result) and 1.0 wavelength (for fully constructive interference, or the brightest possible result), but closer to 1.0 wavelength. Thus, the waves would produce intermediate interference that is closer to fully constructive interference—they would produce a relatively bright spot.

35-3 | Diffraction

In the next section we shall discuss the experiment that first proved that light is a wave. To prepare for that discussion, we must introduce the idea of **diffraction** of waves, a phenomenon that we explore much more fully in Chapter 36. Its essence is this: If a wave encounters a barrier that has an opening of dimensions similar to the wavelength, the part of the wave that passes through the opening will flare (spread) out—will *diffract*—into the region beyond the barrier. The flaring is consistent with the spreading of wavelets in the Huygens construction of Fig. 35-2. Diffraction occurs for waves of all types, not just light waves; Fig. 35-6 shows the diffraction of water waves traveling across the surface of water in a shallow tank.

Figure 35-7a shows the situation schematically for an incident plane wave of wavelength λ encountering a slit that has width $a = 6.0\lambda$ and extends into and out of the page. The part of the wave that passes through the slit flares out on the far side. Figures 35-7b (with $a = 3.0\lambda$) and 35-7c ($a = 1.5\lambda$) illustrate the main feature of diffraction: the narrower the slit, the greater the diffraction.

Diffraction limits geometrical optics, in which we represent an electromagnetic wave with a ray. If we actually try to form a ray by sending light through

FIG. 35-6 Waves produced by an oscillating paddle at the left flare out through an opening in a barrier along the water surface. (*Runk Schoenberger/Grant Heilman Photography*)

FIG. 35-7 Diffraction represented schematically. For a given wavelength λ, the diffraction is more pronounced the smaller the slit width a. The figures show the cases for (a) slit width $a = 6.0\lambda$, (b) slit width $a = 3.0\lambda$, and (c) slit width $a = 1.5\lambda$. In all three cases, the screen and the length of the slit extend well into and out of the page, perpendicular to it.

FIG. 35-8 In Young's interference experiment, incident monochromatic light is diffracted by slit S_0, which then acts as a point source of light that emits semicircular wavefronts. As that light reaches screen B, it is diffracted by slits S_1 and S_2, which then act as two point sources of light. The light waves traveling from slits S_1 and S_2 overlap and undergo interference, forming an interference pattern of maxima and minima on viewing screen C. This figure is a cross section; the screens, slits, and interference pattern extend into and out of the page. Between screens B and C, the semicircular wavefronts centered on S_2 depict the waves that would be there if only S_2 were open. Similarly, those centered on S_1 depict waves that would be there if only S_1 were open.

FIG. 35-9 A photograph of the interference pattern produced by the arrangement shown in Fig. 35-8. (The photograph is a front view of part of screen C.) The alternating maxima and minima are called *interference fringes* (because they resemble the decorative fringe sometimes used on clothing and rugs). *(From Michel Cagnet, Maurice Franzon, and Jean Claude Thierr, Atlas of Optical Phenomena. Springer-Verlag, New York, 1962. Reproduced with permission.)*

a narrow slit, or through a series of narrow slits, diffraction will always defeat our effort because it always causes the light to spread. Indeed, the narrower we make the slits (in the hope of producing a narrower beam), the greater the spreading is. Thus, geometrical optics holds only when slits or other apertures that might be located in the path of light do not have dimensions comparable to or smaller than the wavelength of the light.

35-4 | Young's Interference Experiment

In 1801, Thomas Young experimentally proved that light is a wave, contrary to what most other scientists then thought. He did so by demonstrating that light undergoes interference, as do water waves, sound waves, and waves of all other types. In addition, he was able to measure the average wavelength of sunlight; his value, 570 nm, is impressively close to the modern accepted value of 555 nm. We shall here examine Young's experiment as an example of the interference of light waves.

Figure 35-8 gives the basic arrangement of Young's experiment. Light from a distant monochromatic source illuminates slit S_0 in screen A. The emerging light then spreads via diffraction to illuminate two slits S_1 and S_2 in screen B. Diffraction of the light by these two slits sends overlapping circular waves into the region beyond screen B, where the waves from one slit interfere with the waves from the other slit.

The "snapshot" of Fig. 35-8 depicts the interference of the overlapping waves. However, we cannot see evidence for the interference except where a viewing screen C intercepts the light. Where it does so, points of interference maxima form visible bright rows—called *bright bands, bright fringes,* or (loosely speaking) *maxima*—that extend across the screen (into and out of the page in Fig. 35-8). Dark regions—called *dark bands, dark fringes,* or (loosely speaking) *minima*—result from fully destructive interference and are visible between adjacent pairs of bright fringes. (*Maxima* and *minima* more properly refer to the center of a band.) The pattern of bright and dark fringes on the screen is called an **interference pattern.** Figure 35-9 is a photograph of part of the interference pattern that would be seen by an observer standing to the left of screen C in the arrangement of Fig. 35-8.

Locating the Fringes

Light waves produce fringes in a *Young's double-slit interference experiment,* as it is called, but what exactly determines the locations of the fringes? To answer, we shall use the arrangement in Fig. 35-10a. There, a plane wave of monochromatic light is incident on two slits S_1 and S_2 in screen B; the light diffracts through the slits and produces an interference pattern on screen C. We draw a central axis from the point halfway between the slits to screen C as a reference. We then pick, for discussion, an arbitrary point P on the screen, at angle θ to the central axis. This point intercepts the wave of ray r_1 from the bottom slit and the wave of ray r_2 from the top slit.

These waves are in phase when they pass through the two slits because there they are just portions of the same incident wave. However, once they have passed the slits, the two waves must travel different distances to reach P. We saw a similar situation in Section 17-5 with sound waves and concluded that

> ☞ The phase difference between two waves can change if the waves travel paths of different lengths.

The change in phase difference is due to the *path length difference* ΔL in the paths taken by the waves. Consider two waves initially exactly in phase, traveling along paths with a path length difference ΔL, and then passing through some common point. When ΔL is zero or an integer number of wavelengths, the waves arrive at the common point exactly in phase and they interfere fully constructively there. If that is true for the waves of rays r_1 and r_2 in Fig. 35-10, then point P is part of a bright fringe. When, instead, ΔL is an odd multiple of half a wavelength, the waves arrive at the common point exactly out of phase and they interfere fully destructively there. If that is true for the waves of rays r_1 and r_2, then point P is part of a dark fringe. (And, of course, we can have intermediate situations of interference and thus intermediate illumination at P.) Thus,

> ☞ What appears at each point on the viewing screen in a Young's double-slit interference experiment is determined by the path length difference ΔL of the rays reaching that point.

We can specify where each bright fringe and each dark fringe is located on the screen by giving the angle θ from the central axis to that fringe. To find θ, we must relate it to ΔL. We start with Fig. 35-10a by finding a point b along ray r_1 such that the path length from b to P equals the path length from S_2 to P. Then the path length difference ΔL between the two rays is the distance from S_1 to b.

The relation between this S_1-to-b distance and θ is complicated, but we can simplify it considerably if we arrange for the distance D from the slits to the screen to be much greater than the slit separation d. Then we can approximate rays r_1 and r_2 as being parallel to each other and at angle θ to the central axis (Fig. 35-10b). We can also approximate the triangle formed by S_1, S_2, and b as being a right triangle, and approximate the angle inside that triangle at S_2 as being θ. Then, for that triangle, $\sin\theta = \Delta L/d$ and thus

$$\Delta L = d\sin\theta \qquad \text{(path length difference).} \qquad (35\text{-}12)$$

For a bright fringe, we saw that ΔL must be either zero or an integer number of wavelengths. Using Eq. 35-12, we can write this requirement as

$$\Delta L = d\sin\theta = (\text{integer})(\lambda), \qquad (35\text{-}13)$$

or as

$$d\sin\theta = m\lambda, \qquad \text{for } m = 0, 1, 2, \ldots \qquad \text{(maxima—bright fringes).} \qquad (35\text{-}14)$$

For a dark fringe, ΔL must be an odd multiple of half a wavelength. Again using

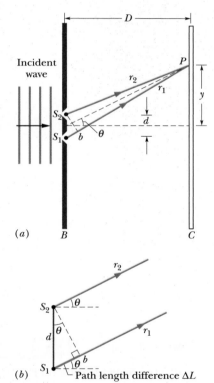

FIG. 35-10 (*a*) Waves from slits S_1 and S_2 (which extend into and out of the page) combine at P, an arbitrary point on screen C at distance y from the central axis. The angle θ serves as a convenient locator for P. (*b*) For $D \gg d$, we can approximate rays r_1 and r_2 as being parallel, at angle θ to the central axis.

Eq. 35-12, we can write this requirement as

$$\Delta L = d \sin \theta = (\text{odd number})(\tfrac{1}{2}\lambda), \qquad (35\text{-}15)$$

or as

$$d \sin \theta = (m + \tfrac{1}{2})\lambda, \qquad \text{for } m = 0, 1, 2, \dots \qquad \text{(minima—dark fringes).} \quad (35\text{-}16)$$

With Eqs. 35-14 and 35-16, we can find the angle θ to any fringe and thus locate that fringe; further, we can use the values of m to label the fringes. For the value and label $m = 0$, Eq. 35-14 tells us that a bright fringe is at $\theta = 0$ and thus on the central axis. This *central maximum* is the point at which waves arriving from the two slits have a path length difference $\Delta L = 0$, hence zero phase difference.

For, say, $m = 2$, Eq. 35-14 tells us that *bright* fringes are at the angle

$$\theta = \sin^{-1}\!\left(\frac{2\lambda}{d}\right)$$

above and below the central axis. Waves from the two slits arrive at these two fringes with $\Delta L = 2\lambda$ and with a phase difference of two wavelengths. These fringes are said to be the *second-order bright fringes* (meaning $m = 2$) or the *second side maxima* (the second maxima to the side of the central maximum), or they are described as being the second bright fringes from the central maximum.

For $m = 1$, Eq. 35-16 tells us that *dark* fringes are at the angle

$$\theta = \sin^{-1}\!\left(\frac{1.5\lambda}{d}\right)$$

above and below the central axis. Waves from the two slits arrive at these two fringes with $\Delta L = 1.5\lambda$ and with a phase difference, in wavelengths, of 1.5. These fringes are called the *second-order dark fringes* or *second minima* because they are the second dark fringes to the side of the central axis. (The first dark fringes, or first minima, are at locations for which $m = 0$ in Eq. 35-16.)

We derived Eqs. 35-14 and 35-16 for the situation $D \gg d$. However, they also apply if we place a converging lens between the slits and the viewing screen and then move the viewing screen closer to the slits, to the focal point of the lens. (The screen is then said to be in the *focal plane* of the lens; that is, it is in the plane perpendicular to the central axis at the focal point.) One property of a converging lens is that it focuses all rays that are parallel to one another to the same point on its focal plane. Thus, the rays that now arrive at any point on the screen (in the focal plane) were exactly parallel (rather than approximately) when they left the slits. They are like the initially parallel rays in Fig. 34-15a that are directed to a point (the focal point) by a lens.

✓ **CHECKPOINT 3** In Fig. 35-10a, what are ΔL (as a multiple of the wavelength) and the phase difference (in wavelengths) for the two rays if point P is (a) a third side maximum and (b) a third minimum?

Sample Problem | 35-2

What is the distance on screen C in Fig. 35-10a between adjacent maxima near the center of the interference pattern? The wavelength λ of the light is 546 nm, the slit separation d is 0.12 mm, and the slit–screen separation D is 55 cm. Assume that θ in Fig. 35-10 is small enough to permit use of the approximations $\sin \theta \approx \tan \theta \approx \theta$, in which θ is expressed in radian measure.

KEY IDEAS (1) First, let us pick a maximum with a low value of m to ensure that it is near the center of the pat-

tern. Then, from the geometry of Fig. 35-10a, the maximum's vertical distance y_m from the center of the pattern is related to its angle θ from the central axis by

$$\tan \theta \approx \theta = \frac{y_m}{D}.$$

(2) From Eq. 35-14, this angle θ for the mth maximum is given by

$$\sin \theta \approx \theta = \frac{m\lambda}{d}.$$

Calculations: If we equate our two expressions for angle θ and then solve for y_m, we find

$$y_m = \frac{m\lambda D}{d}. \qquad (35\text{-}17)$$

For the next maximum as we move away from the pattern's center, we have

$$y_{m+1} = \frac{(m + 1)\lambda D}{d}. \qquad (35\text{-}18)$$

We find the distance between these adjacent maxima by subtracting Eq. 35-17 from Eq. 35-18:

$$\Delta y = y_{m+1} - y_m = \frac{\lambda D}{d}$$

$$= \frac{(546 \times 10^{-9}\ \text{m})(55 \times 10^{-2}\ \text{m})}{0.12 \times 10^{-3}\ \text{m}}$$

$$= 2.50 \times 10^{-3}\ \text{m} \approx 2.5\ \text{mm}. \qquad \text{(Answer)}$$

As long as d and θ in Fig. 35-10a are small, the separation of the interference fringes is independent of m; that is, the fringes are evenly spaced.

Sample Problem   35-3   Build your skill

A double-slit interference pattern is produced on a screen, as in Fig. 35-10; the light is monochromatic at a wavelength of 600 nm. A strip of transparent plastic with index of refraction $n = 1.50$ is to be placed over one of the slits. Its presence changes the interference between light waves from the two slits, causing the interference pattern to be shifted across the screen from the original pattern. Figure 35-11a shows the original locations of the central bright fringe ($m = 0$) and the first bright fringes ($m = 1$) above and below the central fringe. The purpose of the plastic is to shift the pattern upward so that the lower $m = 1$ bright fringe is shifted to the center of the pattern. Should the plastic be placed over the top slit (as arbitrarily drawn in Fig. 35-11b) or the bottom slit, and what thickness L should it have?

KEY IDEA   The interference at a point on the screen depends on the phase difference of the light rays arriving from the two slits. The light rays are in phase at the slits, but their relative phase can shift on the way to the screen due to (1) a difference in the length of the paths they follow and (2) a difference in the number of their internal wavelengths λ_n in the materials through which they pass. Let's check both possibilities.

Path-length difference: Figure 35-11a shows rays r_1 and r_2 along which waves from the two slits travel to reach the lower $m = 1$ bright fringe. Those waves start in phase at the slits but arrive at the fringe with a phase difference of exactly 1 wavelength. To remind ourselves of this main characteristic of the fringe, let us call it the 1λ fringe. The one-wavelength phase difference is due to the one-wavelength path length difference between the rays reaching the fringe; that is, there is exactly one more wavelength along ray r_2 than along r_1.

Figure 35-11b shows the 1λ fringe shifted up to the center of the pattern with the plastic strip over the top slit (we still do not know whether the plastic should be there or over the bottom slit). The figure also shows the new orientations of rays r_1 and r_2 to reach that fringe. There still must be one more wavelength along r_2 than along r_1 (because they still produce the 1λ fringe), but now the path length difference between those rays is

zero, as we can tell from the geometry of Fig. 35-11b. However, r_2 now passes through the plastic.

Internal wavelength: The wavelength λ_n of light in a material with index of refraction n is smaller than the wavelength in vacuum, as given by Eq. 35-8 ($\lambda_n = \lambda/n$). Here, this means that the wavelength of the light is smaller in the plastic than in the air. Thus, the ray that passes through the plastic will have more wavelengths along it than the ray that passes through only air—so we do get the one extra wavelength we need along ray r_2 by placing the plastic over the top slit, as drawn in Fig. 35-11b.

Thickness: To determine the required thickness L of the plastic, we use the procedure in Sample Problem 35-1a. Here, as there, waves that are initially in phase travel equal distances L through different materials (plastic and air). However, here we know the phase difference and require L; so we again use Eq. 35-11,

$$N_2 - N_1 = \frac{L}{\lambda}(n_2 - n_1). \qquad (35\text{-}19)$$

We know that $N_2 - N_1$ is 1 for a phase difference of one wavelength, n_2 is 1.50 for the plastic in front of the top slit, n_1 is 1.00 for the air in front of the bottom slit, and λ is 600×10^{-9} m. Then Eq. 35-19 yields

$$L = \frac{\lambda(N_2 - N_1)}{n_2 - n_1} = \frac{(600 \times 10^{-9}\ \text{m})(1)}{1.50 - 1.00}$$

$$= 1.2 \times 10^{-6}\ \text{m}. \qquad \text{(Answer)}$$

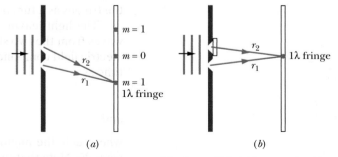

(a)   *(b)*

FIG. 35-11   (a) Arrangement for two-slit interference (not to scale). The locations of three bright fringes (or maxima) are indicated. (b) A strip of plastic covers the top slit. We want the 1λ fringe to be at the center of the pattern.

35-5 | Coherence

For the interference pattern to appear on viewing screen C in Fig. 35-8, the light waves reaching any point P on the screen must have a phase difference that does not vary in time. That is the case in Fig. 35-8 because the waves passing through slits S_1 and S_2 are portions of the single light wave that illuminates the slits. Because the phase difference remains constant, the light from slits S_1 and S_2 is said to be completely **coherent.**

Direct sunlight is partially coherent; that is, sunlight waves intercepted at two points have a constant phase difference only if the points are very close. If you look closely at your fingernail in bright sunlight, you can see a faint interference pattern called *speckle* that causes the nail to appear to be covered with specks. You see this effect because light waves scattering from very close points on the nail are sufficiently coherent to interfere with one another at your eye. The slits in a double-slit experiment, however, are not close enough, and in direct sunlight, the light at the slits would be **incoherent.** To get coherent light, we would have to send the sunlight through a single slit as in Fig. 35-8; because that single slit is small, light that passes through it is coherent. In addition, the smallness of the slit causes the coherent light to spread via diffraction to illuminate both slits in the double-slit experiment.

If we replace the double slits with two similar but independent mono-chromatic light sources, such as two fine incandescent wires, the phase difference between the waves emitted by the sources varies rapidly and randomly. (This occurs because the light is emitted by vast numbers of atoms in the wires, acting randomly and independently for extremely short times—of the order of nanoseconds.) As a result, at any given point on the viewing screen, the inter-ference between the waves from the two sources varies rapidly and randomly between fully constructive and fully destructive. The eye (and most common optical detectors) cannot follow such changes, and no interference pattern can be seen. The fringes disappear, and the screen is seen as being uniformly illuminated.

A *laser* differs from common light sources in that its atoms emit light in a cooperative manner, thereby making the light coherent. Moreover, the light is almost monochromatic, is emitted in a thin beam with little spreading, and can be focused to a width that almost matches the wavelength of the light.

35-6 | Intensity in Double-Slit Interference

Equations 35-14 and 35-16 tell us how to locate the maxima and minima of the double-slit interference pattern on screen C of Fig. 35-10 as a function of the angle θ in that figure. Here we wish to derive an expression for the intensity I of the fringes as a function of θ.

The light leaving the slits is in phase. However, let us assume that the light waves from the two slits are not in phase when they arrive at point P. Instead, the electric field components of those waves at point P are not in phase and vary with time as

$$E_1 = E_0 \sin \omega t \qquad (35\text{-}20)$$

and
$$E_2 = E_0 \sin(\omega t + \phi), \qquad (35\text{-}21)$$

where ω is the angular frequency of the waves and ϕ is the phase constant of wave E_2. Note that the two waves have the same amplitude E_0 and a phase difference of ϕ. Because that phase difference does not vary, the waves are coherent. We shall show that these two waves will combine at P to produce an intensity I

given by

$$I = 4I_0 \cos^2 \tfrac{1}{2} \phi, \tag{35-22}$$

and that

$$\phi = \frac{2\pi d}{\lambda} \sin \theta. \tag{35-23}$$

In Eq. 35-22, I_0 is the intensity of the light that arrives on the screen from one slit when the other slit is temporarily covered. We assume that the slits are so narrow in comparison to the wavelength that this single-slit intensity is essentially uniform over the region of the screen in which we wish to examine the fringes.

Equations 35-22 and 35-23, which together tell us how the intensity I of the fringe pattern varies with the angle θ in Fig. 35-10, necessarily contain information about the location of the maxima and minima. Let us see if we can extract that information to find equations about those locations.

Study of Eq. 35-22 shows that intensity maxima will occur when

$$\tfrac{1}{2} \phi = m\pi, \qquad \text{for } m = 0, 1, 2, \dots. \tag{35-24}$$

If we put this result into Eq. 35-23, we find

$$2m\pi = \frac{2\pi d}{\lambda} \sin \theta, \qquad \text{for } m = 0, 1, 2, \dots$$

or $\qquad d \sin \theta = m\lambda, \qquad$ for $m = 0, 1, 2, \dots$ (maxima), $\tag{35-25}$

which is exactly Eq. 35-14, the expression that we derived earlier for the locations of the maxima.

The minima in the fringe pattern occur when

$$\tfrac{1}{2} \phi = (m + \tfrac{1}{2})\pi, \qquad \text{for } m = 0, 1, 2, \dots. \tag{35-26}$$

If we combine this relation with Eq. 35-23, we are led at once to

$$d \sin \theta = (m + \tfrac{1}{2})\lambda, \qquad \text{for } m = 0, 1, 2, \dots \quad \text{(minima)}, \tag{35-27}$$

which is just Eq. 35-16, the expression we derived earlier for the locations of the fringe minima.

Figure 35-12, which is a plot of Eq. 35-22, shows the intensity of double-slit interference patterns as a function of the phase difference ϕ between the waves at the screen. The horizontal solid line is I_0, the (uniform) intensity on the screen when one of the slits is covered up. Note in Eq. 35-22 and the graph that the intensity I varies from zero at the fringe minima to $4I_0$ at the fringe maxima.

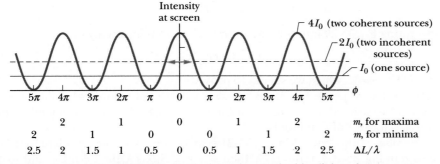

FIG. 35-12 A plot of Eq. 35-22, showing the intensity of a double-slit interference pattern as a function of the phase difference between the waves when they arrive from the two slits. I_0 is the (uniform) intensity that would appear on the screen if one slit were covered. The average intensity of the fringe pattern is $2I_0$, and the *maximum* intensity (for coherent light) is $4I_0$.

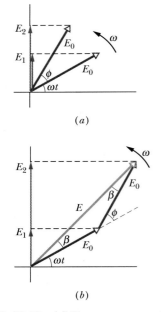

FIG. 35-13 (*a*) Phasors representing, at time *t*, the electric field components given by Eqs. 35-20 and 35-21. Both phasors have magnitude E_0 and rotate with angular speed ω. Their phase difference is ϕ. (*b*) Vector addition of the two phasors gives the phasor representing the resultant wave, with amplitude E and phase constant β.

If the waves from the two sources (slits) were *incoherent*, so that no enduring phase relation existed between them, there would be no fringe pattern and the intensity would have the uniform value $2I_0$ for all points on the screen; the horizontal dashed line in Fig. 35-12 shows this uniform value.

Interference cannot create or destroy energy but merely redistributes it over the screen. Thus, the *average* intensity on the screen must be the same $2I_0$ regardless of whether the sources are coherent. This follows at once from Eq. 35-22; if we substitute $\frac{1}{2}$, the average value of the cosine-squared function, this equation reduces to $I_{avg} = 2I_0$.

Proof of Eqs. 35-22 and 35-23

We shall combine the electric field components E_1 and E_2, given by Eqs. 35-20 and 35-21, respectively, by the method of phasors as is discussed in Section 16-11. In Fig. 35-13*a*, the waves with components E_1 and E_2 are represented by phasors of magnitude E_0 that rotate around the origin at angular speed ω. The values of E_1 and E_2 at any time are the projections of the corresponding phasors on the vertical axis. Figure 35-13*a* shows the phasors and their projections at an arbitrary time *t*. Consistent with Eqs. 35-20 and 35-21, the phasor for E_1 has a rotation angle ωt and the phasor for E_2 has a rotation angle $\omega t + \phi$.

To combine the field components E_1 and E_2 at any point P in Fig. 35-10, we add their phasors vectorially, as shown in Fig. 35-13*b*. The magnitude of the vector sum is the amplitude E of the resultant wave at point P, and that wave has a certain phase constant β. To find the amplitude E in Fig. 35-13*b*, we first note that the two angles marked β are equal because they are opposite equal-length sides of a triangle. From the theorem (for triangles) that an exterior angle (here ϕ, as shown in Fig. 35-13*b*) is equal to the sum of the two opposite interior angles (here that sum is $\beta + \beta$), we see that $\beta = \frac{1}{2}\phi$. Thus, we have

$$E = 2(E_0 \cos \beta)$$
$$= 2E_0 \cos \tfrac{1}{2}\phi. \tag{35-28}$$

If we square each side of this relation, we obtain

$$E^2 = 4E_0^2 \cos^2 \tfrac{1}{2}\phi. \tag{35-29}$$

Now, from Eq. 33-24, we know that the intensity of an electromagnetic wave is proportional to the square of its amplitude. Therefore, the waves we are combining in Fig. 35-13*b*, whose amplitudes are E_0, each has an intensity I_0 that is proportional to E_0^2, and the resultant wave, with amplitude E, has an intensity I that is proportional to E^2. Thus,

$$\frac{I}{I_0} = \frac{E^2}{E_0^2}.$$

Substituting Eq. 35-29 into this equation and rearranging then yield

$$I = 4I_0 \cos^2 \tfrac{1}{2}\phi,$$

which is Eq. 35-22, which we set out to prove.

We still must prove Eq. 35-23, which relates the phase difference ϕ between the waves arriving at any point P on the screen of Fig. 35-10 to the angle θ that serves as a locator of that point.

The phase difference ϕ in Eq. 35-21 is associated with the path length difference S_1b in Fig. 35-10*b*. If S_1b is $\frac{1}{2}\lambda$, then ϕ is π; if S_1b is λ, then ϕ is 2π, and so on. This suggests

$$\left(\begin{matrix} \text{phase} \\ \text{difference} \end{matrix} \right) = \frac{2\pi}{\lambda} \left(\begin{matrix} \text{path length} \\ \text{difference} \end{matrix} \right). \tag{35-30}$$

The path length difference S_1b in Fig. 35-10*b* is $d \sin \theta$; so Eq. 35-30 for the phase difference between the two waves arriving at point P on the screen becomes

$$\phi = \frac{2\pi d}{\lambda} \sin \theta,$$

which is Eq. 35-23, the other equation that we set out to prove to relate ϕ to the angle θ that locates P.

Combining More Than Two Waves

In a more general case, we might want to find the resultant of more than two sinusoidally varying waves at a point. Whatever the number of waves is, our general procedure is this:

1. Construct a series of phasors representing the waves to be combined. Draw them end to end, maintaining the proper phase relations between adjacent phasors.

2. Construct the vector sum of this array. The length of this vector sum gives the amplitude of the resultant phasor. The angle between the vector sum and the first phasor is the phase of the resultant with respect to this first phasor. The projection of this vector-sum phasor on the vertical axis gives the time variation of the resultant wave.

✓ **CHECKPOINT 4**   Each of four pairs of light waves arrives at a certain point on a screen. The waves have the same wavelength. At the arrival point, their amplitudes and phase differences are (a) $2E_0, 6E_0,$ and π rad; (b) $3E_0, 5E_0,$ and π rad; (c) $9E_0, 7E_0,$ and 3π rad; (d) $2E_0, 2E_0,$ and 0 rad. Rank the four pairs according to the intensity of the light at the arrival point, greatest first. (*Hint:* Draw phasors.)

Sample Problem   35-4

Three light waves combine at a certain point where their electric field components are

$$E_1 = E_0 \sin \omega t,$$
$$E_2 = E_0 \sin(\omega t + 60°),$$
$$E_3 = E_0 \sin(\omega t - 30°).$$

Find their resultant component $E(t)$ at that point.

KEY IDEA   The resultant wave is

$$E(t) = E_1(t) + E_2(t) + E_3(t).$$

We can use the method of phasors to find this sum, and we are free to evaluate the phasors at any time t.

Calculations: To simplify the solution, we choose $t = 0$, for which the phasors representing the three waves are shown in Fig. 35-14. We can add these three phasors either directly on a vector-capable calculator or by components. For the component approach, we first write the sum of their horizontal components as

$$\sum E_h = E_0 \cos 0 + E_0 \cos 60° + E_0 \cos(-30°) = 2.37E_0.$$

The sum of their vertical components, which is the value of E at $t = 0$, is

$$\sum E_v = E_0 \sin 0 + E_0 \sin 60° + E_0 \sin(-30°) = 0.366E_0.$$

The resultant wave $E(t)$ thus has an amplitude E_R of

$$E_R = \sqrt{(2.37E_0)^2 + (0.366E_0)^2} = 2.4E_0,$$

and a phase angle β relative to the phasor representing E_1 of

$$\beta = \tan^{-1}\left(\frac{0.366E_0}{2.37E_0}\right) = 8.8°.$$

We can now write, for the resultant wave $E(t)$,

$$E = E_R \sin(\omega t + \beta)$$
$$= 2.4E_0 \sin(\omega t + 8.8°). \qquad \text{(Answer)}$$

Be careful to interpret the angle β correctly in Fig. 35-14: It is the constant angle between E_R and the phasor representing E_1 as the four phasors rotate as a single unit around the origin. The angle between E_R and the horizontal axis in Fig. 35-14 does not remain equal to β.

FIG. 35-14   Three phasors, representing waves with equal amplitudes E_0 and with phase constants 0°, 60°, and −30°, shown at time $t = 0$. The phasors combine to give a resultant phasor with magnitude E_R, at angle β.

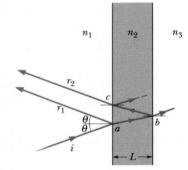

FIG. 35-15 Light waves, represented with ray i, are incident on a thin film of thickness L and index of refraction n_2. Rays r_1 and r_2 represent light waves that have been reflected by the front and back surfaces of the film, respectively. (All three rays are actually nearly perpendicular to the film.) The interference of the waves of r_1 and r_2 with each other depends on their phase difference. The index of refraction n_1 of the medium at the left can differ from the index of refraction n_3 of the medium at the right, but for now we assume that both media are air, with $n_1 = n_3 = 1.0$, which is less than n_2.

35-7 | Interference from Thin Films

The colors we see when sunlight illuminates a soap bubble or an oil slick are caused by the interference of light waves reflected from the front and back surfaces of a thin transparent film. The thickness of the soap or oil film is typically of the order of magnitude of the wavelength of the (visible) light involved. (Greater thicknesses spoil the coherence of the light needed to produce the colors due to interference.)

Figure 35-15 shows a thin transparent film of uniform thickness L and index of refraction n_2, illuminated by bright light of wavelength λ from a distant point source. For now, we assume that air lies on both sides of the film and thus that $n_1 = n_3$ in Fig. 35-15. For simplicity, we also assume that the light rays are almost perpendicular to the film ($\theta \approx 0$). We are interested in whether the film is bright or dark to an observer viewing it almost perpendicularly. (Since the film is brightly illuminated, how could it possibly be dark? You will see.)

The incident light, represented by ray i, intercepts the front (left) surface of the film at point a and undergoes both reflection and refraction there. The reflected ray r_1 is intercepted by the observer's eye. The refracted light crosses the film to point b on the back surface, where it undergoes both reflection and refraction. The light reflected at b crosses back through the film to point c, where it undergoes both reflection and refraction. The light refracted at c, represented by ray r_2, is intercepted by the observer's eye.

If the light waves of rays r_1 and r_2 are exactly in phase at the eye, they produce an interference maximum and region ac on the film is bright to the observer. If they are exactly out of phase, they produce an interference minimum and region ac is dark to the observer, *even though it is illuminated.* If there is some intermediate phase difference, there are intermediate interference and brightness.

Thus, the key to what the observer sees is the phase difference between the waves of rays r_1 and r_2. Both rays are derived from the same ray i, but the path involved in producing r_2 involves light traveling twice across the film (a to b, and then b to c), whereas the path involved in producing r_1 involves no travel through the film. Because θ is about zero, we approximate the path length difference between the waves of r_1 and r_2 as $2L$. However, to find the phase difference between the waves, we cannot just find the number of wavelengths λ that is equivalent to a path length difference of $2L$. This simple approach is impossible for two reasons: (1) the path length difference occurs in a medium other than air, and (2) reflections are involved, which can change the phase.

> ☞ The phase difference between two waves can change if one or both are reflected.

Before we continue our discussion of interference from thin films, we must discuss changes in phase that are caused by reflections.

Reflection Phase Shifts

Refraction at an interface never causes a phase change—but reflection can, depending on the indexes of refraction on the two sides of the interface. Figure 35-16 shows what happens when reflection causes a phase change, using as an example pulses on a denser string (along which pulse travel is relatively slow) and a lighter string (along which pulse travel is relatively fast).

When a pulse traveling relatively slowly along the denser string in Fig. 35-16a reaches the interface with the lighter string, the pulse is partially transmitted and partially reflected, with no change in orientation. For light, this situation corresponds to the incident wave traveling in the medium of greater index of refraction n (recall that greater n means slower speed). In that case, the wave that is reflected at the interface does not undergo a change in phase; that is, its *reflection phase shift* is zero.

When a pulse traveling more quickly along the lighter string in Fig. 35-16b reaches the interface with the denser string, the pulse is again partially transmit-

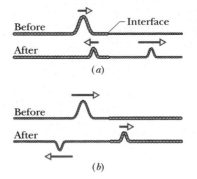

FIG. 35-16 Phase changes when a pulse is reflected at the interface between two stretched strings of different linear densities. The wave speed is greater in the lighter string. (a) The incident pulse is in the denser string. (b) The incident pulse is in the lighter string. Only here is there a phase change, and only in the reflected wave.

ted and partially reflected. The transmitted pulse again has the same orientation as the incident pulse, but now the reflected pulse is inverted. For a sinusoidal wave, such an inversion involves a phase change of π rad, or half a wavelength. For light, this situation corresponds to the incident wave traveling in the medium of lesser index of refraction (with greater speed). In that case, the wave that is reflected at the interface undergoes a phase shift of π rad, or half a wavelength.

We can summarize these results for light in terms of the index of refraction of the medium off which (or from which) the light reflects:

Reflection	Reflection phase shift
Off lower index	0
Off higher index	0.5 wavelength

This might be remembered as "higher means half."

Equations for Thin-Film Interference

In this chapter we have now seen three ways in which the phase difference between two waves can change:

1. by reflection

2. by the waves traveling along paths of different lengths

3. by the waves traveling through media of different indexes of refraction

When light reflects from a thin film, producing the waves of rays r_1 and r_2 shown in Fig. 35-15, all three ways are involved. Let us consider them one by one.

We first reexamine the two reflections in Fig. 35-15. At point a on the front interface, the incident wave (in air) reflects from the medium having the higher of the two indexes of refraction; so the wave of reflected ray r_1 has its phase shifted by 0.5 wavelength. At point b on the back interface, the incident wave reflects from the medium (air) having the lower of the two indexes of refraction; so the wave reflected there is not shifted in phase by the reflection, and thus neither is the portion of it that exits the film as ray r_2. We can organize this information with the first line in Table 35-1, which refers to the simplified drawing in Fig. 35-17 for a thin film in air. So far, as a result of the reflection phase shifts, the waves of r_1 and r_2 have a phase difference of 0.5 wavelength and thus are exactly out of phase.

Now we must consider the path length difference $2L$ that occurs because the wave of ray r_2 crosses the film twice. (This difference $2L$ is shown on the second line in Table 35-1.) If the waves of r_1 and r_2 are to be exactly in phase so that they produce fully constructive interference, the path length $2L$ must cause an additional phase difference of 0.5, 1.5, 2.5, . . . wavelengths. Only then will the net phase difference be an integer number of wavelengths. Thus, for a bright film, we must have

$$2L = \frac{\text{odd number}}{2} \times \text{wavelength} \quad \text{(in-phase waves).} \quad (35\text{-}31)$$

The wavelength we need here is the wavelength λ_{n2} of the light in the medium containing path length $2L$—that is, in the medium with index of refraction n_2. Thus, we can rewrite Eq. 35-31 as

$$2L = \frac{\text{odd number}}{2} \times \lambda_{n2} \quad \text{(in-phase waves).} \quad (35\text{-}32)$$

If, instead, the waves are to be exactly out of phase so that there is fully destructive interference, the path length $2L$ must cause either no additional phase difference or a phase difference of 1, 2, 3, . . . wavelengths. Only then will the net phase difference be an odd number of half-wavelengths. For a dark film,

TABLE 35-1

An Organizing Table for Thin-Film Interference in Air (Fig. 35-17)[a]

	r_1	r_2
Reflection phase shifts	0.5 wavelength	0
Path length difference	2L	
Index in which path length difference occurs	n_2	
In phase[a]: $2L = \dfrac{\text{odd number}}{2} \times \dfrac{\lambda}{n_2}$		
Out of phase[a]: $2L = \text{integer} \times \dfrac{\lambda}{n_2}$		

[a]Valid for $n_2 > n_1$ and $n_2 > n_3$.

FIG. 35-17   Reflections from a thin film in air.

FIG. 35-18 The reflection of light from a soapy water film spanning a vertical loop. The top portion is so thin that the light reflected there undergoes destructive interference, making that portion dark. Colored interference fringes, or bands, decorate the rest of the film but are marred by circulation of liquid within the film as the liquid is gradually pulled downward by gravitation. *(Richard Megna/Fundamental Photographs)*

we must have

$$2L = \text{integer} \times \text{wavelength}, \tag{35-33}$$

where, again, the wavelength is the wavelength λ_{n2} in the medium containing $2L$. Thus, this time we have

$$2L = \text{integer} \times \lambda_{n2} \quad \text{(out-of-phase waves).} \tag{35-34}$$

Now we can use Eq. 35-8 ($\lambda_n = \lambda/n$) to write the wavelength of the wave of ray r_2 inside the film as

$$\lambda_{n2} = \frac{\lambda}{n_2}, \tag{35-35}$$

where λ is the wavelength of the incident light in vacuum (and approximately also in air). Substituting Eq. 35-35 into Eq. 35-32 and replacing "odd number/2" with $(m + \frac{1}{2})$ give us

$$2L = (m + \tfrac{1}{2})\frac{\lambda}{n_2}, \quad \text{for } m = 0, 1, 2, \dots \quad \text{(maxima—bright film in air).} \tag{35-36}$$

Similarly, with m replacing "integer," Eq. 35-34 yields

$$2L = m\frac{\lambda}{n_2}, \quad \text{for } m = 0, 1, 2, \dots \quad \text{(minima—dark film in air).} \tag{35-37}$$

For a given film thickness L, Eqs. 35-36 and 35-37 tell us the wavelengths of light for which the film appears bright and dark, respectively, one wavelength for each value of m. Intermediate wavelengths give intermediate brightnesses. For a given wavelength λ, Eqs. 35-36 and 35-37 tell us the thicknesses of the films that appear bright and dark in that light, respectively, one thickness for each value of m. Intermediate thicknesses give intermediate brightnesses.

Film Thickness Much Less Than λ

A special situation arises when a film is so thin that L is much less than λ, say, $L < 0.1\lambda$. Then the path length difference $2L$ can be neglected, and the phase difference between r_1 and r_2 is due *only* to reflection phase shifts. If the film of Fig. 35-17, where the reflections cause a phase difference of 0.5 wavelength, has thickness $L < 0.1\lambda$, then r_1 and r_2 are exactly out of phase, and thus the film is dark, regardless of the wavelength and intensity of the light. This special situation corresponds to $m = 0$ in Eq. 35-37. We shall count *any* thickness $L < 0.1\lambda$ as being the least thickness specified by Eq. 35-37 to make the film of Fig. 35-17 dark. (Every such thickness will correspond to $m = 0$.) The next greater thickness that will make the film dark is that corresponding to $m = 1$.

Figure 35-18 shows a vertical soap film whose thickness increases from top to bottom because gravitation has caused the film to slump. Bright white light illuminates the film. However, the top portion is so thin that it is dark. In the (somewhat thicker) middle we see fringes, or bands, whose color depends primarily on the wavelength at which reflected light undergoes fully constructive interference for a particular thickness. Toward the (thickest) bottom the fringes become progressively narrower and the colors begin to overlap and fade.

Color Shifting by Butterflies, Inks, and Paints

A surface that displays colors due to thin-film interference is said to be *iridescent* because the tints of the colors change as you change your view of the surface. The iridescence of the top surface of a *Morpho* butterfly wing is due to thin-film interference of light reflected by thin terraces of transparent cuticle-like material

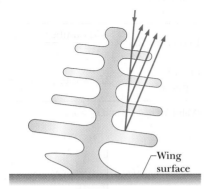

FIG. 35-19 Reflecting structure extending up from a Morpho butterfly wing. Reflections from the top surfaces of the transparent "terraces" give an interference color to the wing.

on the wing (Fig. 35-19). These terraces are arranged like wide, flat branches on a tree-like structure that extends perpendicular to the wing.

Suppose you look directly down on these terraces as white light shines directly down on the wing. Then the light reflected back up to you from the terraces undergoes fully constructive interference in the blue-green region of the visible spectrum. Light in the yellow and red regions, at the opposite end of the spectrum, is weaker because it undergoes only intermediate interference. Thus, the top surface of the wing looks blue-green to you.

If you intercept light that reflects from the wing in some other direction, the light has traveled along a slanted path through the terraces. Then the wavelength at which there is fully constructive interference is somewhat different from that for light reflected directly upward. Thus, if the wing moves in your view so that the angle at which you view it changes, the color at which the wing is brightest changes somewhat, producing the iridescence of the wing.

The color-shifting inks and paints used on paper currencies, cars, guitars, and other objects function in almost the same way as the color-shifting wing on a *Morpho* butterfly. Figure 35-20a shows a cross section of the ink layer used on some currencies. The color shifting is due to thin multilayered flakes suspended in regular ink. Figure 35-20b shows a cross section through one of the flakes. Light penetrating the regular ink above the flake travels through thin layers of chromium (Cr), magnesium fluoride (MgF_2), and aluminum (Al). The Cr layers function as weak mirrors, the Al layer functions as a better mirror, and the MgF_2 layers function like soap films. The result is that light reflected upward from each boundary between layers passes back through the regular ink and then undergoes interference at an observer's eye. Which color undergoes fully constructive interference depends on the thickness L of the MgF_2 layers. In U.S. currency printed with color-shifting inks, the value of L is designed to give fully constructive interference for red or red-yellow light when the observer looks directly down on the currency. When the observer tilts the currency and thus each flake, the light reaching the observer from the flakes undergoes constructive interference for green light. Thus, by changing the angle of view, the observer can shift the color. Other countries use other designs of thin-film flakes to achieve different shifts in the colors on their currencies.

FIG. 35-20 (a) Color-shifting ink on a paper currency consists of multilayered thin-film flakes suspended in regular ink. (b) Cross section of one of the flakes. Light penetrates the five layers, reflecting from each boundary. The color that results from the interference of these reflected light waves is determined by the thickness L of the magnesium fluoride layers.

PROBLEM-SOLVING TACTICS

Tactic 1: Thin-Film Equations Some students believe that Eq. 35-36 gives the maxima and Eq. 35-37 gives the minima for *all* thin-film situations. This is not true. These relations were derived only for the situation in which $n_2 > n_1$ and $n_2 > n_3$ in Fig. 35-15 and for the situation of Fig. 35-17, in which a thin film lies in air.

The appropriate equations for other relative values of the indexes of refraction can be derived by following the reasoning of this section and constructing new versions of Table 35-1. In each case you will end up with Eqs. 35-36 and 35-37, but sometimes Eq. 35-36 will give the minima and Eq. 35-37 will give the maxima—the opposite of what we found here. Which equation gives which depends on whether the reflections at the two interfaces give the same reflection phase shift.

✓ **CHECKPOINT 5** The figure shows four situations in which light reflects perpendicularly from a thin film of thickness L, with indexes of refraction as given. (a) For which situations does reflection at the film interfaces cause a zero phase difference for the two reflected rays? (b) For which situations will the film be dark if the path length difference $2L$ causes a phase difference of 0.5 wavelength?

White light, with a uniform intensity across the visible wavelength range of 400 to 690 nm, is perpendicularly incident on a water film, of index of refraction $n_2 = 1.33$ and thickness $L = 320$ nm, that is suspended in air. At what wavelength λ is the light reflected by the film brightest to an observer?

KEY IDEA The reflected light from the film is brightest at the wavelengths λ for which the reflected rays are in phase with one another. The equation relating these wavelengths λ to the given film thickness L and film index of refraction n_2 is either Eq. 35-36 or Eq. 35-37, depending on the reflection phase shifts for this particular film.

Calculations: To determine which equation is needed, we should fill out an organizing table like Table 35-1. However, because there is air on both sides of the water film, the situation here is exactly like that in Fig. 35-17, and thus the table would be exactly like Table 35-1. Then from Table 35-1, we see that the reflected rays are

in phase (and thus the film is brightest) when

$$2L = \frac{\text{odd number}}{2} \times \frac{\lambda}{n_2},$$

which leads to Eq. 35-36:

$$2L = (m + \tfrac{1}{2})\frac{\lambda}{n_2}.$$

Solving for λ and substituting for L and n_2, we find

$$\lambda = \frac{2n_2 L}{m + \tfrac{1}{2}} = \frac{(2)(1.33)(320 \text{ nm})}{m + \tfrac{1}{2}} = \frac{851 \text{ nm}}{m + \tfrac{1}{2}}.$$

For $m = 0$, this gives us $\lambda = 1700$ nm, which is in the infrared region. For $m = 1$, we find $\lambda = 567$ nm, which is yellow-green light, near the middle of the visible spectrum. For $m = 2$, $\lambda = 340$ nm, which is in the ultraviolet region. Thus, the wavelength at which the light seen by the observer is brightest is

$$\lambda = 567 \text{ nm}. \qquad \text{(Answer)}$$

In Fig. 35-21, a glass lens is coated on one side with a thin film of magnesium fluoride (MgF_2) to reduce reflection from the lens surface. The index of refraction of MgF_2 is 1.38; that of the glass is 1.50. What is the least coating thickness that eliminates (via interference) the reflections at the middle of the visible spectrum ($\lambda = 550$ nm)? Assume that the light is approximately perpendicular to the lens surface.

KEY IDEA Reflection is eliminated if the film thickness L is such that light waves reflected from the two film interfaces are exactly out of phase. The equation relating L to the given wavelength λ and the index of refraction n_2 of the thin film is either Eq. 35-36 or Eq. 35-37, depending on the reflection phase shifts at the interfaces.

Calculations: To determine which equation is needed, we fill out an organizing table like Table 35-1. At the first interface, the incident light is in air, which has a lesser index of refraction than the MgF_2 (the thin film). Thus, we fill in 0.5 wavelength under r_1 in our organizing table (meaning that the waves of ray r_1 are shifted by 0.5λ at the first interface). At the second interface, the incident light is in the MgF_2, which has a lesser index of refraction than the glass on the other side of the interface. Thus, we fill in 0.5 wavelength under r_2 in our table.

Because both reflections cause the same phase shift, they tend to put the waves of r_1 and r_2 in phase. Since we want those waves to be *out of phase*, their path length difference $2L$ must be an odd number of half-wavelengths:

$$2L = \frac{\text{odd number}}{2} \times \frac{\lambda}{n_2}.$$

This leads to Eq. 35-36. Solving that equation for L then gives us the film thicknesses that will eliminate reflection from the lens and coating:

$$L = (m + \tfrac{1}{2})\frac{\lambda}{2n_2}, \qquad \text{for } m = 0, 1, 2, \ldots. \quad (35\text{-}38)$$

We want the least thickness for the coating—that is, the smallest value of L. Thus, we choose $m = 0$, the smallest possible value of m. Substituting it and the given data in Eq. 35-38, we obtain

$$L = \frac{\lambda}{4n_2} = \frac{550 \text{ nm}}{(4)(1.38)} = 99.6 \text{ nm}. \quad \text{(Answer)}$$

FIG. 35-21 Unwanted reflections from glass can be suppressed (at a chosen wavelength) by coating the glass with a thin transparent film of magnesium fluoride of the properly chosen thickness.

Sample Problem 35-7 **Build your skill**

Figure 35-22a shows a transparent plastic block with a thin wedge of air at the right. (The wedge thickness is exaggerated in the figure.) A broad beam of red light, with wavelength $\lambda = 632.8$ nm, is directed downward through the top of the block (at an incidence angle of 0°). Some of the light that passes into the plastic is reflected back up from the top and bottom surfaces of the wedge, which acts as a thin film (of air) with a thickness that varies uniformly and gradually from L_L at the left-hand end to L_R at the right-hand end. (The plastic layers above and below the wedge of air are too thick to act as thin films.) An observer looking down on the block sees an interference pattern consisting of six dark fringes and five bright red fringes along the wedge. What is the change in thickness $\Delta L \, (= L_R - L_L)$ along the wedge?

KEY IDEAS (1) The brightness at any point along the left–right length of the air wedge is due to the interference of the waves reflected at the top and bottom interfaces of the wedge. (2) The variation of brightness in the pattern of bright and dark fringes is due to the variation in the thickness of the wedge. In some regions, the thickness puts the reflected waves in phase and thus produces a bright reflection (a bright red fringe). In other regions, the thickness puts the reflected waves out of phase and thus produces no reflection (a dark fringe).

Organizing the reflections: Because the observer sees more dark fringes than bright fringes, we can assume that a dark fringe is produced at both the left and right ends of the wedge. Thus, the interference pattern is that shown in Fig. 35-22b.

We can represent the reflection of light at the top and bottom interfaces of the wedge, at any point along its length, with Fig. 35-22c, in which L is the wedge thickness at that point. Let us apply this figure to the left end of the wedge, where the reflections give a dark fringe.

We know that, for a dark fringe, the waves of rays r_1 and r_2 in Fig. 35-22c must be out of phase. We also know that the equation relating the film thickness L to the light's wavelength λ and the film's index of refraction n_2 is either Eq. 35-36 or Eq. 35-37, depending on the reflection phase shifts. To determine which equation gives a dark fringe at the left end of the wedge, we should fill out an organizing table like Table 35-1.

At the top interface of the wedge, the incident light is in the plastic, which has a greater n than the air beneath that interface. So, we fill in 0 under r_1 in our organizing table. At the bottom interface of the wedge, the incident light is in air, which has a lesser n than the plastic beneath that interface. So we fill in 0.5 wavelength under r_2. Thus the reflections alone tend to put the waves of r_1 and r_2 out of phase.

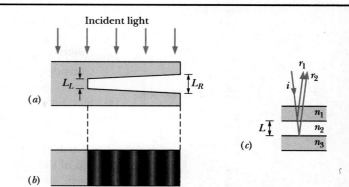

FIG. 35-22 (a) Red light is incident on a thin, air-filled wedge in the side of a transparent plastic block. The thickness of the wedge is L_L at the left end and L_R at the right end. (b) The view from above the block: an interference pattern of six dark fringes and five bright red fringes lies over the region of the wedge. (c) A representation of the incident ray i, reflected rays r_1 and r_2, and thickness L of the wedge anywhere along the length of the wedge.

Reflections at left end: Because the waves are out of phase at the left end of the air wedge, the path length difference $2L$ at that end of the wedge must be given by

$$2L = \text{integer} \times \frac{\lambda}{n_2},$$

which leads to Eq. 35-37:

$$2L = m\frac{\lambda}{n_2}, \qquad \text{for } m = 0, 1, 2, \ldots . \quad (35\text{-}39)$$

Reflections at right end: Equation 35-39 holds not only for the left end of the wedge but also for any point along the wedge where a dark fringe is observed, including the right end, with a different integer value of m for each fringe. The least value of m is associated with the least thickness of the wedge where a dark fringe is observed. Progressively greater values of m are associated with progressively greater thicknesses of the wedge where a dark fringe is observed. Let m_L be the value at the left end. Then the value at the right end must be $m_L + 5$ because, from Fig. 35-22b, the right end is located at the fifth dark fringe from the left end.

Thickness difference: To find ΔL, we first solve Eq. 35-39 twice—once for the thickness L_L at the left end and once for the thickness L_R at the right end:

$$L_L = (m_L)\frac{\lambda}{2n_2}, \qquad L_R = (m_L + 5)\frac{\lambda}{2n_2}. \quad (35\text{-}40)$$

We can now subtract L_L from L_R and substitute $n_2 = 1.00$ for the air within the wedge and $\lambda = 632.8 \times 10^{-9}$ m:

$$\Delta L = L_R - L_L = \frac{(m_L + 5)\lambda}{2n_2} - \frac{m_L\lambda}{2n_2} = \frac{5}{2}\frac{\lambda}{n_2}$$

$$= 1.58 \times 10^{-6} \text{ m.} \qquad \text{(Answer)}$$

FIG. 35-23 Michelson's interferometer, showing the path of light originating at point P of an extended source S. Mirror M splits the light into two beams, which reflect from mirrors M_1 and M_2 back to M and then to telescope T. In the telescope an observer sees a pattern of interference fringes.

35-8 | Michelson's Interferometer

An **interferometer** is a device that can be used to measure lengths or changes in length with great accuracy by means of interference fringes. We describe the form originally devised and built by A. A. Michelson in 1881.

Consider light that leaves point P on extended source S in Fig. 35-23 and encounters *beam splitter M*. A beam splitter is a mirror that transmits half the incident light and reflects the other half. In the figure we have assumed, for convenience, that this mirror possesses negligible thickness. At M the light thus divides into two waves. One proceeds by transmission toward mirror M_1 at the end of one arm of the instrument; the other proceeds by reflection toward mirror M_2 at the end of the other arm. The waves are entirely reflected at these mirrors and are sent back along their directions of incidence, each wave eventually entering telescope T. What the observer sees is a pattern of curved or approximately straight interference fringes; in the latter case the fringes resemble the stripes on a zebra.

The path length difference for the two waves when they recombine at the telescope is $2d_2 - 2d_1$, and anything that changes this path length difference will cause a change in the phase difference between these two waves at the eye. As an example, if mirror M_2 is moved by a distance $\frac{1}{2}\lambda$, the path length difference is changed by λ and the fringe pattern is shifted by one fringe (as if each dark stripe on a zebra had moved to where the adjacent dark stripe had been). Similarly, moving mirror M_2 by $\frac{1}{4}\lambda$ causes a shift by half a fringe (each dark zebra stripe shifts to where the adjacent white stripe was).

A shift in the fringe pattern can also be caused by the insertion of a thin transparent material into the optical path of one of the mirrors—say, M_1. If the material has thickness L and index of refraction n, then the number of wavelengths along the light's to-and-fro path through the material is, from Eq. 35-9,

$$N_m = \frac{2L}{\lambda_n} = \frac{2Ln}{\lambda}. \tag{35-41}$$

The number of wavelengths in the same thickness $2L$ of air before the insertion of the material is

$$N_a = \frac{2L}{\lambda}. \tag{35-42}$$

When the material is inserted, the light returned from mirror M_1 undergoes a phase change (in terms of wavelengths) of

$$N_m - N_a = \frac{2Ln}{\lambda} - \frac{2L}{\lambda} = \frac{2L}{\lambda}(n - 1). \tag{35-43}$$

For each phase change of one wavelength, the fringe pattern is shifted by one fringe. Thus, by counting the number of fringes through which the material causes the pattern to shift, and substituting that number for $N_m - N_a$ in Eq. 35-43, you can determine the thickness L of the material in terms of λ.

By such techniques the lengths of objects can be expressed in terms of the wavelengths of light. In Michelson's day, the standard of length—the meter—was chosen by international agreement to be the distance between two fine scratches on a certain metal bar preserved at Sèvres, near Paris. Michelson was able to show, using his interferometer, that the standard meter was equivalent to 1 553 163.5 wavelengths of a certain monochromatic red light emitted from a light source containing cadmium. For this careful measurement, Michelson received the 1907 Nobel Prize in physics. His work laid the foundation for the eventual abandonment (in 1961) of the meter bar as a standard of length and for the redefinition of the meter in terms of the wavelength of light. By 1983, even this wavelength standard was not precise enough to meet the growing requirements of science and technology, and it was replaced with a new standard based on a defined value for the speed of light.

REVIEW & SUMMARY

Huygens' Principle The three-dimensional transmission of waves, including light, may often be predicted by *Huygens' principle,* which states that all points on a wavefront serve as point sources of spherical secondary wavelets. After a time t, the new position of the wavefront will be that of a surface tangent to these secondary wavelets.

The law of refraction can be derived from Huygens' principle by assuming that the index of refraction of any medium is $n = c/v$, in which v is the speed of light in the medium and c is the speed of light in vacuum.

Wavelength and Index of Refraction The wavelength λ_n of light in a medium depends on the index of refraction n of the medium:

$$\lambda_n = \frac{\lambda}{n}, \tag{35-8}$$

in which λ is the wavelength of the light in vacuum. Because of this dependency, the phase difference between two waves can change if they pass through different materials with different indexes of refraction.

Young's Experiment In **Young's interference experiment,** light passing through a single slit falls on two slits in a screen. The light leaving these slits flares out (by diffraction), and interference occurs in the region beyond the screen. A fringe pattern, due to the interference, forms on a viewing screen.

The light intensity at any point on the viewing screen depends in part on the difference in the path lengths from the slits to that point. If this difference is an integer number of wavelengths, the waves interfere constructively and an intensity maximum results. If it is an odd number of half-wavelengths, there is destructive interference and an intensity minimum occurs. The conditions for maximum and minimum intensity are

$$d \sin \theta = m\lambda, \qquad \text{for } m = 0, 1, 2, \dots$$
$$\text{(maxima—bright fringes)}, \tag{35-14}$$

$$d \sin \theta = (m + \tfrac{1}{2})\lambda, \qquad \text{for } m = 0, 1, 2, \dots$$
$$\text{(minima—dark fringes)}, \tag{35-16}$$

where θ is the angle the light path makes with a central axis and d is the slit separation.

Coherence If two light waves that meet at a point are to interfere perceptibly, the phase difference between them must remain constant with time; that is, the waves must be **coherent.** When two coherent waves meet, the resulting intensity may be found by using phasors.

Intensity in Two-Slit Interference In Young's interference experiment, two waves, each with intensity I_0, yield a resultant wave of intensity I at the viewing screen, with

$$I = 4I_0 \cos^2 \tfrac{1}{2}\phi, \qquad \text{where } \phi = \frac{2\pi d}{\lambda} \sin \theta. \tag{35-22, 35-23}$$

Equations 35-14 and 35-16, which identify the positions of the fringe maxima and minima, are contained within this relation.

Thin-Film Interference When light is incident on a thin transparent film, the light waves reflected from the front and back surfaces interfere. For near-normal incidence the wavelength conditions for maximum and minimum intensity of the light reflected from a *film in air* are

$$2L = (m + \tfrac{1}{2}) \frac{\lambda}{n_2}, \qquad \text{for } m = 0, 1, 2, \dots$$
$$\text{(maxima—bright film in air)}, \tag{35-36}$$

$$2L = m \frac{\lambda}{n_2}, \qquad \text{for } m = 0, 1, 2, \dots$$
$$\text{(minima—dark film in air)}, \tag{35-37}$$

where n_2 is the index of refraction of the film, L is its thickness, and λ is the wavelength of the light in air.

If the light incident at an interface between media with different indexes of refraction is in the medium with the smaller index of refraction, the reflection causes a phase change of π rad, or half a wavelength, in the reflected wave. Otherwise, there is no phase change due to the reflection. Refraction at an interface does not cause a phase shift.

The Michelson Interferometer In *Michelson's interferometer* a light wave is split into two beams that, after traversing paths of different lengths, are recombined so they interfere and form a fringe pattern. Varying the path length of one of the beams allows distances to be accurately expressed in terms of wavelengths of light, by counting the number of fringes through which the pattern shifts because of the change.

QUESTIONS

1 Figure 35-24 shows two light rays that are initially exactly in phase and that reflect from several glass surfaces. Neglect the slight slant in the path of the light in the second arrangement. (a) What is the path length difference of the rays? In wavelengths λ, (b) what should that path length difference equal if the rays are to be exactly out of phase when they emerge, and (c) what is the smallest value of d that will allow that final phase difference?

FIG. 35-24 Question 1.

2 In Fig. 35-25, three pulses of light—a, b, and c—of the same wavelength are sent through layers of plastic having the given indexes of refraction and along the paths indicated. Rank the pulses according to their travel time through the plastic layers, greatest first.

FIG. 35-25 Question 2.

3 Light travels along the length of a 1500-nm-long nano-structure. When a peak of the wave is at one end of the nanostructure, is there a peak or a valley at the other end if the wavelength is (a) 500 nm and (b) 1000 nm?

4 (a) If you move from one bright fringe in a two-slit interference pattern to the next one farther out, (b) does the path length difference ΔL increase or decrease and (c) by how much does it change, in wavelengths λ?

5 Does the spacing between fringes in a two-slit interference pattern increase, decrease, or stay the same if (a) the slit separation is increased, (b) the color of the light is switched from red to blue, and (c) the whole apparatus is submerged in cooking sherry? (d) If the slits are illuminated with white light, then at any side maximum, does the blue component or the red component peak closer to the central maximum?

6 Figure 35-26 shows two rays of light, of wavelength 600 nm, that reflect from glass surfaces separated by 150 nm. The rays are initially in phase. (a) What is the path length difference of the rays? (b) When they have cleared the reflection region, are the rays exactly in phase, exactly out of phase, or in some intermediate state?

FIG. 35-26   Question 6.

7 Is there an interference maximum, a minimum, an intermediate state closer to a maximum, or an intermediate state closer to a minimum at point P in Fig. 35-10 if the path length difference of the two rays is (a) 2.2λ, (b) 3.5λ, (c) 1.8λ, and (d) 1.0λ? For each situation, give the value of m associated with the maximum or minimum involved.

8 Figure 35-27a gives intensity I versus position x on the viewing screen for the central portion of a two-slit interference pattern. The other parts of the figure give phasor diagrams for the electric field components of the waves arriving at the screen from the two slits (as in Fig. 35-13a). Which numbered points on the screen best correspond to which phasor diagram?

FIG. 35-27   Question 8.

9 Figure 35-28 shows two sources S_1 and S_2 that emit radio waves of wavelength λ in all directions. The sources are exactly in phase and are separated by a distance equal to 1.5λ.

The vertical broken line is the perpendicular bisector of the distance between the sources. (a) If we start at the indicated start point and travel along path 1, does the interference produce a maximum all along the path, a minimum all along the path, or alternating maxima and minima? Repeat for (b) path 2 and (c) path 3.

FIG. 35-28   Question 9.

10 Figure 35-29 shows the transmission of light through a thin film in air by a perpendicular beam (tilted in the figure for clarity). (a) Did ray r_3 undergo a phase shift due to reflection? (b) In wavelengths, what is the reflection phase shift for ray r_4? (c) If the film thickness is L, what is the path length difference between rays r_3 and r_4?

FIG. 35-29   Question 10.

11 Figure 35-30 shows four situations in which light reflects perpendicularly from a thin film of thickness L sandwiched between much thicker materials. The indexes of refraction are given. In which situations does Eq. 35-36 correspond to the reflections yielding maxima (that is, a bright film)?

FIG. 35-30   Question 11.

12 Figure 35-31a shows the cross section of a vertical thin film whose width increases downward because gravitation causes slumping. Figure 35-31b is a face-on view of the film, showing four bright (red) interference fringes that result when the film is illuminated with a perpendicular beam of red light. Points in the cross section corresponding to the bright fringes are labeled. In terms of the wavelength of the light inside the film, what is the difference in film thickness between (a) points a and b and (b) points b and d?

FIG. 35-31   Question 12.

PROBLEMS

GO   Tutoring problem available (at instructor's discretion) in *WileyPLUS* and WebAssign

SSM   Worked-out solution available in Student Solutions Manual   **WWW**   Worked-out solution is at ———

• – •••   Number of dots indicates level of problem difficulty   **ILW**   Interactive solution is at ———

Additional information available in *The Flying Circus of Physics* and at flyingcircusofphysics.com

http://www.wiley.com/college/halliday

sec. 35-2   Light as a Wave

•1   The speed of yellow light (from a sodium lamp) in a certain liquid is measured to be 1.92×10^8 m/s. What is the index of refraction of this liquid for the light?

•2   In Fig. 35-32a, a beam of light in material 1 is incident on a boundary at an angle of 30°. The extent to which the light is bent due to refraction depends, in part, on the index of refraction n_2 of material 2. Figure 35-32b gives the angle of refraction θ_2 versus n_2 for a range of possible n_2 values, from $n_a = 1.30$ to $n_b = 1.90$. What is the speed of light in material 1?

      (a)   (b)

FIG. 35-32   Problem 2.

•3   How much faster, in meters per second, does light travel in sapphire than in diamond? See Table 33-1.

•4   The wavelength of yellow sodium light in air is 589 nm. (a) What is its frequency? (b) What is its wavelength in glass whose index of refraction is 1.52? (c) From the results of (a) and (b), find its speed in this glass.

•5   In Fig. 35-4, assume that two waves of light in air, of wavelength 400 nm, are initially in phase. One travels through a glass layer of index of refraction $n_1 = 1.60$ and thickness L. The other travels through an equally thick plastic layer of index of refraction $n_2 = 1.50$. (a) What is the smallest value L should have if the waves are to end up with a phase difference of 5.65 rad? (b) If the waves arrive at some common point with the same amplitude, is their interference fully constructive, fully destructive, intermediate but closer to fully constructive, or intermediate but closer to fully destructive? **SSM**

•6   In Fig. 35-33, a light wave along ray r_1 reflects once from a mirror and a light wave along ray r_2 reflects twice from that same mirror and once from a tiny mirror at distance L from the bigger mirror. (Neglect the slight tilt of the rays.) The waves have wavelength λ and are initially exactly out of phase. What are the (a) smallest, (b) second smallest, and (c) third smallest values of L/λ that result in the final waves being exactly in phase?

FIG. 35-33

Problems 6 and 7.

•7   In Fig. 35-33, a light wave along ray r_1 reflects once from a mirror and a light wave along ray r_2 reflects twice from that same mirror and once from a tiny mirror at distance L from the bigger mirror. (Neglect the slight tilt of the rays.) The waves have wavelength 620 nm and are initially in phase. (a) What is the smallest value of L that puts the final light waves exactly out of phase? (b) With the tiny mirror initially at that value of L, how far must it be moved away from the bigger mirror to again put the final waves out of phase?

•8   In Fig. 35-34, two light pulses are sent through layers of plastic with thicknesses of either L or $2L$ as shown and indexes of refraction $n_1 = 1.55$, $n_2 = 1.70$, $n_3 = 1.60$, $n_4 = 1.45$, $n_5 = 1.59$, $n_6 = 1.65$, and $n_7 = 1.50$. (a) Which pulse travels through the plastic in less time? (b) What multiple of L/c gives the difference in the traversal times of the pulses?

FIG. 35-34   Problem 8.

••9   Suppose that the two waves in Fig. 35-4 have wavelength $\lambda = 500$ nm in air. What multiple of λ gives their phase difference when they emerge if (a) $n_1 = 1.50, n_2 = 1.60$, and $L = 8.50\ \mu$m; (b) $n_1 = 1.62, n_2 = 1.72$, and $L = 8.50\ \mu$m; and (c) $n_1 = 1.59, n_2 = 1.79$, and $L = 3.25\ \mu$m? (d) Suppose that in each of these three situations the waves arrive at a common point (with the same amplitude) after emerging. Rank the situations according to the brightness the waves produce at the common point.

••10   In Fig. 35-35, two light rays go through different paths by reflecting from the various flat surfaces shown. The light waves have a wavelength of 420.0 nm and are initially in phase. What are the (a) smallest and (b) second smallest value of distance L that will put the waves exactly out of phase as they emerge from the region?

••11   In Fig. 35-4, assume that the two light waves, of wavelength 620 nm in air, are initially out of phase by π rad. The indexes of refraction of the media are $n_1 = 1.45$ and $n_2 = 1.65$. What are the (a) smallest and (b) second smallest value of L that will put the waves exactly in phase once they pass through the two media?

FIG. 35-35   Problems 10 and 94.

••12   In Fig. 35-36, a light ray is incident at angle $\theta_1 = 50°$ on a series of five transparent layers with parallel boundaries. For layers 1 and 3, $L_1 = 20$ μm, $L_3 = 25$ μm, $n_1 = 1.6$, and $n_3 = 1.45$. (a) At what angle does the light emerge back into air at the right? (b) How much time does the light take to travel through layer 3?

FIG. 35-36 Problem 12.

••13 Two waves of light in air, of wavelength $\lambda = 600.0$ nm, are initially in phase. They then travel through plastic layers as shown in Fig. 35-37, with $L_1 = 4.00$ μm, $L_2 = 3.50$ μm, $n_1 = 1.40$, and $n_2 = 1.60$. (a) What multiple of λ gives their phase difference after they both have emerged from the layers? (b) If the waves later arrive at some common point with the same amplitude, is their interference fully constructive, fully destructive, intermediate but closer to fully constructive, or intermediate but closer to fully destructive? **ILW**

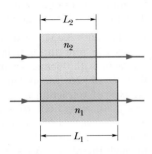

FIG. 35-37 Problem 13.

sec. 35-4 Young's Interference Experiment

•14 Monochromatic green light, of wavelength 550 nm, illuminates two parallel narrow slits 7.70 μm apart. Calculate the angular deviation (θ in Fig. 35-10) of the third-order (for $m = 3$) bright fringe (a) in radians and (b) in degrees.

•15 In Fig. 35-38, two radio-frequency point sources S_1 and S_2, separated by distance $d = 2.0$ m, are radiating in phase with $\lambda = 0.50$ m. A detector moves in a large circular path around the two sources in a plane containing them. How many maxima does it detect? **SSM**

FIG. 35-38 Problems 15 and 24.

•16 In a double-slit arrangement the slits are separated by a distance equal to 100 times the wavelength of the light passing through the slits. (a) What is the angular separation in radians between the central maximum and an adjacent maximum? (b) What is the distance between these maxima on a screen 50.0 cm from the slits?

•17 A double-slit arrangement produces interference fringes for sodium light ($\lambda = 589$ nm) that have an angular separation of 3.50×10^{-3} rad. For what wavelength would the angular separation be 10.0% greater? **SSM**

•18 A double-slit arrangement produces interference fringes for sodium light ($\lambda = 589$ nm) that are 0.20° apart. What is the angular fringe separation if the entire arrangement is immersed in water ($n = 1.33$)?

•19 Suppose that Young's experiment is performed with blue-green light of wavelength 500 nm. The slits are 1.20 mm apart, and the viewing screen is 5.40 m from the slits. How far apart are the bright fringes near the center of the interference pattern? **SSM ILW**

•20 In the two-slit experiment of Fig. 35-10, let angle θ be 20.0°, the slit separation be 4.24 μm, and the wavelength be $\lambda = 500$ nm. (a) What multiple of λ gives the phase difference between the waves of rays r_1 and r_2 when they arrive at point P on the distant screen? (b) What is the phase difference in radians? (c) Determine where in the interference pattern point P lies by giving the maximum or minimum on which it lies, or the maximum and minimum between which it lies.

••21 In Fig. 35-39, sources A and B emit long-range radio waves of wavelength 400 m, with the phase of the emission from A ahead of that from source B by 90°. The distance r_A from A to detector D is greater than the corresponding distance r_B by 100 m. What is the phase difference of the waves at D?

FIG. 35-39 Problem 21.

••22 Sunlight is used in a double-slit interference experiment. The fourth-order maximum for a wavelength of 450 nm occurs at an angle of $\theta = 90°$. Thus, it is on the verge of being eliminated from the pattern because θ cannot exceed 90° in Eq. 35-14. (a) What range of wavelengths in the visible range (400 nm to 700 nm) are not present in the third-order maxima? To eliminate all of the visible light in the fourth-order maximum, (b) should the slit separation be increased or decreased and (c) what least change in separation is needed?

••23 In Fig. 35-40, two isotropic point sources of light (S_1 and S_2) are separated by distance 2.70 μm along a y axis and emit in phase at wavelength 900 nm and at the same amplitude. A light detector is located at point P at coordinate x_P on the x axis. What is the greatest value of x_P at which the detected light is minimum due to destructive interference? **GO**

FIG. 35-40 Problems 23, 28, and 122.

••24 In Fig. 35-38, two isotropic point sources S_1 and S_2 emit identical light waves in phase at wavelength λ. The sources lie at separation d on an x axis, and a light detector is moved in a circle of large radius around the midpoint between them. It detects 30 points of zero intensity, including two on the x axis, one of them to the left of the sources and the other to the right of the sources. What is the value of d/λ?

••25 In a double-slit experiment, the distance between slits is 5.0 mm and the slits are 1.0 m from the screen. Two interference patterns can be seen on the screen: one due to light of wavelength 480 nm, and the other due to light of wavelength 600 nm. What is the separation on the screen between the third-order ($m = 3$) bright fringes of the two interference patterns?

••26 In Fig. 35-41, two isotropic point sources S_1 and S_2 emit light in phase at wavelength λ and at the same amplitude. The sources are separated by distance $2d = 6.00\lambda$. They lie on an axis that is parallel to an x axis, which runs along a viewing screen at distance $D = 20.0\lambda$. The origin lies on the perpendicular bisector between the sources. The figure shows two rays reaching point P on the screen, at position x_P. (a) At what value of x_P do the rays have the minimum possible phase difference? (b) What multiple of λ gives that minimum

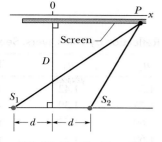

FIG. 35-41   Problem 26.

phase difference? (c) At what value of x_P do the rays have the maximum possible phase difference? What multiple of λ gives (d) that maximum phase difference and (e) the phase difference when $x_P = 6.00\lambda$? (f) When $x_P = 6.00\lambda$, is the resulting intensity at point P maximum, minimum, intermediate but closer to maximum, or intermediate but closer to minimum?

•••27   A thin flake of mica ($n = 1.58$) is used to cover one slit of a double-slit interference arrangement. The central point on the viewing screen is now occupied by what had been the seventh bright side fringe ($m = 7$). If $\lambda = 550$ nm, what is the thickness of the mica?

•••28   Figure 35-40 shows two isotropic point sources of light (S_1 and S_2) that emit in phase at wavelength 400 nm and at the same amplitude. A detection point P is shown on an x axis that extends through source S_1. The phase difference ϕ between the light arriving at point P from the two sources is to be measured as P is moved along the x axis from $x = 0$ out to $x = +\infty$. The results out to $x_s = 10 \times 10^{-7}$ m are given in Fig. 35-42. On the way out to $+\infty$, what is the greatest value of x at which the light arriving at P from S_1 is exactly out of phase with the light arriving at P from S_2?

FIG. 35-42   Problem 28.

sec. 35-6   Intensity in Double-Slit Interference

•29   Two waves of the same frequency have amplitudes 1.00 and 2.00. They interfere at a point where their phase difference is 60.0°. What is the resultant amplitude? SSM

•30   Find the sum y of the following quantities:
$$y_1 = 10 \sin \omega t \quad \text{and} \quad y_2 = 8.0 \sin(\omega t + 30°).$$

••31   Three electromagnetic waves travel through a certain point P along an x axis. They are polarized parallel to a y axis, with the following variations in their amplitudes. Find their resultant at P. GO

$$E_1 = (10.0\ \mu\text{V/m}) \sin[(2.0 \times 10^{14}\ \text{rad/s})t]$$
$$E_2 = (5.00\ \mu\text{V/m}) \sin[(2.0 \times 10^{14}\ \text{rad/s})t + 45.0°]$$
$$E_3 = (5.00\ \mu\text{V/m}) \sin[(2.0 \times 10^{14}\ \text{rad/s})t - 45.0°]$$

••32   In the double-slit experiment of Fig. 35-10, the electric fields of the waves arriving at point P are given by
$$E_1 = (2.00\ \mu\text{V/m}) \sin[(1.26 \times 10^{15})t]$$
$$E_2 = (2.00\ \mu\text{V/m}) \sin[(1.26 \times 10^{15})t + 39.6\ \text{rad}],$$
where time t is in seconds. (a) What is the amplitude of the resultant electric field at point P? (b) What is the ratio of the intensity I_P at point P to the intensity I_{cen} at the center of the interference pattern? (c) Describe where point P is in the interference pattern by giving the maximum or minimum on which it lies, or the maximum and minimum between which it lies. In a phasor diagram of the electric fields, (d) at what rate would the phasors rotate around the origin and (e) what is the angle between the phasors?

••33   Add the quantities $y_1 = 10 \sin \omega t$, $y_2 = 15 \sin(\omega t + 30°)$, and $y_3 = 5.0 \sin(\omega t - 45°)$ using the phasor method. ILW

••34   In the double-slit experiment of Fig. 35-10, the viewing screen is at distance $D = 4.00$ m, point P lies at distance $y = 20.5$ cm from the center of the pattern, the slit separation d is $4.50\ \mu$m, and the wavelength λ is 580 nm. (a) Determine where point P is in the interference pattern by giving the maximum or minimum on which it lies, or the maximum and minimum between which it lies. (b) What is the ratio of the intensity I_P at point P to the intensity I_{cen} at the center of the pattern?

sec. 35-7   Interference from Thin Films

•35   The rhinestones in costume jewelry are glass with index of refraction 1.50. To make them more reflective, they are often coated with a layer of silicon monoxide of index of refraction 2.00. What is the minimum coating thickness needed to ensure that light of wavelength 560 nm and of perpendicular incidence will be reflected from the two surfaces of the coating with fully constructive interference?

•36   White light is sent downward onto a horizontal thin film that is sandwiched between two materials. The indexes of refraction are 1.80 for the top material, 1.70 for the thin film, and 1.50 for the bottom material. The film thickness is 5.00×10^{-7} m. Of the visible wavelengths (400 to 700 nm) that result in fully constructive interference at an observer above the film, which is the (a) longer and (b) shorter wavelength? The materials and film are then heated so that the film thickness increases. (c) Does the light resulting in fully constructive interference shift toward longer or shorter wavelengths?

•37   Light of wavelength 624 nm is incident perpendicularly on a soap film ($n = 1.33$) suspended in air. What are the (a) least and (b) second least thicknesses of the film for which the reflections from the film undergo fully constructive interference? ILW

•38   A 600-nm-thick soap film ($n = 1.40$) in air is illuminated with white light in a direction perpendicular to the film. For how many different wavelengths in the 300 to 700 nm range is there (a) fully constructive interference and (b) fully destructive interference in the reflected light?

•39   We wish to coat flat glass ($n = 1.50$) with a transparent material ($n = 1.25$) so that reflection of light at wavelength 600 nm is eliminated by interference. What minimum thickness can the coating have to do this? SSM

••40 A thin film of acetone (n = 1.25) coats a thick glass plate (n = 1.50). White light is incident normal to the film. In the reflections, fully destructive interference occurs at 600 nm and fully constructive interference at 700 nm. Calculate the thickness of the acetone film.

••41 through 52 *Reflection by thin layers.* In Fig. 35-43, light is incident perpendicularly on a thin layer of material 2 that lies between (thicker) materials 1 and 3. (The rays are tilted only for clarity.) The waves of rays r_1 and r_2 interfere, and here we consider the type of interference to be either maximum (max) or minimum (min). For this situation, each problem in Table 35-2 refers to the indexes of refraction n_1, n_2, and n_3, the type of interference, the thin-layer thickness L in nanometers, and the wavelength λ in nanometers of the light as measured in air. Where λ is missing, give the wavelength that is in the visible range. Where L is missing, give the second least thickness or the third least thickness as indicated. **SSM** 41, 47 **GO** 43, 51

••53 A disabled tanker leaks kerosene (n = 1.20) into the Persian Gulf, creating a large slick on top of the water (n = 1.30). (a) If you are looking straight down from an airplane, while the Sun is overhead, at a region of the slick where its thickness is 460 nm, for which wavelength(s) of visible light is the reflection brightest because of constructive interference? (b) If you are scuba diving directly under this same region of the slick, for which wavelength(s) of visible light is the transmitted intensity strongest?

••54 A thin film, with a thickness of 272.7 nm and with air on both sides, is illuminated with a beam of white light. The beam is perpendicular to the film and consists of the full range of wavelengths for the visible spectrum. In the light reflected by the film, light with a wavelength of 600.0 nm undergoes fully constructive interference. At what wavelength does the reflected light undergo fully destructive interference? (*Hint:* You must make a reasonable assumption about the index of refraction.)

••55 The reflection of perpendicularly incident white light by a soap film in air has an interference maximum at 600 nm and a minimum at 450 nm, with no minimum in be-

TABLE 35-2

Problems 41 through 52: Reflection by Thin Layers. See the setup for these problems.

	n_1	n_2	n_3	Type	L	λ
41	1.50	1.34	1.42	min	380	
42	1.32	1.75	1.39	max	325	
43	1.55	1.60	1.33	max	3rd	612
44	1.55	1.60	1.33	max	285	
45	1.60	1.40	1.80	min	200	
46	1.40	1.46	1.75	min	2nd	482
47	1.40	1.46	1.75	min	210	
48	1.50	1.34	1.42	max	2nd	587
49	1.68	1.59	1.50	min	2nd	342
50	1.68	1.59	1.50	min	415	
51	1.32	1.75	1.39	max	3rd	382
52	1.60	1.40	1.80	max	2nd	632

FIG. 35-43 Problems 41 through 52.

tween. If n = 1.33 for the film, what is the film thickness, assumed uniform?

••56 A plane wave of monochromatic light is incident normally on a uniform thin film of oil that covers a glass plate. The wavelength of the source can be varied continuously. Fully destructive interference of the reflected light is observed for wavelengths of 500 and 700 nm and for no wavelengths in between. If the index of refraction of the oil is 1.30 and that of the glass is 1.50, find the thickness of the oil film.

••57 through 68 *Transmission through thin layers.* In Fig. 35-44, light is incident perpendicularly on a thin layer of material 2 that lies between (thicker) materials 1 and 3. (The rays are tilted only for clarity.) Part of the light ends up in material 3 as ray r_3

FIG. 35-44 Problems 57 through 68.

TABLE 35-3

Problems 57 through 68: Transmission Through Thin Layers. See the setup for these problems.

	n_1	n_2	n_3	Type	L	λ
57	1.50	1.34	1.42	min	2nd	587
58	1.68	1.59	1.50	max	2nd	342
59	1.40	1.46	1.75	max	2nd	482
60	1.32	1.75	1.39	min	3rd	382
61	1.60	1.40	1.80	min	2nd	632
62	1.55	1.60	1.33	min	3rd	612
63	1.68	1.59	1.50	max	415	
64	1.50	1.34	1.42	max	380	
65	1.32	1.75	1.39	min	325	
66	1.40	1.46	1.75	max	210	
67	1.55	1.60	1.33	min	285	
68	1.60	1.40	1.80	max	200	

(the light does not reflect inside material 2) and r_4 (the light reflects twice inside material 2). The waves of r_3 and r_4 interfere, and here we consider the type of interference to be either maximum (max) or minimum (min). For this situation, each problem in Table 35-3 refers to the indexes of refraction n_1, n_2, and n_3, the type of interference, the thin-layer thickness L in nanometers, and the wavelength λ in nanometers of the light as measured in air. Where λ is missing, give the wavelength that is in the visible range. Where L is missing, give the second least thickness or the third least thickness as indicated. **SSM** 63 **GO** 61

••69 In Fig. 35-45, a broad beam of light of wavelength 630 nm is incident at 90° on a thin, wedge-shaped film with index of refraction 1.50. An observer intercepting the light transmitted by the film sees 10 bright and 9 dark fringes along the length of the film. By how much does the film thickness change over this length? **GO**

FIG. 35-45 Problem 69.

••70 Two rectangular glass plates ($n = 1.60$) are in contact along one edge and are separated along the opposite edge (Fig. 35-46). Light with a wavelength of 600 nm is incident perpendicularly onto the top plate. The air between the plates acts as a thin film. Nine dark fringes and eight bright fringes are observed from above the top plate. If the distance between the two plates along the separated edges is increased by 600 nm, how many dark fringes will there then be across the top plate?

FIG. 35-46 Problems 70, 71, 72–74, 106, 107, and 113.

••71 In Fig. 35-46, a broad beam of light of wavelength 683 nm is sent directly downward through the top plate of a pair of glass plates. The plates are 120 mm long, touch at the left end, and are separated by 48.0 μm at the right end. The air between the plates acts as a thin film. How many bright fringes will be seen by an observer looking down through the top plate? **SSM**

••72 In Fig. 35-46, a broad beam of light of wavelength 620 nm is sent directly downward through the top plate of a pair of glass plates touching at the left end. The air between the plates acts as a thin film, and an interference pattern can be seen from above the plates. Initially, a dark fringe lies at the left end, a bright fringe lies at the right end, and nine dark fringes lie between those two end fringes. The plates are then very gradually squeezed together at a constant rate to decrease the angle between them. As a result, the fringe at the right side changes between being bright to being dark every 15.0 s. (a) At what rate is the spacing between the plates at the right end being changed? (b) By how much has the spacing there changed when both left and right ends have a dark fringe and there are five dark fringes between them?

••73 In Fig. 35-46, two microscope slides touch at one end and are separated at the other end. When light of wavelength 500 nm shines vertically down on the slides, an overhead observer sees an interference pattern on the slides with the dark fringes separated by 1.2 mm. What is the angle between the slides?

••74 In Fig. 35-46, a broad beam of monochromatic light is directed perpendicularly through two glass plates that are held together at one end to create a wedge of air between them. An observer intercepting light reflected from the wedge of air, which acts as a thin film, sees 4001 dark fringes along the length of the wedge. When the air between the plates is evacuated, only 4000 dark fringes are seen. Calculate to six significant figures the index of refraction of air from these data.

••75 Figure 35-47a shows a lens with radius of curvature R lying on a flat glass plate and illuminated from above by light with wavelength λ. Figure 35-47b (a photograph taken from above the lens) shows that circular interference fringes (called *Newton's rings*) appear, associated with the variable thickness d of the air film between the lens and the plate. Find the radii r of the interference maxima assuming $r/R \ll 1$. **SSM ILW**

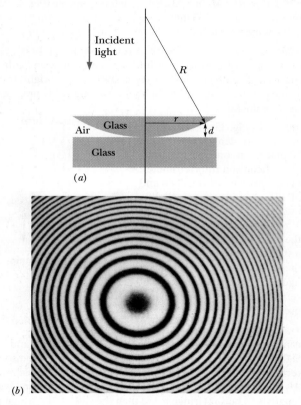

FIG. 35-47 Problems 75 and 77. *(Courtesy Bausch & Lomb)*

••76 In a Newton's rings experiment (see Problem 75), the radius of curvature R of the lens is 5.0 m and the lens diameter is 20 mm. (a) How many bright rings are produced? Assume that $\lambda = 589$ nm. (b) How many bright rings would be produced if the arrangement were immersed in water ($n = 1.33$)?

••77 A Newton's rings apparatus is to be used to determine the radius of curvature of a lens (see Fig. 35-47 and Problem 75). The radii of the nth and $(n + 20)$th bright rings are measured and found to be 0.162 and 0.368 cm, respectively, in light of wavelength 546 nm. Calculate the radius of curvature of the lower surface of the lens.

•••78 A thin film of liquid is held in a horizontal circular ring, with air on both sides of the film. A beam of light at wavelength 550 nm is directed perpendicularly onto the film, and the intensity I of its reflection is monitored. Figure 35-48 gives intensity I as a function of time t; the horizontal scale is set by $t_s = 20.0$ s. The intensity changes because of evaporation from the two sides of the film. Assume that the film is flat and has parallel sides, a radius of 1.80 cm, and an index of refraction of 1.40. Also assume that the film's volume decreases at a constant rate. Find that rate.

FIG. 35-48   Problem 78.

sec. 35-8 Michelson's Interferometer

•79 If mirror M_2 in a Michelson interferometer (Fig. 35-23) is moved through 0.233 mm, a shift of 792 bright fringes occurs. What is the wavelength of the light producing the fringe pattern?

•80 A thin film with index of refraction $n = 1.40$ is placed in one arm of a Michelson interferometer, perpendicular to the optical path. If this causes a shift of 7.0 bright fringes of the pattern produced by light of wavelength 589 nm, what is the film thickness?

••81 In Fig. 35-49, an airtight chamber of length $d = 5.0$ cm is placed in one of the arms of a Michelson interferometer. (The glass window on each end of the chamber has negligible thickness.) Light of wavelength $\lambda = 500$ nm is used. Evacuating the air from the chamber causes a shift of 60 bright fringes. From these data and to six significant figures, find the index of refraction of air at atmospheric pressure. **SSM WWW**

FIG. 35-49   Problem 81.

••82 The element sodium can emit light at two wavelengths, $\lambda_1 = 589.10$ nm and $\lambda_2 = 589.59$ nm. Light from sodium is being used in a Michelson interferometer (Fig. 35-23). Through what distance must mirror M_2 be moved if the shift in the fringe pattern for one wavelength is to be 1.00 fringe more than the shift in the fringe pattern for the other wavelength?

Additional Problems

83 Ocean waves moving at a speed of 4.0 m/s are approaching a beach at angle $\theta_1 = 30°$ to the normal, as shown from above in Fig. 35-50. Suppose the water depth changes abruptly at a certain distance from the beach and the wave speed there drops to 3.0 m/s. (a) Close to the beach, what is the angle θ_2 between the direc-

FIG. 35-50   Problem 83.

tion of wave motion and the normal? (Assume the same law of refraction as for light.) (b) Explain why most waves come in normal to a shore even though at large distances they approach at a variety of angles.

84 Figure 35-51a shows two light rays that are initially in phase as they travel upward through a block of plastic, with wavelength 400 nm as measured in air. Light ray r_1 exits directly into air. However, before light ray r_2 exits into air, it travels through a liquid in a hollow cylinder within the plastic. Initially the height L_{liq} of the liquid is 40.0 μm, but then the liquid begins to evaporate. Let ϕ be the phase difference between rays r_1 and r_2 once they both exit into the air. Figure 35-51b shows ϕ versus the liquid's height L_{liq} until the liquid disappears, with ϕ given in terms of wavelength and the horizontal scale set by $L_s = 40.00$ μm. What are (a) the index of refraction of the plastic and (b) the index of refraction of the liquid?

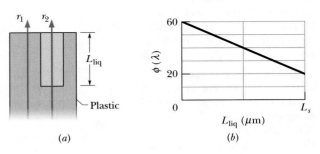

FIG. 35-51   Problem 84.

85 Two light rays, initially in phase and with a wavelength of 500 nm, go through different paths by reflecting from the various mirrors shown in Fig. 35-52. (Such a reflection does not itself produce a phase shift.) (a) What least value of distance d will put the rays exactly out of phase when they emerge from the region? (Ignore the slight tilt of the path for ray 2.) (b) Repeat the question assuming that the entire apparatus is immersed in a protein solution with an index of refraction of 1.38. **GO**

FIG. 35-52   Problem 85.

86 In Figure 35-53, two isotropic point sources S_1 and S_2 emit light in phase at wavelength λ and at the same amplitude. The sources are separated by distance $d = 6.00\lambda$ on an x axis. A viewing screen is at distance $D = 20.0\lambda$ from S_2 and parallel to the y axis. The figure shows two rays reaching point P on the screen, at height y_P. (a) At what value of y_P do the rays have the minimum possible phase difference? (b) What multiple of λ gives that minimum phase difference? (c) At what value of y_P do the rays have the maximum possible phase difference? What multiple of λ gives

FIG. 35-53   Problem 86.

(d) that maximum phase difference and (e) the phase difference when $y_P = d$? (f) When $y_P = d$, is the resulting intensity at point P maximum, minimum, intermediate but closer to maximum, or intermediate but closer to minimum?

87 In Fig. 35-54, a microwave transmitter at height a above the water level of a wide lake transmits microwaves of wavelength λ toward a receiver on the opposite shore, a distance x above the water level. The microwaves reflecting from the water interfere with the microwaves arriving directly from the transmitter. Assuming that the lake width D is much greater than a and x, and that $\lambda \geq a$, find an expression that gives the values of x for which the signal at the receiver is maximum. (*Hint:* Does the reflection cause a phase change?) SSM

FIG. 35-54 Problem 87.

88 In Fig. 35-55, two isotropic point sources S_1 and S_2 emit light at wavelength $\lambda = 400$ nm. Source S_1 is located at $y = 640$ nm; source S_2 is located at $y = -640$ nm. At point P_1 (at $x = 720$ nm), the wave from S_2 arrives ahead of the wave from S_1 by a phase difference of 0.600π rad. (a) What multiple of λ gives the phase difference between the waves from the two sources as the waves arrive at point P_2, which is located at $y = 720$ nm. (The figure is not drawn to scale.) (b) If the waves arrive at P_2 with equal amplitudes, is the interference there fully constructive, fully destructive, intermediate but closer to fully constructive, or intermediate but closer to fully destructive?

FIG. 35-55
Problem 88.

89 A double-slit arrangement produces bright interference fringes for sodium light ($\lambda = 589$ nm) that are angularly separated by $0.30°$ near the center of the pattern. What is the angular fringe separation if the entire arrangement is immersed in water, which has an index of refraction of 1.33? SSM

90 In Fig. 35-56a, the waves along rays 1 and 2 are initially in phase, with the same wavelength λ in air. Ray 2 goes through a material with length L and index of refraction n. The rays are then reflected by mirrors to a common point P on a screen. Suppose that we can vary n from $n = 1.0$ to $n = 2.5$. Suppose also that, from $n = 1.0$ to $n = n_s = 1.5$, the intensity I of the

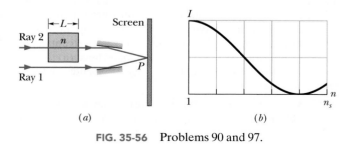

FIG. 35-56 Problems 90 and 97.

light at point P varies with n as given in Fig. 35-56b. At what values of n greater than 1.4 is intensity I (a) maximum and (b) zero? (c) What multiple of λ gives the phase difference between the rays at point P when $n = 2.0$?

91 In a phasor diagram for the waves at any point on the viewing screen for the two-slit experiment in Fig. 35-10, the phasor of the resultant wave rotates $60.0°$ in 2.50×10^{-16} s. What is the wavelength of the light?

92 Light of wavelength 700.0 nm is sent along a route of length 2000 nm. The route is then filled with a medium having an index of refraction of 1.400. In degrees, by how much does the medium phase-shift the light? Give (a) the full shift and (b) the equivalent shift that has a value less than $360°$.

93 Two parallel slits are illuminated with monochromatic light of wavelength 500 nm. An interference pattern is formed on a screen some distance from the slits, and the fourth dark band is located 1.68 cm from the central bright band on the screen. (a) What is the path length difference corresponding to the fourth dark band? (b) What is the distance on the screen between the central bright band and the first bright band on either side of the central band? (*Hint:* The angle to the fourth dark band and the angle to the first bright band are small enough that $\tan \theta \approx \sin \theta$.) SSM

94 In two experiments, light is to be sent along the two paths shown in Fig. 35-35 by reflecting it from the various flat surfaces shown. In the first experiment, rays 1 and 2 are initially in phase and have a wavelength of 620.0 nm. In the second experiment, rays 1 and 2 are initially in phase and have a wavelength of 496.0 nm. What least value of distance L is required such that the 620.0 nm waves emerge from the region exactly in phase but the 496.0 nm waves emerge exactly out of phase?

95 Find the slit separation of a double-slit arrangement that will produce interference fringes 0.018 rad apart on a distant screen when the light has wavelength $\lambda = 589$ nm.

96 A thin film suspended in air is 0.410 μm thick and is illuminated with white light incident perpendicularly on its surface. The index of refraction of the film is 1.50. At what wavelength will visible light that is reflected from the two surfaces of the film undergo fully constructive interference?

97 In Fig. 35-56a, the waves along rays 1 and 2 are initially in phase, with the same wavelength λ in air. Ray 2 goes through a material with length L and index of refraction n. The rays are then reflected by mirrors to a common point P on a screen. Suppose that we can vary L from 0 to 2400 nm. Suppose also that, from $L = 0$ to $L_s = 900$ nm, the intensity I

FIG. 35-57 Problem 97.

of the light at point P varies with L as given in Fig. 35-57. At what values of L greater than L_s is intensity I (a) maximum and (b) zero? (c) What multiple of λ gives the phase difference between ray 1 and ray 2 at common point P when $L = 1200$ nm? SSM

98 A camera lens with index of refraction greater than 1.30 is coated with a thin transparent film of index of refraction 1.25 to eliminate by interference the reflection of light at wavelength λ that is incident perpendicularly on the lens. What multiple of λ gives the minimum film thickness needed?

99 If the distance between the first and tenth minima of a double-slit pattern is 18.0 mm and the slits are separated by 0.150 mm with the screen 50.0 cm from the slits, what is the wavelength of the light used? SSM

100 What is the speed in fused quartz of light of wavelength 550 nm? (See Fig. 33-18.)

101 In Sample Problem 35-6, assume that the coating eliminates the reflection of light of wavelength 550 nm at normal incidence. By what percentages is reflection diminished by the coating at (a) 450 nm and (b) 650 nm?

102 Laser light of wavelength 632.8 nm passes through a double-slit arrangement at the front of a lecture room, reflects off a mirror 20.0 m away at the back of the room, and then produces an interference pattern on a screen at the front of the room. The distance between adjacent bright fringes is 10.0 cm. (a) What is the slit separation? (b) What is at the center of the pattern when the lecturer places a thin cellophane sheet over one slit, thereby increasing by 2.50 the number of wavelengths along the path that includes the cellophane?

103 Light of wavelength λ is used in a Michelson interferometer. Let x be the position of the movable mirror, with $x = 0$ when the arms have equal lengths $d_2 = d_1$. Write an expression for the intensity of the observed light as a function of x, letting I_m be the maximum intensity. SSM

104 A sheet of glass having an index of refraction of 1.40 is to be coated with a film of material having an index of refraction of 1.55 in order that green light with a wavelength of 525 nm (in air) will be preferentially transmitted via constructive interference. (a) What is the minimum thickness of the film that will achieve the result? (b) Why are other parts of the visible spectrum not also preferentially transmitted? (c) Will the transmission of any colors be sharply reduced? If so, which colors?

105 One slit of a double-slit arrangement is covered by a thin glass plate with index of refraction 1.4, and the other by a thin glass plate with index of refraction 1.7. The point on the screen at which the central maximum fell before the glass plates were inserted is now occupied by what had been the $m = 5$ bright fringe. Assuming that $\lambda = 480$ nm and that the plates have the same thickness t, find t.

106 In Fig. 35-46, two glass plates are held together at one end to form a wedge of air that acts as a thin film. A broad beam of light of wavelength 480 nm is directed through the plates, perpendicular to the first plate. An observer intercepting light reflected from the plates sees on the plates an interference pattern that is due to the wedge of air. How much thicker is the wedge at the sixteenth bright fringe than it is at the sixth bright fringe, counting from where the plates touch?

107 A broad beam of light of wavelength 600 nm is sent directly downward through the glass plate ($n = 1.50$) in Fig. 35-46. That plate and a plastic plate ($n = 1.20$) form a thin wedge of air that acts as a thin film. An observer looking

FIG. 35-58 Problem 107.

down through the top plate sees the fringe pattern shown in Fig. 35-58, with dark fringes centered on the ends. (a) What is the thickness of the wedge at the right end? (b) How many dark fringes will the observer see if the air between the plates is replaced with water ($n = 1.33$)?

108 Sodium light ($\lambda = 589$ nm) illuminates two slits separated by $d = 2.0$ mm. The slit–screen distance D is 40 mm. What percentage error is made by using Eq. 35-14 to locate the $m = 10$ bright fringe on the screen rather than using the exact path length difference?

109 Figure 35-59 shows an optical fiber in which a central plastic core of index of refraction $n_1 = 1.58$ is surrounded by a plastic sheath of index of refraction $n_2 = 1.53$. Light can travel along different paths within the central core, leading to different travel times through the fiber. This causes an initially short pulse of light to spread as it travels along

FIG. 35-59 Problem 109.

the fiber, resulting in information loss. Consider light that travels directly along the central axis of the fiber and light that is repeatedly reflected at the critical angle along the core–sheath interface, reflecting from side to side as it travels down the central core. If the fiber length is 300 m, what is the difference in the travel times along these two routes?

110 When an electron moves through a medium at a speed exceeding the speed of light in that medium, the electron radiates electromagnetic energy (the *Cerenkov effect*). What minimum speed must an electron have in a liquid with index of refraction 1.54 in order to radiate?

111 Point sources S_1 and S_2 radiate in phase at wavelength 400 nm and at the same amplitude. The sources are located on an x axis at $x = 6.5$ μm and $x = -6.0$ μm, respectively. (a) Determine the phase difference (in radians) at the origin between the radiation from S_1 and the radiation from S_2. (b) Suppose a slab of transparent material with thickness 1.5 μm and index of refraction $n = 1.5$ is placed between $x = 0$ and $x = 1.5$ μm. What then is the phase difference (in radians) at the origin between the radiation from S_1 and the radiation from S_2?

112 The second dark band in a double-slit interference pattern is 1.2 cm from the central maximum of the pattern. The separation between the two slits is equal to 800 wavelengths of the monochromatic light incident (perpendicularly) on the slits. What is the distance between the plane of the slits and the viewing screen?

113 In Fig. 35-46, two glass plates ($n = 1.60$) form a wedge, and a fluid ($n = 1.50$) fills the wedge-shaped space. At the left end the plates touch; at the right, they are separated by 580 nm. Light with a wavelength (in air) of 580 nm shines downward on the assembly, and an observer intercepts light sent back upward. Does a dark or bright band lie at (a) the left end and (b) the right end? (c) How many dark bands are there along the plates?

114 A thin film ($n = 1.25$) is deposited on a glass plate ($n = 1.40$) and illuminated with light of wavelength 550 nm (measured in air). The light beam is perpendicular to the plate. What is the minimum (nonzero) thickness for the film that will (a) maximally transmit and (b) maximally reflect the light?

115 A light beam with a wavelength of 600 nm in air passes through film 1 ($n_1 = 1.2$) of thickness 1.0 μm, then through film 2 (air) of thickness 1.5 μm, and finally through film 3 ($n_3 = 1.8$) of thickness 1.0 μm. The beam is perpendicular to the film surfaces, which are parallel to one another. (a) Which film does the light cross in the least time, and (b) what is that least time? (c) What is the total number of wavelengths (at any instant) across all three films together?

116 Two light rays, initially in phase and having wavelength $\lambda = 6.00 \times 10^{-7}$ m, pass through different plastic layers of the same thickness, 7.00×10^{-6} m. The indexes of refraction are 1.65 for one layer and 1.49 for the other. (a) What *least* multiple of λ gives the phase difference between the rays when they emerge? (b) If those two rays then reach a common point with the same amplitude, does the interference result in complete darkness, maximum brightness, intermediate illumination but closer to complete darkness, or intermediate illumination but closer to maximum brightness? If the two rays are, instead, initially exactly out of phase, what are the answers to (c) part (a) and (d) part (b)?

117 In a double-slit interference experiment, the slit separation is 2.00 μm, the light wavelength is 500 nm, and the separation between the slits and the screen is 4.00 m. (a) What is the angle between the center and the third side bright fringe? If we decrease the light frequency to 90.0% of its initial value, (b) does the third side bright fringe move along the screen toward or away from the pattern's center and (c) how far does it move?

118 A plane monochromatic light wave in air is perpendicularly incident on a thin film of oil that covers a glass plate. The wavelength of the source may be varied continuously. Fully destructive interference of the reflected light occurs for wavelengths 500 and 700 nm and for no wavelength in between. The index of refraction of the glass is 1.50. Show that the index of refraction of the oil must be less than 1.50.

119 Figure 35-60 shows the design of a Texas arcade game. Four laser pistols are pointed toward the center of an array of plastic layers where a clay armadillo is the target. The indexes of refraction of the layers are $n_1 = 1.55$, $n_2 = 1.70$, $n_3 = 1.45$, $n_4 = 1.60$, $n_5 = 1.45$, $n_6 = 1.61$, $n_7 = 1.59$, $n_8 = 1.70$, and $n_9 = 1.60$. The layer thicknesses are either 2.00 mm or 4.00 mm, as drawn. What is the travel time through the layers for the laser burst from (a) pistol 1, (b) pistol 2, (c) pistol 3, and (d) pistol 4? (e) If the pistols are fired simultaneously, which laser burst hits the target first?

120 In Fig. 35-10, let the angle θ of the two rays be 20.0°, slit separation d be 58.00 μm, and wavelength λ be 500.9 nm. (a) What multiple of λ gives the phase difference of the two rays when they reach a common point on a distant screen? (b) Does their interference result in complete darkness, maximum brightness, intermediate illumination but closer to complete darkness, or intermediate illumination but closer to maximum brightness?

121 Slits of unequal widths are used in a double-slit arrangement to produce an interference pattern on a distant screen. If only the narrower slit 1 is illuminated (the wider slit 2 is covered), the light reaching the center of the pattern has amplitude E_0 and intensity I_0. If only slit 2 is illuminated, the amplitude of the light reaching the pattern center is $2E_0$. When both slits are illuminated and a two-slit interference pattern is on the screen, what is the intensity $I(\theta)$ in the pattern as a function of angle θ? Answer in the style of Eqs. 35-22 and 35-23.

122 Figure 35-40 shows two point sources S_1 and S_2 that emit light at wavelength $\lambda = 500$ nm and with the same amplitude. The emissions are isotropic and in phase, and the separation between the sources is $d = 2.00$ μm. At any point P on the x axis, the wave from S_1 and the wave from S_2 interfere. When P is very far away ($x \approx \infty$), what are (a) the phase difference between the waves arriving from S_1 and S_2 and (b) the type of interference they produce (approximately fully constructive or fully destructive)? (c) As we then move P along the x axis toward S_1, does the phase difference between the waves from S_1 and S_2 increase or decrease? (d)–(o) Fill in Table 35-4 for the given phase differences by determining the type of interference and the x coordinate at which the interference occurs.

TABLE 35-4

Problem 122: Parts (d) through (o)

Phase Difference	Type	Position x
0	(d)	(e)
0.500λ	(f)	(g)
1.00λ	(h)	(i)
1.50λ	(j)	(k)
2.00λ	(l)	(m)
2.50λ	(n)	(o)

123 (a) Use the result of Problem 75 to show that, in a Newton's rings experiment, the difference in radius between adjacent bright rings (maxima) is given by

$$\Delta r = r_{m+1} - r_m \approx \tfrac{1}{2}\sqrt{\lambda R/m},$$

assuming $m \gg 1$. (b) Now show that the *area* between adjacent bright rings is given by $A = \pi\lambda R$, assuming $m \gg 1$. Note that this area is independent of m.

FIG. 35-60 Problem 119.

36 Diffraction

Many surfaces (such as shirts and pants) have color because dye molecules absorb visible light of certain wavelengths, allowing the rest of the light to reflect. Some surfaces (such as iridescent butterfly wings) have color because thin-film interference removes certain wavelengths and allows the rest of the light to be reflected. However, the arresting blue skin of this mandrill baboon is due to neither dye molecules nor thin films.

What makes the face of a mandrill baboon blue?

The answer is in this chapter.

Timothy Laman/National Geographic/Getty Images, Inc.

36-1 | WHAT IS PHYSICS?

One focus of physics in the study of light is to understand and put to use the diffraction of light as it passes through a narrow slit or (as we shall discuss) past either a narrow obstacle or an edge. We touched on this phenomenon in Chapter 35 when we looked at how light flared—diffracted—through the slits in Young's experiment. Diffraction through a given slit is more complicated than simple flaring, however, because the light also interferes with itself and produces an interference pattern. It is because of such complications that light is rich with application opportunities. Even though the diffraction of light as it passes through a slit or past an obstacle seems awfully academic, countless engineers and scientists make their living using this physics, and the total worth of diffraction applications worldwide is probably incalculable.

Before we can discuss some of these applications, we first must discuss why diffraction is due to the wave nature of light.

36-2 | Diffraction and the Wave Theory of Light

In Chapter 35 we defined diffraction rather loosely as the flaring of light as it emerges from a narrow slit. More than just flaring occurs, however, because the light produces an interference pattern called a **diffraction pattern.** For example, when monochromatic light from a distant source (or a laser) passes through a narrow slit and is then intercepted by a viewing screen, the light produces on the screen a diffraction pattern like that in Fig. 36-1. This pattern consists of a broad and intense (very bright) central maximum plus a number of narrower and less intense maxima (called **secondary** or **side** maxima) to both sides. In between the maxima are minima.

Such a pattern would be totally unexpected in geometrical optics: If light traveled in straight lines as rays, then the slit would allow some of those rays through to form a sharp rendition of the slit on the viewing screen. As in Chapter 35, we must conclude that geometrical optics is only an approximation.

Diffraction is not limited to situations when light passes through a narrow opening (such as a slit or pinhole). It also occurs when light passes an edge, such as the edges of the razor blade whose diffraction pattern is shown in Fig. 36-2. Note the lines of maxima and minima that run approximately parallel to the edges, at both the inside edges of the blade and the outside edges. As the light passes, say, the vertical edge at the left, it flares left and right and undergoes interference, producing the pattern along the left edge. The rightmost portion of that pattern actually lies behind the blade, within what would be the blade's shadow if geometrical optics prevailed.

You encounter a common example of diffraction when you look at a clear blue sky and see tiny specks and hairlike structures floating in your view. These *floaters,* as they are called, are produced when light passes the edges of tiny deposits in the vitreous humor, the transparent material filling most of the eyeball. What you are seeing when a floater is in your field of vision is the diffraction pattern produced on the retina by one of these deposits. If you sight through a pinhole in a piece of cardboard so as to make the light entering your eye approximately a plane wave, you can distinguish individual maxima and minima in the patterns.

Diffraction is a wave effect. That is, it occurs because light is a wave and it occurs with other types of waves as well. For example, you have probably seen diffraction in action at football games. When a cheerleader near the playing field yells up at several thousand noisy fans, the yell can hardly be heard because the sound waves diffract when they pass through the narrow opening of the

FIG. 36-1 This diffraction pattern appeared on a viewing screen when light that had passed through a narrow vertical slit reached the screen. Diffraction caused the light to flare out perpendicular to the long sides of the slit. That flaring produced an interference pattern consisting of a broad central maximum plus less intense and narrower secondary (or side) maxima, with minima between them. *(Ken Kay/Fundamental Photographs)*

FIG. 36-2 The diffraction pattern produced by a razor blade in monochromatic light. Note the lines of alternating maximum and minimum intensity. *(Ken Kay/Fundamental Photographs)*

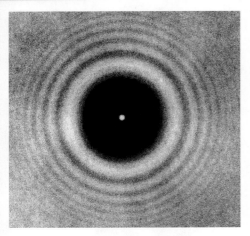

FIG. 36-3 A photograph of the diffraction pattern of a disk. Note the concentric diffraction rings and the Fresnel bright spot at the center of the pattern. This experiment is essentially identical to that arranged by the committee testing Fresnel's theories, because both the sphere they used and the disk used here have a cross section with a circular edge. *(From Michel Cagnet, Maurice Franzon, and Jean Thierr, Atlas of Optical Phenomena. Springer-Verlag, New York, 1962. Reproduced with permission.)*

cheerleader's mouth. This flaring leaves little of the waves traveling toward the fans in front of the cheerleader. To offset the diffraction, the cheerleader can yell through a megaphone. The sound waves then emerge from the much wider opening at the end of the megaphone. The flaring is thus reduced, and much more of the sound reaches the fans in front of the cheerleader.

The Fresnel Bright Spot

Diffraction finds a ready explanation in the wave theory of light. However, this theory, originally advanced in the late 1600s by Huygens and used 123 years later by Young to explain double-slit interference, was very slow in being adopted, largely because it ran counter to Newton's theory that light was a stream of particles.

Newton's view was the prevailing view in French scientific circles of the early 19th century, when Augustin Fresnel was a young military engineer. Fresnel, who believed in the wave theory of light, submitted a paper to the French Academy of Sciences describing his experiments with light and his wave-theory explanations of them.

In 1819, the Academy, dominated by supporters of Newton and thinking to challenge the wave point of view, organized a prize competition for an essay on the subject of diffraction. Fresnel won. The Newtonians, however, were neither converted nor silenced. One of them, S. D. Poisson, pointed out the "strange result" that if Fresnel's theories were correct, then light waves should flare into the shadow region of a sphere as they pass the edge of the sphere, producing a bright spot at the center of the shadow. The prize committee arranged a test of Poisson's prediction and discovered that the predicted *Fresnel bright spot*, as we call it today, was indeed there (Fig. 36-3). Nothing builds confidence in a theory so much as having one of its unexpected and counterintuitive predictions verified by experiment.

36-3 | Diffraction by a Single Slit: Locating the Minima

Let us now examine the diffraction pattern of plane waves of light of wavelength λ that are diffracted by a single long, narrow slit of width a in an otherwise opaque screen B, as shown in cross section in Fig. 36-4. (In that figure, the slit's length extends into and out of the page, and the incoming wavefronts are parallel to screen B.) When the diffracted light reaches viewing screen C, waves from different points within the slit undergo interference and produce a diffraction pattern of bright and dark fringes (interference maxima and minima) on the screen. To locate the fringes, we shall use a procedure somewhat similar to the one we used to locate the fringes in a two-slit interference pattern. However, diffraction is more mathematically challenging, and here we shall be able to find equations for only the dark fringes.

Before we do that, however, we can justify the central bright fringe seen in Fig. 36-1 by noting that the Huygens wavelets from all points in the slit travel about the same distance to reach the center of the pattern and thus are in phase there. As for the other bright fringes, we can say only that they are approximately halfway between adjacent dark fringes.

To find the dark fringes, we shall use a clever (and simplifying) strategy that involves pairing up all the rays coming through the slit and then finding what conditions cause the wavelets of the rays in each pair to cancel each other. We apply this strategy in Fig. 36-4 to locate the first dark fringe, at point P_1. First, we mentally divide the slit into two *zones* of equal widths $a/2$. Then we extend to P_1 a light ray r_1 from the top point of the top zone and a light ray r_2 from the

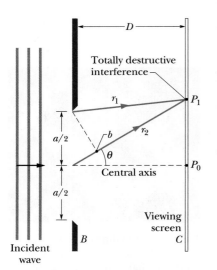

FIG. 36-4 Waves from the top points of two zones of width $a/2$ undergo fully destructive interference at point P_1 on viewing screen C.

top point of the bottom zone. A central axis is drawn from the center of the slit to screen C, and P_1 is located at an angle θ to that axis.

The wavelets of the pair of rays r_1 and r_2 are in phase within the slit because they originate from the same wavefront passing through the slit, along the width of the slit. However, to produce the first dark fringe they must be out of phase by $\lambda/2$ when they reach P_1; this phase difference is due to their path length difference, with the path traveled by the wavelet of r_2 to reach P_1 being longer than the path traveled by the wavelet of r_1. To display this path length difference, we find a point b on ray r_2 such that the path length from b to P_1 matches the path length of ray r_1. Then the path length difference between the two rays is the distance from the center of the slit to b.

When viewing screen C is near screen B, as in Fig. 36-4, the diffraction pattern on C is difficult to describe mathematically. However, we can simplify the mathematics considerably if we arrange for the screen separation D to be much larger than the slit width a. Then we can approximate rays r_1 and r_2 as being parallel, at angle θ to the central axis (Fig. 36-5). We can also approximate the triangle formed by point b, the top point of the slit, and the center point of the slit as being a right triangle, and one of the angles inside that triangle as being θ. The path length difference between rays r_1 and r_2 (which is still the distance from the center of the slit to point b) is then equal to $(a/2)\sin\theta$.

We can repeat this analysis for any other pair of rays originating at corresponding points in the two zones (say, at the midpoints of the zones) and extending to point P_1. Each such pair of rays has the same path length difference $(a/2)\sin\theta$. Setting this common path length difference equal to $\lambda/2$ (our condition for the first dark fringe), we have

$$\frac{a}{2}\sin\theta = \frac{\lambda}{2},$$

which gives us

$$a\sin\theta = \lambda \quad \text{(first minimum)}. \quad (36\text{-}1)$$

Given slit width a and wavelength λ, Eq. 36-1 tells us the angle θ of the first dark fringe above and (by symmetry) below the central axis.

Note that if we begin with $a > \lambda$ and then narrow the slit while holding the wavelength constant, we increase the angle at which the first dark fringes appear; that is, the extent of the diffraction (the extent of the flaring and the width of the pattern) is *greater* for a *narrower* slit. When we have reduced the slit width to the wavelength (that is, $a = \lambda$), the angle of the first dark fringes is 90°. Since the first dark fringes mark the two edges of the central bright fringe, that bright fringe must then cover the entire viewing screen.

We find the second dark fringes above and below the central axis as we found the first dark fringes, except that we now divide the slit into *four* zones of equal widths $a/4$, as shown in Fig. 36-6a. We then extend rays $r_1, r_2, r_3,$ and r_4 from the top points of the zones to point P_2, the location of the second dark fringe above the central axis. To produce that fringe, the path length difference between r_1 and r_2, that between r_2 and r_3, and that between r_3 and r_4 must all be equal to $\lambda/2$.

For $D \gg a$, we can approximate these four rays as being parallel, at angle θ to the central axis. To display their path length differences, we extend a perpendicular line through each adjacent pair of rays, as shown in Fig. 36-6b, to form a series of right triangles, each of which has a path length difference as one side. We see from the top triangle that the path length difference between r_1 and r_2 is $(a/4)\sin\theta$. Similarly, from the bottom triangle, the path length difference between r_3 and r_4 is also $(a/4)\sin\theta$. In fact, the path length difference for any two rays that originate at corresponding points in two adjacent zones is $(a/4)\sin\theta$. Since in each such case the path length difference is equal to $\lambda/2$, we have

$$\frac{a}{4}\sin\theta = \frac{\lambda}{2},$$

FIG. 36-5 For $D \gg a$, we can approximate rays r_1 and r_2 as being parallel, at angle θ to the central axis.

(a)

(b)

FIG. 36-6 (a) Waves from the top points of four zones of width $a/4$ undergo fully destructive interference at point P_2. (b) For $D \gg a$, we can approximate rays $r_1, r_2, r_3,$ and r_4 as being parallel, at angle θ to the central axis.

which gives us

$$a \sin \theta = 2\lambda \qquad \text{(second minimum).} \qquad (36\text{-}2)$$

We could now continue to locate dark fringes in the diffraction pattern by splitting up the slit into more zones of equal width. We would always choose an even number of zones so that the zones (and their waves) could be paired as we have been doing. We would find that the dark fringes above and below the central axis can be located with the general equation

$$a \sin \theta = m\lambda, \qquad \text{for } m = 1, 2, 3, \ldots \quad \text{(minima—dark fringes).} \qquad (36\text{-}3)$$

You can remember this result in the following way. Draw a triangle like the one in Fig. 36-5, but for the full slit width a, and note that the path length difference between the top and bottom rays equals $a \sin \theta$. Thus, Eq. 36-3 says:

➤ In a single-slit diffraction experiment, dark fringes are produced where the path length differences ($a \sin \theta$) between the top and bottom rays are equal to $\lambda, 2\lambda, 3\lambda, \ldots$.

This may seem to be wrong because the waves of those two particular rays will be exactly in phase with each other when their path length difference is an integer number of wavelengths. However, they each will still be part of a pair of waves that are exactly out of phase with each other; thus, *each* wave will be canceled by some other wave, resulting in darkness.

Equations 36-1, 36-2, and 36-3 are derived for the case of $D \gg a$. However, they also apply if we place a converging lens between the slit and the viewing screen and then move the screen in so that it coincides with the focal plane of the lens. The lens ensures that rays which now reach any point on the screen are *exactly* parallel (rather than approximately) back at the slit. They are like the initially parallel rays of Fig. 34-15a that are directed to the focal point by a converging lens.

✓ **CHECKPOINT 1**   We produce a diffraction pattern on a viewing screen by means of a long narrow slit illuminated by blue light. Does the pattern expand away from the bright center (the maxima and minima shift away from the center) or contract toward it if we (a) switch to yellow light or (b) decrease the slit width?

Sample Problem 36-1

A slit of width a is illuminated by white light.

(a) For what value of a will the first minimum for red light of wavelength $\lambda = 650$ nm appear at $\theta = 15°$?

KEY IDEA Diffraction occurs separately for each wavelength in the range of wavelengths passing through the slit, with the locations of the minima for each wavelength given by Eq. 36-3 ($a \sin \theta = m\lambda$).

Calculation: When we set $m = 1$ (for the first minimum) and substitute the given values of θ and λ, Eq. 36-3 yields

$$a = \frac{m\lambda}{\sin \theta} = \frac{(1)(650 \text{ nm})}{\sin 15°}$$
$$= 2511 \text{ nm} \approx 2.5 \ \mu\text{m}. \qquad \text{(Answer)}$$

For the incident light to flare out that much ($\pm 15°$ to the first minima) the slit has to be very fine indeed—in this case, a mere four times the wavelength. For comparison, note that a fine human hair may be about 100 μm in diameter.

(b) What is the wavelength λ' of the light whose first side diffraction maximum is at $15°$, thus coinciding with the first minimum for the red light?

KEY IDEA The first side maximum for any wavelength is about halfway between the first and second minima for that wavelength.

Calculations: Those first and second minima can be located with Eq. 36-3 by setting $m = 1$ and $m = 2$, respectively. Thus, the first side maximum can be located *approximately* by setting $m = 1.5$. Then Eq. 36-3 becomes

$$a \sin \theta = 1.5\lambda'.$$

Solving for λ' and substituting known data yield

$$\lambda' = \frac{a \sin \theta}{1.5} = \frac{(2511 \text{ nm})(\sin 15°)}{1.5}$$
$$= 430 \text{ nm}. \qquad \text{(Answer)}$$

Light of this wavelength is violet. From the two equations we used, can you see that the first side maximum for light of wavelength 430 nm will always coincide with the first minimum for light of wavelength 650 nm, no matter what the slit width is? However, the angle θ at which this overlap occurs does depend on slit width. If the slit is relatively narrow, the angle will be relatively large, and conversely.

36-4 | Intensity in Single-Slit Diffraction, Qualitatively

In Section 36-3 we saw how to find the positions of the minima and the maxima in a single-slit diffraction pattern. Now we turn to a more general problem: find an expression for the intensity I of the pattern as a function of θ, the angular position of a point on a viewing screen.

To do this, we divide the slit of Fig. 36-4 into N zones of equal widths Δx small enough that we can assume each zone acts as a source of Huygens wavelets. We wish to superimpose the wavelets arriving at an arbitrary point P on the viewing screen, at angle θ to the central axis, so that we can determine the amplitude E_θ of the electric component of the resultant wave at P. The intensity of the light at P is then proportional to the square of that amplitude.

To find E_θ, we need the phase relationships among the arriving wavelets. The phase difference between wavelets from adjacent zones is given by

$$\begin{pmatrix} \text{phase} \\ \text{difference} \end{pmatrix} = \left(\frac{2\pi}{\lambda}\right)\begin{pmatrix} \text{path length} \\ \text{difference} \end{pmatrix}.$$

For point P at angle θ, the path length difference between wavelets from adjacent zones is $\Delta x \sin \theta$; so the phase difference $\Delta\phi$ between wavelets from adjacent zones is

$$\Delta\phi = \left(\frac{2\pi}{\lambda}\right)(\Delta x \sin \theta). \qquad (36\text{-}4)$$

We assume that the wavelets arriving at P all have the same amplitude ΔE. To find the amplitude E_θ of the resultant wave at P, we add the amplitude ΔE via phasors. To do this, we construct a diagram of N phasors, one corresponding to the wavelet from each zone in the slit.

For point P_0 at $\theta = 0$ on the central axis of Fig. 36-4, Eq. 36-4 tells us that the phase difference $\Delta\phi$ between the wavelets is zero; that is, the wavelets all arrive in phase. Figure 36-7a is the corresponding phasor diagram; adjacent phasors represent wavelets from adjacent zones and are arranged head to tail. Because there is zero phase difference between the wavelets, there is zero angle between each pair of adjacent phasors. The amplitude E_θ of the net wave at P_0 is the vector sum of these phasors. This arrangement of the phasors turns out to be the one that gives the greatest value for the amplitude E_θ. We call this value E_m; that is, E_m is the value of E_θ for $\theta = 0$.

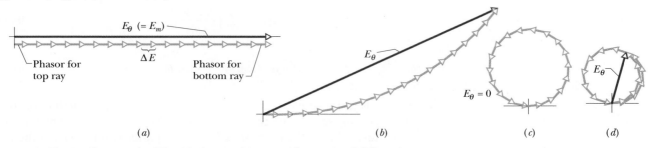

FIG. 36-7 Phasor diagrams for $N = 18$ phasors, corresponding to the division of a single slit into 18 zones. Resultant amplitudes E_θ are shown for (a) the central maximum at $\theta = 0$, (b) a point on the screen lying at a small angle θ to the central axis, (c) the first minimum, and (d) the first side maximum.

We next consider a point P that is at a small angle θ to the central axis. Equation 36-4 now tells us that the phase difference $\Delta\phi$ between wavelets from adjacent zones is no longer zero. Figure 36-7b shows the corresponding phasor diagram; as before, the phasors are arranged head to tail, but now there is an angle $\Delta\phi$ between adjacent phasors. The amplitude E_θ at this new point is still the vector sum of the phasors, but it is smaller than that in Fig. 36-7a, which means that the intensity of the light is less at this new point P than at P_0.

If we continue to increase θ, the angle $\Delta\phi$ between adjacent phasors increases, and eventually the chain of phasors curls completely around so that the head of the last phasor just reaches the tail of the first phasor (Fig. 36-7c). The amplitude E_θ is now zero, which means that the intensity of the light is also zero. We have reached the first minimum, or dark fringe, in the diffraction pattern. The first and last phasors now have a phase difference of 2π rad, which means that the path length difference between the top and bottom rays through the slit equals one wavelength. Recall that this is the condition we determined for the first diffraction minimum.

As we continue to increase θ, the angle $\Delta\phi$ between adjacent phasors continues to increase, the chain of phasors begins to wrap back on itself, and the resulting coil begins to shrink. Amplitude E_θ now increases until it reaches a maximum value in the arrangement shown in Fig. 36-7d. This arrangement corresponds to the first side maximum in the diffraction pattern.

If we increase θ a bit more, the resulting shrinkage of the coil decreases E_θ, which means that the intensity also decreases. When θ is increased enough, the head of the last phasor again meets the tail of the first phasor. We have then reached the second minimum.

We could continue this qualitative method of determining the maxima and minima of the diffraction pattern but, instead, we shall now turn to a quantitative method.

> ✓**CHECKPOINT 2**   The figures represent, in smoother form (with more phasors) than Fig. 36-7, the phasor diagrams for two points of a diffraction pattern that are on opposite sides of a certain diffraction maximum. (a) Which maximum is it? (b) What is the approximate value of m (in Eq. 36-3) that corresponds to this maximum?
>
>
>
> (a) (b)

36-5 | Intensity in Single-Slit Diffraction, Quantitatively

Equation 36-3 tells us how to locate the minima of the single-slit diffraction pattern on screen C of Fig. 36-4 as a function of the angle θ in that figure. Here we wish to derive an expression for the intensity $I(\theta)$ of the pattern as a function of θ. We state, and shall prove below, that the intensity is given by

$$I(\theta) = I_m \left(\frac{\sin\alpha}{\alpha} \right)^2, \tag{36-5}$$

where

$$\alpha = \tfrac{1}{2}\phi = \frac{\pi a}{\lambda} \sin\theta. \tag{36-6}$$

The symbol α is just a convenient connection between the angle θ that locates a point on the viewing screen and the light intensity $I(\theta)$ at that point. The intensity I_m is the greatest value of the intensities $I(\theta)$ in the pattern and occurs at the central maximum (where $\theta = 0$), and ϕ is the phase difference (in radians) between the top and bottom rays from the slit of width a.

Study of Eq. 36-5 shows that intensity minima will occur where

$$\alpha = m\pi, \qquad \text{for } m = 1, 2, 3, \ldots . \tag{36-7}$$

If we put this result into Eq. 36-6, we find

$$m\pi = \frac{\pi a}{\lambda} \sin \theta, \qquad \text{for } m = 1, 2, 3, \ldots,$$

or $\qquad a \sin \theta = m\lambda, \qquad \text{for } m = 1, 2, 3, \ldots \qquad \text{(minima—dark fringes),} \qquad (36\text{-}8)$

which is exactly Eq. 36-3, the expression that we derived earlier for the location of the minima.

Figure 36-8 shows plots of the intensity of a single-slit diffraction pattern, calculated with Eqs. 36-5 and 36-6 for three slit widths: $a = \lambda$, $a = 5\lambda$, and $a = 10\lambda$. Note that as the slit width increases (relative to the wavelength), the width of the *central diffraction maximum* (the central hill-like region of the graphs) decreases; that is, the light undergoes less flaring by the slit. The secondary maxima also decrease in width (and become weaker). In the limit of slit width a being much greater than wavelength λ, the secondary maxima due to the slit disappear; we then no longer have single-slit diffraction (but we still have diffraction due to the edges of the wide slit, like that produced by the edges of the razor blade in Fig. 36-2).

Proof of Eqs. 36-5 and 36-6

The arc of phasors in Fig. 36-9 represents the wavelets that reach an arbitrary point P on the viewing screen of Fig. 36-4, corresponding to a particular small angle θ. The amplitude E_θ of the resultant wave at P is the vector sum of these phasors. If we divide the slit of Fig. 36-4 into infinitesimal zones of width Δx, the arc of phasors in Fig. 36-9 approaches the arc of a circle; we call its radius R as indicated in that figure. The length of the arc must be E_m, the amplitude at the center of the diffraction pattern, because if we straightened out the arc we would have the phasor arrangement of Fig. 36-7a (shown lightly in Fig. 36-9).

The angle ϕ in the lower part of Fig. 36-9 is the difference in phase between the infinitesimal vectors at the left and right ends of arc E_m. From the geometry, ϕ is also the angle between the two radii marked R in Fig. 36-9. The dashed line in that figure, which bisects ϕ, then forms two congruent right triangles. From either triangle we can write

$$\sin \tfrac{1}{2}\phi = \frac{E_\theta}{2R}. \qquad (36\text{-}9)$$

In radian measure, ϕ is (with E_m considered to be a circular arc)

$$\phi = \frac{E_m}{R}.$$

Solving this equation for R and substituting in Eq. 36-9 lead to

$$E_\theta = \frac{E_m}{\tfrac{1}{2}\phi} \sin \tfrac{1}{2}\phi. \qquad (36\text{-}10)$$

In Section 33-5 we saw that the intensity of an electromagnetic wave is proportional to the square of the amplitude of its electric field. Here, this means that the maximum intensity I_m (which occurs at the center of the diffraction pattern) is proportional to E_m^2 and the intensity $I(\theta)$ at angle θ is proportional to E_θ^2. Thus, we may write

$$\frac{I(\theta)}{I_m} = \frac{E_\theta^2}{E_m^2}. \qquad (36\text{-}11)$$

Substituting for E_θ with Eq. 36-10 and then substituting $\alpha = \tfrac{1}{2}\phi$, we are led to the following expression for the intensity as a function of θ:

$$I(\theta) = I_m \left(\frac{\sin \alpha}{\alpha} \right)^2.$$

This is exactly Eq. 36-5, one of the two equations we set out to prove.

(a)

(b)

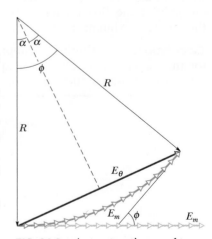
(c)

FIG. 36-8 The relative intensity in single-slit diffraction for three values of the ratio a/λ. The wider the slit is, the narrower is the central diffraction maximum.

FIG. 36-9 A construction used to calculate the intensity in single-slit diffraction. The situation shown corresponds to that of Fig. 36-7b.

The second equation we wish to prove relates α to θ. The phase difference ϕ between the rays from the top and bottom of the entire slit may be related to a path length difference with Eq. 36-4; it tells us that

$$\phi = \left(\frac{2\pi}{\lambda}\right)(a \sin \theta),$$

where a is the sum of the widths Δx of the infinitesimal zones. However, $\phi = 2\alpha$, so this equation reduces to Eq. 36-6.

✓ **CHECKPOINT 3**   Two wavelengths, 650 and 430 nm, are used separately in a single-slit diffraction experiment. The figure shows the results as graphs of intensity I versus angle θ for the two diffraction patterns. If both wavelengths are then used simultaneously, what color will be seen in the combined diffraction pattern at (a) angle A and (b) angle B?

Sample Problem   **36-2**

Find the intensities of the first three secondary maxima (side maxima) in the single-slit diffraction pattern of Fig. 36-1, measured as a percentage of the intensity of the central maximum.

KEY IDEAS   The secondary maxima lie approximately halfway between the minima, whose angular locations are given by Eq. 36-7 ($\alpha = m\pi$). The locations of the secondary maxima are then given (approximately) by

$$a = (m + \tfrac{1}{2})\pi, \qquad \text{for } m = 1, 2, 3, \ldots,$$

with α in radian measure. We can relate the intensity I at any point in the diffraction pattern to the intensity I_m of the central maximum via Eq. 36-5.

Calculations: Substituting the approximate values of α for the secondary maxima into Eq. 36-5 to obtain the relative intensities at those maxima, we get

$$\frac{I}{I_m} = \left(\frac{\sin \alpha}{\alpha}\right)^2 = \left(\frac{\sin(m + \tfrac{1}{2})\pi}{(m + \tfrac{1}{2})\pi}\right)^2, \quad \text{for } m = 1, 2, 3, \ldots.$$

The first of the secondary maxima occurs for $m = 1$, and its relative intensity is

$$\frac{I_1}{I_m} = \left(\frac{\sin(1 + \tfrac{1}{2})\pi}{(1 + \tfrac{1}{2})\pi}\right)^2 = \left(\frac{\sin 1.5\pi}{1.5\pi}\right)^2$$

$$= 4.50 \times 10^{-2} \approx 4.5\%. \qquad \text{(Answer)}$$

For $m = 2$ and $m = 3$ we find that

$$\frac{I_2}{I_m} = 1.6\% \quad \text{and} \quad \frac{I_3}{I_m} = 0.83\%. \quad \text{(Answer)}$$

Successive secondary maxima decrease rapidly in intensity. Figure 36-1 was deliberately overexposed to reveal them.

36-6 | Diffraction by a Circular Aperture

Here we consider diffraction by a circular aperture—that is, a circular opening, such as a circular lens, through which light can pass. Figure 36-10 shows the image of a distant point source of light (a star, for instance) formed on photographic film placed in the focal plane of a converging lens. This image is not a point, as geometrical optics would suggest, but a circular disk surrounded by several progressively fainter secondary rings. Comparison with Fig. 36-1 leaves little doubt that we are dealing with a diffraction phenomenon. Here, however, the aperture is a circle of diameter d rather than a rectangular slit.

The (complex) analysis of such patterns shows that the first minimum for the diffraction pattern of a circular aperture of diameter d is located by

$$\sin \theta = 1.22 \frac{\lambda}{d} \qquad \text{(first minimum—circular aperture).} \qquad (36\text{-}12)$$

The angle θ here is the angle from the central axis to any point on that (circular) minimum. Compare this with Eq. 36-1,

$$\sin \theta = \frac{\lambda}{a} \qquad \text{(first minimum—single slit),} \qquad (36\text{-}13)$$

which locates the first minimum for a long narrow slit of width a. The main difference is the factor 1.22, which enters because of the circular shape of the aperture.

Resolvability

The fact that lens images are diffraction patterns is important when we wish to *resolve* (distinguish) two distant point objects whose angular separation is small. Figure 36-11 shows, in three different cases, the visual appearance and corresponding intensity pattern for two distant point objects (stars, say) with small angular separation. In Figure 36-11a, the objects are not resolved because of diffraction; that is, their diffraction patterns (mainly their central maxima) overlap so much that the two objects cannot be distinguished from a single point object. In Fig. 36-11b the objects are barely resolved, and in Fig. 36-11c they are fully resolved.

In Fig. 36-11b the angular separation of the two point sources is such that the central maximum of the diffraction pattern of one source is centered on the first minimum of the diffraction pattern of the other, a condition called **Rayleigh's criterion** for resolvability. From Eq. 36-12, two objects that are barely resolvable by this criterion must have an angular separation θ_R of

$$\theta_R = \sin^{-1} \frac{1.22\lambda}{d}.$$

Since the angles are small, we can replace $\sin \theta_R$ with θ_R expressed in radians:

$$\theta_R = 1.22 \frac{\lambda}{d} \qquad \text{(Rayleigh's criterion).} \qquad (36\text{-}14)$$

Applying Rayleigh's criterion for resolvability to human vision is only an approximation because visual resolvability depends on many factors, such as the relative brightness of the sources and their surroundings, turbulence in the air

FIG. 36-10 The diffraction pattern of a circular aperture. Note the central maximum and the circular secondary maxima. The figure has been overexposed to bring out these secondary maxima, which are much less intense than the central maximum. *(From Michel Cagnet, Maurice Franzon, and Jean Thierr, Atlas of Optical Phenomena. Springer-Verlag, New York, 1962. Reproduced with permission.)*

FIG. 36-11 At the top, the images of two point sources (stars) formed by a converging lens. At the bottom, representations of the image intensities. In (a) the angular separation of the sources is too small for them to be distinguished, in (b) they can be marginally distinguished, and in (c) they are clearly distinguished. Rayleigh's criterion is satisfied in (b), with the central maximum of one diffraction pattern coinciding with the first minimum of the other.

(a)   (b)   (c)

FIG. 36-12 The pointillistic painting *The Seine at Herblay* by Maximilien Luce consists of thousands of colored dots. With the viewer very close to the canvas, the dots and their true colors are visible. At normal viewing distances, the dots are irresolvable and thus blend. *(Maxilien Luce, The Seine at Herblay, 1890. Musee d'Orsay, Paris, France. Photo by Erich Lessing/Art Resource)*

between the sources and the observer, and the functioning of the observer's visual system. Experimental results show that the least angular separation that can actually be resolved by a person is generally somewhat greater than the value given by Eq. 36-14. However, for calculations here, we shall take Eq. 36-14 as being a precise criterion: If the angular separation θ between the sources is greater than θ_R, we can visually resolve the sources; if it is less, we cannot.

Rayleigh's criterion can explain the arresting illusions of color in the style of painting known as pointillism (Fig. 36-12). In this style, a painting is made not with brush strokes in the usual sense but rather with a myriad of small colored dots. One fascinating aspect of a pointillistic painting is that when you change your distance from it, the colors shift in subtle, almost subconscious ways. This color shifting has to do with whether you can resolve the colored dots. When you stand close enough to the painting, the angular separations θ of adjacent dots are greater than θ_R and thus the dots can be seen individually. Their colors are the true colors of the paints used. However, when you stand far enough from the painting, the angular separations θ are less than θ_R and the dots cannot be seen individually. The resulting blend of colors coming into your eye from any group of dots can then cause your brain to "make up" a color for that group—a color that may not actually exist in the group. In this way, a pointillistic painter uses your visual system to create the colors of the art.

When we wish to use a lens instead of our visual system to resolve objects of small angular separation, it is desirable to make the diffraction pattern as small as possible. According to Eq. 36-14, this can be done either by increasing the lens diameter or by using light of a shorter wavelength. For this reason ultraviolet light is often used with microscopes; because of its shorter wavelength, it permits finer detail to be examined than would be possible for the same microscope operated with visible light.

✓ **CHECKPOINT 4**   Suppose that you can barely resolve two red dots because of diffraction by the pupil of your eye. If we increase the general illumination around you so that the pupil decreases in diameter, does the resolvability of the dots improve or diminish? Consider only diffraction. (You might experiment to check your answer.)

Sample Problem   36-3

Figure 36-13a is a representation of the colored dots on a pointillistic painting. Assume that the average center-to-center separation of the dots is $D = 2.0$ mm. Also assume that the diameter of the pupil of your eye is $d = 1.5$ mm and that the least angular separation between dots you can resolve is set only by Rayleigh's criterion. What is the least viewing distance from which you cannot distinguish any dots on the painting?

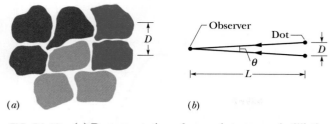

(a)   (b)

FIG. 36-13 (a) Representation of some dots on a pointillistic painting, showing an average center-to-center separation D. (b) The arrangement of separation D between two dots, their angular separation θ, and the viewing distance L.

KEY IDEA Consider any two adjacent dots that you can distinguish when you are close to the painting. As you move away, you continue to distinguish the dots until their angular separation θ (in your view) has decreased to the angle given by Rayleigh's criterion:

$$\theta_R = 1.22 \frac{\lambda}{d}. \qquad (36\text{-}15)$$

Calculations: Figure 36-13b shows, from the side, the angular separation θ of the dots, their center-to-center separation D, and your distance L from them. Because

D/L is small, angle θ is also small and we can make the approximation

$$\theta = \frac{D}{L}. \qquad (36\text{-}16)$$

Setting θ of Eq. 36-16 equal to θ_R of Eq. 36-15 and solving for L, we then have

$$L = \frac{Dd}{1.22\lambda}. \qquad (36\text{-}17)$$

Equation 36-17 tells us that L is larger for smaller λ. Thus, as you move away from the painting, adjacent red dots (long wavelengths) become indistinguishable before adjacent blue dots do. To find the least distance L at which *no* colored dots are distinguishable, we substitute $\lambda = 400$ nm (blue or violet light) into Eq. 36-17:

$$L = \frac{(2.0 \times 10^{-3} \text{ m})(1.5 \times 10^{-3} \text{ m})}{(1.22)(400 \times 10^{-9} \text{ m})} = 6.1 \text{ m}.$$

(Answer)

At this or a greater distance, the color you perceive at any given spot on the painting is a blended color that may not actually exist there.

Sample Problem   36-4

A circular converging lens, with diameter $d = 32$ mm and focal length $f = 24$ cm, forms images of distant point objects in the focal plane of the lens. The wavelength is $\lambda = 550$ nm.

(a) Considering diffraction by the lens, what angular separation must two distant point objects have to satisfy Rayleigh's criterion?

KEY IDEA Figure 36-14 shows two distant point objects P_1 and P_2, the lens, and a viewing screen in the focal plane of the lens. It also shows, on the right, plots of light intensity I versus position on the screen for the central maxima of the images formed by the lens. Note that the angular separation θ_o of the objects equals the angular separation θ_i of the images. Thus, if the images are to satisfy Rayleigh's criterion for resolvability, the angular separations on both sides of the lens must be given by Eq. 36-14 (assuming small angles).

Calculations: From Eq. 36-14, we obtain

$$\theta_o = \theta_i = \theta_R = 1.22 \frac{\lambda}{d}$$

$$= \frac{(1.22)(550 \times 10^{-9} \text{ m})}{32 \times 10^{-3} \text{ m}} = 2.1 \times 10^{-5} \text{ rad.}$$

(Answer)

At this angular separation, each central maximum in the two intensity curves of Fig. 36-14 is centered on the first minimum of the other curve.

(b) What is the separation Δx of the centers of the *images* in the focal plane? (That is, what is the separation of the *central* peaks in the two intensity-versus-position curves?)

Calculations: From either triangle between the lens and the screen in Fig. 36-14, we see that tan $\theta_i/2 = \Delta x/2f$. Rearranging this equation and making the approximation tan $\theta \approx \theta$, we find

$$\Delta x = f\theta_i, \qquad (36\text{-}18)$$

where θ_i is in radian measure. Substituting known data then yields

$$\Delta x = (0.24 \text{ m})(2.1 \times 10^{-5} \text{ rad}) = 5.0 \ \mu\text{m}. \quad \text{(Answer)}$$

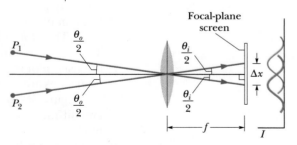

FIG. 36-14 Light from two distant point objects P_1 and P_2 passes through a converging lens and forms images on a viewing screen in the focal plane of the lens. Only one representative ray from each object is shown. The images are not points but diffraction patterns, with intensities approximately as plotted at the right. The angular separation of the objects is θ_o and that of the images is θ_i; the central maxima of the images have a separation Δx.

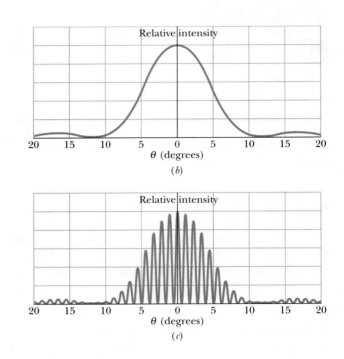

FIG. 36-15 (*a*) The intensity plot to be expected in a double-slit interference experiment with vanishingly narrow slits. (*b*) The intensity plot for diffraction by a typical slit of width *a* (not vanishingly narrow). (*c*) The intensity plot to be expected for two slits of width *a*. The curve of (*b*) acts as an envelope, limiting the intensity of the double-slit fringes in (*a*). Note that the first minima of the diffraction pattern of (*b*) eliminate the double-slit fringes that would occur near 12° in (*c*).

36-7 | Diffraction by a Double Slit

In the double-slit experiments of Chapter 35, we implicitly assumed that the slits were much narrower than the wavelength of the light illuminating them; that is, $a \ll \lambda$. For such narrow slits, the central maximum of the diffraction pattern of either slit covers the entire viewing screen. Moreover, the interference of light from the two slits produces bright fringes with approximately the same intensity (Fig. 35-12).

In practice with visible light, however, the condition $a \ll \lambda$ is often not met. For relatively wide slits, the interference of light from two slits produces bright fringes that do not all have the same intensity. That is, the intensities of the fringes produced by double-slit interference (as discussed in Chapter 35) are modified by diffraction of the light passing through each slit (as discussed in this chapter).

As an example, the intensity plot of Fig. 36-15*a* suggests the double-slit interference pattern that would occur if the slits were infinitely narrow (and thus $a \ll \lambda$); all the bright interference fringes would have the same intensity. The intensity plot of Fig. 36-15*b* is that for diffraction by a single actual slit; the diffraction pattern has a broad central maximum and weaker secondary maxima at ±17°. The plot of Fig. 36-15*c* suggests the interference pattern for two actual slits. That plot was constructed by using the curve of Fig. 36-15*b* as an *envelope* on the intensity plot in Fig. 36-15*a*. The positions of the fringes are not changed; only the intensities are affected.

Figure 36-16*a* shows an actual pattern in which both double-slit interference and diffraction are evident. If one slit is covered, the single-slit diffraction pattern of Fig. 36-16*b* results. Note the correspondence between Figs. 36-16*a* and 36-15*c*, and between Figs. 36-16*b* and 36-15*b*. In comparing these figures, bear in mind that Fig. 36-16 has been deliberately overexposed to bring out the faint secondary maxima and that two secondary maxima (rather than one) are shown.

With diffraction effects taken into account, the intensity of a double-slit interference pattern is given by

$$I(\theta) = I_m (\cos^2 \beta) \left(\frac{\sin \alpha}{\alpha} \right)^2 \qquad \text{(double slit),} \qquad (36\text{-}19)$$

FIG. 36-16 (*a*) Interference fringes for an actual double-slit system; compare with Fig. 36-15*c*. (*b*) The diffraction pattern of a single slit; compare with Fig. 36-15*b*. *(From Michel Cagnet, Maurice Franzon, and Jean Thierr, Atlas of Optical Phenomena. Springer-Verlag, New York, 1962. Reproduced with permission.)*

in which

$$\beta = \frac{\pi d}{\lambda} \sin \theta \tag{36-20}$$

and

$$\alpha = \frac{\pi a}{\lambda} \sin \theta. \tag{36-21}$$

Here *d* is the distance between the centers of the slits and *a* is the slit width. Note carefully that the right side of Eq. 36-19 is the product of I_m and two factors. (1) The *interference factor* $\cos^2 \beta$ is due to the interference between two slits with slit separation *d* (as given by Eqs. 35-22 and 35-23). (2) The *diffraction factor* $[(\sin \alpha)/\alpha]^2$ is due to diffraction by a single slit of width *a* (as given by Eqs. 36-5 and 36-6).

Let us check these factors. If we let $a \to 0$ in Eq. 36-21, for example, then $\alpha \to 0$ and $(\sin \alpha)/\alpha \to 1$. Equation 36-19 then reduces, as it must, to an equation describing the interference pattern for a pair of vanishingly narrow slits with slit separation *d*. Similarly, putting $d = 0$ in Eq. 36-20 is equivalent physically to causing the two slits to merge into a single slit of width *a*. Then Eq. 36-20 yields $\beta = 0$ and $\cos^2 \beta = 1$. In this case Eq. 36-19 reduces, as it must, to an equation describing the diffraction pattern for a single slit of width *a*.

The double-slit pattern described by Eq. 36-19 and displayed in Fig. 36-16*a* combines interference and diffraction in an intimate way. Both are superposition effects, in that they result from the combining of waves with different phases at a given point. If the combining waves originate from a small number of elementary coherent sources—as in a double-slit experiment with $a \ll \lambda$—we call the process *interference*. If the combining waves originate in a single wavefront—as in a single-slit experiment—we call the process *diffraction*. This distinction between interference and diffraction (which is somewhat arbitrary and not always adhered to) is a convenient one, but we should not forget that both are superposition effects and usually both are present simultaneously (as in Fig. 36-16*a*).

Sample Problem   36-5   Build your skill

In a double-slit experiment, the wavelength λ of the light source is 405 nm, the slit separation *d* is 19.44 μm, and the slit width *a* is 4.050 μm. Consider the interference of the light from the two slits and also the diffraction of the light through each slit.

(a) How many bright interference fringes are within the central peak of the diffraction envelope?

KEY IDEAS We first analyze the two basic mechanisms responsible for the optical pattern produced in the experiment:

1. *Single-slit diffraction:* The limits of the central peak are the first minima in the diffraction pattern due to

either slit individually. (See Fig. 36-15.) The angular locations of those minima are given by Eq. 36-3 ($a \sin \theta = m\lambda$). Here let us rewrite this equation as $a \sin \theta = m_1\lambda$, with the subscript 1 referring to the one-slit diffraction. For the first minima in the diffraction pattern, we substitute $m_1 = 1$, obtaining

$$a \sin \theta = \lambda. \tag{36-22}$$

2. *Double-slit interference:* The angular locations of the bright fringes of the double-slit interference pattern are given by Eq. 35-14, which we can write as

$$d \sin \theta = m_2\lambda, \qquad \text{for } m_2 = 0, 1, 2, \dots . \tag{36-23}$$

Here the subscript 2 refers to the double-slit interference.

Calculations: We can locate the first diffraction minimum within the double-slit fringe pattern by dividing Eq. 36-23 by Eq. 36-22 and solving for m_2. By doing so and then substituting the given data, we obtain

$$m_2 = \frac{d}{a} = \frac{19.44 \ \mu m}{4.050 \ \mu m} = 4.8.$$

This tells us that the bright interference fringe for $m_2 = 4$ fits into the central peak of the one-slit diffraction pattern, but the fringe for $m_2 = 5$ does not fit. Within the central diffraction peak we have the central bright fringe ($m_2 = 0$), and four bright fringes (up to $m_2 = 4$) on each side of it. Thus, a total of nine bright fringes of the double-slit interference pattern are within the central peak of the diffraction envelope. The bright fringes to one side of the central bright fringe are shown in Fig. 36-17.

(b) How many bright fringes are within either of the first side peaks of the diffraction envelope?

KEY IDEA The outer limits of the first side diffraction peaks are the second diffraction minima, each of which is at the angle θ given by $a \sin \theta = m_1 \lambda$ with $m_1 = 2$:

$$a \sin \theta = 2\lambda. \qquad (36\text{-}24)$$

Calculation: Dividing Eq. 36-23 by Eq. 36-24, we find

$$m_2 = \frac{2d}{a} = \frac{(2)(19.44 \ \mu m)}{4.050 \ \mu m} = 9.6.$$

FIG. 36-17 One side of the intensity plot for a two-slit interference experiment. The inset shows (vertically expanded) the plot within the first and second side peaks of the diffraction envelope.

This tells us that the second diffraction minimum occurs just before the bright interference fringe for $m_2 = 10$ in Eq. 36-23. Within either first side diffraction peak we have the fringes from $m_2 = 5$ to $m_2 = 9$, for a total of five bright fringes of the double-slit interference pattern (shown in the inset of Fig. 36-17). However, if the $m_2 = 5$ bright fringe, which is almost eliminated by the first diffraction minimum, is considered too dim to count, then only four bright fringes are in the first side diffraction peak.

36-8 | Diffraction Gratings

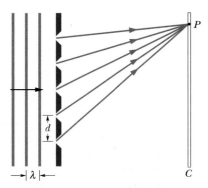

FIG. 36-18 An idealized diffraction grating, consisting of only five rulings, that produces an interference pattern on a distant viewing screen C.

One of the most useful tools in the study of light and of objects that emit and absorb light is the **diffraction grating.** This device is somewhat like the double-slit arrangement of Fig. 35-10 but has a much greater number N of slits, often called *rulings,* perhaps as many as several thousand per millimeter. An idealized grating consisting of only five slits is represented in Fig. 36-18. When monochromatic light is sent through the slits, it forms narrow interference fringes that can be analyzed to determine the wavelength of the light. (Diffraction gratings can also be opaque surfaces with narrow parallel grooves arranged like the slits in Fig. 36-18. Light then scatters back from the grooves to form interference fringes rather than being transmitted through open slits.)

With monochromatic light incident on a diffraction grating, if we gradually increase the number of slits from two to a large number N, the intensity plot changes from the typical double-slit plot of Fig. 36-15c to a much more complicated one and then eventually to a simple graph like that shown in Fig. 36-19a. The pattern you would see on a viewing screen using monochromatic red light from, say, a helium–neon laser is shown in Fig. 36-19b. The maxima are now very narrow (and so are called *lines*); they are separated by relatively wide dark regions.

We use a familiar procedure to find the locations of the bright lines on the viewing screen. We first assume that the screen is far enough from the grating so that the rays reaching a particular point P on the screen are approximately parallel when they leave the grating (Fig. 36-20). Then we apply to each pair of adjacent rulings the same reasoning we used for double-slit interference. The separation d between rulings is called the *grating spacing.* (If N rulings occupy a total width w, then $d = w/N$.) The path length difference between adjacent rays is again $d \sin \theta$ (Fig. 36-20), where θ is the angle from the central axis of the grating (and of the diffraction pattern) to point P. A line will be located at P if the path length difference between adjacent rays is an integer number of wavelengths—that is, if

$$d \sin \theta = m\lambda, \quad \text{for } m = 0, 1, 2, \ldots \quad \text{(maxima—lines)}, \qquad (36\text{-}25)$$

where λ is the wavelength of the light. Each integer m represents a different line; hence these integers can be used to label the lines, as in Fig. 36-19. The integers are then called the *order numbers*, and the lines are called the zeroth-order line (the central line, with $m = 0$), the first-order line ($m = 1$), the second-order line ($m = 2$), and so on.

If we rewrite Eq. 36-25 as $\theta = \sin^{-1}(m\lambda/d)$, we see that, for a given diffraction grating, the angle from the central axis to any line (say, the third-order line) depends on the wavelength of the light being used. Thus, when light of an unknown wavelength is sent through a diffraction grating, measurements of the angles to the higher-order lines can be used in Eq. 36-25 to determine the wavelength. Even light of several unknown wavelengths can be distinguished and identified in this way. We cannot do that with the double-slit arrangement of Section 35-4, even though the same equation and wavelength dependence apply there. In double-slit interference, the bright fringes due to different wavelengths overlap too much to be distinguished.

Width of the Lines

A grating's ability to resolve (separate) lines of different wavelengths depends on the width of the lines. We shall here derive an expression for the *half-width* of the central line (the line for which $m = 0$) and then state an expression for the half-widths of the higher-order lines. We define the **half-width** of the central line as being the angle $\Delta\theta_{hw}$ from the center of the line at $\theta = 0$ outward to where the line effectively ends and darkness effectively begins with the first minimum (Fig. 36-21). At such a minimum, the N rays from the N slits of the grating cancel one another. (The actual width of the central line is, of course, $2(\Delta\theta_{hw})$, but line widths are usually compared via half-widths.)

In Section 36-3 we were also concerned with the cancellation of a great many rays, there due to diffraction through a single slit. We obtained Eq. 36-3, which, because of the similarity of the two situations, we can use to find the first minimum here. It tells us that the first minimum occurs where the path length difference between the top and bottom rays equals λ. For single-slit diffraction, this difference is $a \sin \theta$. For a grating of N rulings, each separated from the next by distance d, the distance between the top and bottom rulings is Nd (Fig. 36-22),

FIG. 36-19   (*a*) The intensity plot produced by a diffraction grating with a great many rulings consists of narrow peaks, here labeled with their order numbers m. (*b*) The corresponding bright fringes seen on the screen are called lines and are here also labeled with order numbers m.

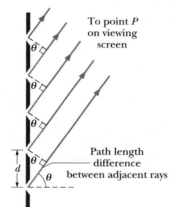

FIG. 36-20   The rays from the rulings in a diffraction grating to a distant point P are approximately parallel. The path length difference between each two adjacent rays is $d \sin \theta$, where θ is measured as shown. (The rulings extend into and out of the page.)

FIG. 36-21   The half-width $\Delta\theta_{hw}$ of the central line is measured from the center of that line to the adjacent minimum on a plot of I versus θ like Fig. 36-19a.

FIG. 36-22   The top and bottom rulings of a diffraction grating of N rulings are separated by Nd. The top and bottom rays passing through these rulings have a path length difference of $Nd \sin \Delta\theta_{hw}$, where $\Delta\theta_{hw}$ is the angle to the first minimum. (The angle is here greatly exaggerated for clarity.)

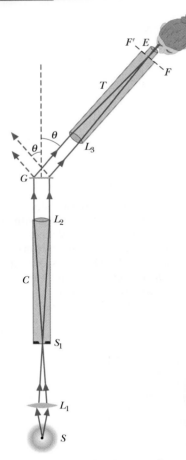

FIG. 36-23 A simple type of grating spectroscope used to analyze the wavelengths of light emitted by source S.

and so the path length difference between the top and bottom rays here is $Nd \sin \Delta\theta_{hw}$. Thus, the first minimum occurs where

$$Nd \sin \Delta\theta_{hw} = \lambda. \tag{36-26}$$

Because $\Delta\theta_{hw}$ is small, $\sin \Delta\theta_{hw} = \Delta\theta_{hw}$ (in radian measure). Substituting this in Eq. 36-26 gives the half-width of the central line as

$$\Delta\theta_{hw} = \frac{\lambda}{Nd} \qquad \text{(half-width of central line).} \tag{36-27}$$

We state without proof that the half-width of any other line depends on its location relative to the central axis and is

$$\Delta\theta_{hw} = \frac{\lambda}{Nd \cos \theta} \qquad \text{(half-width of line at } \theta\text{).} \tag{36-28}$$

Note that for light of a given wavelength λ and a given ruling separation d, the widths of the lines decrease with an increase in the number N of rulings. Thus, of two diffraction gratings, the grating with the larger value of N is better able to distinguish between wavelengths because its diffraction lines are narrower and so produce less overlap.

Grating Spectroscope

Diffraction gratings are widely used to determine the wavelengths that are emitted by sources of light ranging from lamps to stars. Figure 36-23 shows a simple *grating spectroscope* in which a grating is used for this purpose. Light from source S is focused by lens L_1 on a vertical slit S_1 placed in the focal plane of lens L_2. The light emerging from tube C (called a *collimator*) is a plane wave and is incident perpendicularly on grating G, where it is diffracted into a diffraction pattern, with the $m = 0$ order diffracted at angle $\theta = 0$ along the central axis of the grating.

We can view the diffraction pattern that would appear on a viewing screen at any angle θ simply by orienting telescope T in Fig. 36-23 to that angle. Lens L_3 of the telescope then focuses the light diffracted at angle θ (and at slightly smaller and larger angles) onto a focal plane FF' within the telescope. When we look through eyepiece E, we see a magnified view of this focused image.

By changing the angle θ of the telescope, we can examine the entire diffraction pattern. For any order number other than $m = 0$, the original light is spread out according to wavelength (or color) so that we can determine, with Eq. 36-25, just what wavelengths are being emitted by the source. If the source emits discrete wavelengths, what we see as we rotate the telescope horizontally through the angles corresponding to an order m is a vertical line of color for each wavelength, with the shorter-wavelength line at a smaller angle θ than the longer-wavelength line.

For example, the light emitted by a hydrogen lamp, which contains hydrogen gas, has four discrete wavelengths in the visible range. If our eyes intercept this light directly, it appears to be white. If, instead, we view it through a grating spectroscope, we can distinguish, in several orders, the lines of the four colors corresponding to these visible wavelengths. (Such lines are called *emission lines*.) Four orders are represented in Fig. 36-24. In the central order ($m = 0$), the lines

FIG. 36-24 The zeroth, first, second, and fourth orders of the visible emission lines from hydrogen. Note that the lines are farther apart at greater angles. (They are also dimmer and wider, although that is not shown here.)

FIG. 36-25   The visible emission lines of cadmium, as seen through a grating spectroscope. *(Department of Physics, Imperial College/Science Photo Library/Photo Researchers)*

corresponding to all four wavelengths are superimposed, giving a single white line at $\theta = 0$. The colors are separated in the higher orders.

The third order is not shown in Fig. 36-24 for the sake of clarity; it actually overlaps the second and fourth orders. The fourth-order red line is missing because it is not formed by the grating used here. That is, when we attempt to solve Eq. 36-25 for the angle θ for the red wavelength when $m = 4$, we find that $\sin \theta$ is greater than unity, which is not possible. The fourth order is then said to be *incomplete* for this grating; it might not be incomplete for a grating with greater spacing d, which will spread the lines less than in Fig. 36-24. Figure 36-25 is a photograph of the visible emission lines produced by cadmium.

Optically Variable Graphics

Holograms are made by having laser light scatter from an object onto an emulsion. Once the hologram is developed, an image of the object can be created by illuminating the hologram with the same type of laser light. The image is arresting because, unlike common photographs, it has depth, and you can change your perspective of the object by changing the angle at which you view the hologram.

Holograms were thought to be an ideal anticounterfeiting measure for credit cards and other types of personal cards. However, they have several disadvantages. (1) A holographic image can be sharp when viewed in laser light, which is coherent and incident from a single direction. However, the image is murky ("milky") when viewed in the normal light of a store. (Such *diffuse light* is incoherent and incident from many directions.) Thus a store clerk is unlikely to examine a display on a credit card closely enough to see whether it is a legitimate hologram. (2) A hologram can be easily counterfeited because it is a photograph of an actual object. A counterfeiter merely makes a model of that object, then makes a hologram of it, and then attaches the hologram to a counterfeit credit card.

Most credit cards and many identification cards now carry *optically variable graphics* (OVG), which produce an image via the diffraction of diffuse light by gratings embedded in the device (Fig. 36-26). The gratings send out hundreds or even thousands of different orders. Someone viewing the card intercepts some of these orders, and the combined light creates a virtual image that is part of, say, a credit-card logo. For example, in Fig. 36-27a, gratings at point *a* produce a certain image when the viewer is at orientation *A*, and in Fig. 36-27b, gratings at point *b* produce a different image when the viewer is at orientation *B*. These images are bright and sharp because the gratings have been designed to be viewed in diffuse light.

An OVG is very difficult to design because optical engineers must work backwards from a graphic, such as a given logo. The engineers must determine the grating properties across the OVG if a certain image is to be seen from one set of viewing angles and a different image is to be seen from a different set of viewing angles. Such work requires sophisticated programming on computers. Once designed, the OVG structure is so complicated that counterfeiting it is extremely difficult.

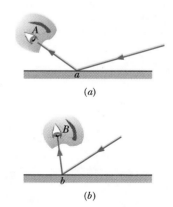

FIG. 36-26   An identification card with OVG. *(Pascal Goetgheluck/Photo Researchers)*

FIG. 36-27   (*a*) Gratings at point *a* on the surface of an OVG device send light to a viewer at orientation *A*, creating a certain virtual image. (*b*) Gratings at point *b* send light to the viewer at orientation *B*, creating a different virtual image.

✓**CHECKPOINT 5**   The figure shows lines of different orders produced by a diffraction grating in monochromatic red light. (a) Is the center of the pattern to the left or right? (b) In monochromatic green light, are the half-widths of the lines produced in the same orders greater than, less than, or the same as the half-widths of the lines shown?

The fine rulings, each 0.5 μm wide, on a compact disc function as a diffraction grating. When a small source of white light illuminates a disc, the diffracted light forms colored "lanes" that are the composite of the diffraction patterns from the rulings.
(Kristen Brochmann/Fundamental Photographs)

36-9 | Gratings: Dispersion and Resolving Power

Dispersion

To be useful in distinguishing wavelengths that are close to each other (as in a grating spectroscope), a grating must spread apart the diffraction lines associated with the various wavelengths. This spreading, called **dispersion,** is defined as

$$D = \frac{\Delta\theta}{\Delta\lambda} \quad \text{(dispersion defined).} \tag{36-29}$$

Here $\Delta\theta$ is the angular separation of two lines whose wavelengths differ by $\Delta\lambda$. The greater D is, the greater is the distance between two emission lines whose wavelengths differ by $\Delta\lambda$. We show below that the dispersion of a grating at angle θ is given by

$$D = \frac{m}{d\cos\theta} \quad \text{(dispersion of a grating).} \tag{36-30}$$

Thus, to achieve higher dispersion we must use a grating of smaller grating spacing d and work in a higher-order m. Note that the dispersion does not depend on the number of rulings N in the grating. The SI unit for D is the degree per meter or the radian per meter.

Resolving Power

To *resolve* lines whose wavelengths are close together (that is, to make the lines distinguishable), the line should also be as narrow as possible. Expressed otherwise, the grating should have a high **resolving power R,** defined as

$$R = \frac{\lambda_{\text{avg}}}{\Delta\lambda} \quad \text{(resolving power defined).} \tag{36-31}$$

Here λ_{avg} is the mean wavelength of two emission lines that can barely be recognized as separate, and $\Delta\lambda$ is the wavelength difference between them. The greater R is, the closer two emission lines can be and still be resolved. We shall show below that the resolving power of a grating is given by the simple expression

$$R = Nm \quad \text{(resolving power of a grating).} \tag{36-32}$$

To achieve high resolving power, we must use many rulings (large N).

Proof of Eq. 36-30

Let us start with Eq. 36-25, the expression for the locations of the lines in the diffraction pattern of a grating:

$$d\sin\theta = m\lambda.$$

Let us regard θ and λ as variables and take differentials of this equation. We find

$$d(\cos\theta)\,d\theta = m\,d\lambda.$$

For small enough angles, we can write these differentials as small differences, obtaining

$$d(\cos\theta)\,\Delta\theta = m\,\Delta\lambda \tag{36-33}$$

or

$$\frac{\Delta\theta}{\Delta\lambda} = \frac{m}{d\cos\theta}.$$

The ratio on the left is simply D (see Eq. 36-29), and so we have indeed derived Eq. 36-30.

TABLE 36-1

Three Gratings[a]

Grating	N	d (nm)	θ	D (°/μm)	R
A	10 000	2540	13.4°	23.2	10 000
B	20 000	2540	13.4°	23.2	20 000
C	10 000	1360	25.5°	46.3	10 000

[a]Data are for $\lambda = 589$ nm and $m = 1$.

Proof of Eq. 36-32

We start with Eq. 36-33, which was derived from Eq. 36-25, the expression for the locations of the lines in the diffraction pattern formed by a grating. Here $\Delta\lambda$ is the small wavelength difference between two waves that are diffracted by the grating, and $\Delta\theta$ is the angular separation between them in the diffraction pattern. If $\Delta\theta$ is to be the smallest angle that will permit the two lines to be resolved, it must (by Rayleigh's criterion) be equal to the half-width of each line, which is given by Eq. 36-28:

$$\Delta\theta_{hw} = \frac{\lambda}{Nd \cos \theta}.$$

If we substitute $\Delta\theta_{hw}$ as given here for $\Delta\theta$ in Eq. 36-33, we find that

$$\frac{\lambda}{N} = m \, \Delta\lambda,$$

from which it readily follows that

$$R = \frac{\lambda}{\Delta\lambda} = Nm.$$

This is Eq. 36-32, which we set out to derive.

Dispersion and Resolving Power Compared

The resolving power of a grating must not be confused with its dispersion. Table 36-1 shows the characteristics of three gratings, all illuminated with light of wavelength $\lambda = 589$ nm, whose diffracted light is viewed in the first order ($m = 1$ in Eq. 36-25). You should verify that the values of D and R as given in the table can be calculated with Eqs. 36-30 and 36-32, respectively. (In the calculations for D, you will need to convert radians per meter to degrees per micrometer.)

For the conditions noted in Table 36-1, gratings A and B have the same *dispersion D* and A and C have the same *resolving power R*.

Figure 36-28 shows the intensity patterns (also called *line shapes*) that would be produced by these gratings for two lines of wavelengths λ_1 and λ_2, in the vicinity of $\lambda = 589$ nm. Grating B, with the higher resolving power, produces narrower lines and thus is capable of distinguishing lines that are much closer together in wavelength than those in the figure. Grating C, with the higher dispersion, produces the greater angular separation between the lines.

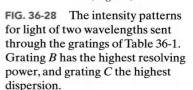

FIG. 36-28 The intensity patterns for light of two wavelengths sent through the gratings of Table 36-1. Grating B has the highest resolving power, and grating C the highest dispersion.

Sample Problem | **36-6**

A diffraction grating has 1.26×10^4 rulings uniformly spaced over width $w = 25.4$ mm. It is illuminated at normal incidence by yellow light from a sodium vapor lamp. This light contains two closely spaced emission lines (known as the sodium doublet) of wavelengths 589.00 nm and 589.59 nm.

(a) At what angle does the first-order maximum occur (on either side of the center of the diffraction pattern) for the wavelength of 589.00 nm?

KEY IDEA The maxima produced by the diffraction grating can be determined with Eq. 36-25 ($d \sin \theta = m\lambda$).

Calculations: The grating spacing d is

$$d = \frac{w}{N} = \frac{25.4 \times 10^{-3} \text{ m}}{1.26 \times 10^4}$$

$$= 2.016 \times 10^{-6} \text{ m} = 2016 \text{ nm}.$$

The first-order maximum corresponds to $m = 1$. Substituting these values for d and m into Eq. 36-25 leads to

$$\theta = \sin^{-1}\frac{m\lambda}{d} = \sin^{-1}\frac{(1)(589.00 \text{ nm})}{2016 \text{ nm}}$$

$$= 16.99° \approx 17.0°. \qquad \text{(Answer)}$$

(b) Using the dispersion of the grating, calculate the angular separation between the two lines in the first order.

KEY IDEAS (1) The angular separation $\Delta\theta$ between the two lines in the first order depends on their wavelength difference $\Delta\lambda$ and the dispersion D of the grating, according to Eq. 36-29 ($D = \Delta\theta/\Delta\lambda$). (2) The dispersion D depends on the angle θ at which it is to be evaluated.

Calculations: We can assume that, in the first order, the two sodium lines occur close enough to each other for us to evaluate D at the angle $\theta = 16.99°$ we found in part (a) for one of those lines. Then Eq. 36-30 gives the dispersion as

$$D = \frac{m}{d\cos\theta} = \frac{1}{(2016 \text{ nm})(\cos 16.99°)}$$

$$= 5.187 \times 10^{-4} \text{ rad/nm}.$$

From Eq. 36-29 and with $\Delta\lambda$ in nanometers, we then have

$$\Delta\theta = D\,\Delta\lambda = (5.187 \times 10^{-4} \text{ rad/nm})(589.59 - 589.00)$$

$$= 3.06 \times 10^{-4} \text{ rad} = 0.0175°. \qquad \text{(Answer)}$$

You can show that this result depends on the grating spacing d but not on the number of rulings there are in the grating.

(c) What is the least number of rulings a grating can have and still be able to resolve the sodium doublet in the first order?

KEY IDEAS (1) The resolving power of a grating in any order m is physically set by the number of rulings N in the grating according to Eq. 36-32 ($R = Nm$). (2) The smallest wavelength difference $\Delta\lambda$ that can be resolved depends on the average wavelength involved and on the resolving power R of the grating, according to Eq. 36-31 ($R = \lambda_{avg}/\Delta\lambda$).

Calculation: For the sodium doublet to be barely resolved, $\Delta\lambda$ must be their wavelength separation of 0.59 nm, and λ_{avg} must be their average wavelength of 589.30 nm. Thus, we find that the smallest number of rulings for a grating to resolve the sodium doublet is

$$N = \frac{R}{m} = \frac{\lambda_{avg}}{m\,\Delta\lambda}$$

$$= \frac{589.30 \text{ nm}}{(1)(0.59 \text{ nm})} = 999 \text{ rulings}. \qquad \text{(Answer)}$$

36-10 | Diffraction by Organized Layers

Here we examine two examples of optical interference due to the diffraction of light by organized layers. First, we discuss x-ray diffraction by organized layers of atoms in crystals, an effect known for about 100 years. Then we discuss visible-light diffraction by organized layers in organic materials, including the skin of mandrill baboons, which is a current subject of research.

X-Ray Diffraction

X rays are electromagnetic radiation whose wavelengths are of the order of 1 Å ($= 10^{-10}$ m). Compare this with a wavelength of 550 nm ($= 5.5 \times 10^{-7}$ m) at the center of the visible spectrum. Figure 36-29 shows that x rays are produced when electrons escaping from a heated filament F are accelerated by a potential difference V and strike a metal target T.

A standard optical diffraction grating cannot be used to discriminate between different wavelengths in the x-ray wavelength range. For $\lambda = 1$ Å ($= 0.1$ nm) and $d = 3000$ nm, for example, Eq. 36-25 shows that the first-order maximum occurs at

$$\theta = \sin^{-1}\frac{m\lambda}{d} = \sin^{-1}\frac{(1)(0.1 \text{ nm})}{3000 \text{ nm}} = 0.0019°.$$

FIG. 36-29 X rays are generated when electrons leaving heated filament F are accelerated through a potential difference V and strike a metal target T. The "window" W in the evacuated chamber C is transparent to x rays.

This is too close to the central maximum to be practical. A grating with $d \approx \lambda$ is desirable, but, because x-ray wavelengths are about equal to atomic diameters, such gratings cannot be constructed mechanically.

In 1912, it occurred to German physicist Max von Laue that a crystalline solid, which consists of a regular array of atoms, might form a natural three-dimensional "diffraction grating" for x-rays. The idea is that, in a crystal such as sodium chloride (NaCl), a basic unit of atoms (called the *unit cell*) repeats itself throughout the array. Figure 36-30*a* represents a section through a crystal of NaCl and identifies this basic unit. The unit cell is a cube measuring a_0 on each side.

When an x-ray beam enters a crystal such as NaCl, x rays are *scattered*—that is, redirected—in all directions by the crystal structure. In some directions the scattered waves undergo destructive interference, resulting in intensity minima; in other directions the interference is constructive, resulting in intensity maxima. This process of scattering and interference is a form of diffraction, although it is unlike the diffraction of light traveling through a slit or past an edge as we discussed earlier.

Although the process of diffraction of x rays by a crystal is complicated, the maxima turn out to be in directions *as if* the x rays were reflected by a family of parallel *reflecting planes* (or *crystal planes*) that extend through the atoms within the crystal and that contain regular arrays of the atoms. (The x rays are not actually reflected; we use these fictional planes only to simplify the analysis of the actual diffraction process.)

Figure 36-30*b* shows three reflecting planes (part of a family containing many parallel planes) with *interplanar spacing d,* from which the incident rays shown are said to reflect. Rays 1, 2, and 3 reflect from the first, second, and third planes, respectively. At each reflection the angle of incidence and the angle of reflection are represented with θ. Contrary to the custom in optics, these angles are defined relative to the *surface* of the reflecting plane rather than a normal to that surface. For the situation of Fig. 36-30*b*, the interplanar spacing happens to be equal to the unit cell dimension a_0.

Figure 36-30*c* shows an edge-on view of reflection from an adjacent pair of planes. The waves of rays 1 and 2 arrive at the crystal in phase. After they are reflected, they must again be in phase because the reflections and the reflecting planes have been defined solely to explain the intensity maxima in the diffraction of x rays by a crystal. Unlike light rays, the x rays do not refract upon entering the crystal; moreover, we do not define an index of refraction for this situation. Thus, the relative phase between the waves of rays 1 and 2 as they leave the crystal is set solely by their path length difference. For these rays to be in phase, the path length difference must be equal to an integer multiple of the wavelength λ of the x rays.

By drawing the dashed perpendiculars in Fig. 36-30*c*, we find that the path length difference is $2d \sin \theta$. In fact, this is true for any pair of adjacent planes in the family of planes represented in Fig. 36-30*b*. Thus, we have, as the criterion for intensity maxima for x-ray diffraction,

$$2d \sin \theta = m\lambda, \qquad \text{for } m = 1, 2, 3, \ldots \quad \text{(Bragg's law)}, \qquad (36\text{-}34)$$

where m is the order number of an intensity maximum. Equation 36-34 is called **Bragg's law** after British physicist W. L. Bragg, who first derived it. (He and his father shared the 1915 Nobel Prize in physics for their use of x rays to study the structures of crystals.) The angle of incidence and reflection in Eq. 36-34 is called a *Bragg angle*.

Regardless of the angle at which x rays enter a crystal, there is always a family of planes from which they can be said to reflect so that we can apply Bragg's law. In Fig. 36-30*d*, notice that the crystal structure has the same orientation as it does in Fig. 36-30*a*, but the angle at which the beam enters the structure differs from that shown in Fig. 36-30*b*. This new angle requires a new family of reflecting

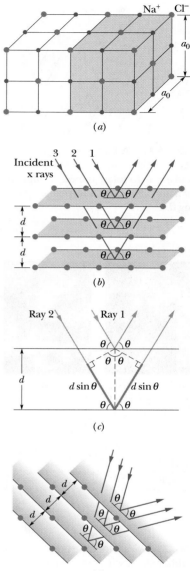

FIG. 36-30 (*a*) The cubic structure of NaCl, showing the sodium and chlorine ions and a unit cell (shaded). (*b*) Incident x rays undergo diffraction by the structure of (*a*). The x rays are diffracted as if they were reflected by a family of parallel planes, with the angle of reflection equal to the angle of incidence, both angles measured relative to the planes (not relative to a normal as in optics). (*c*) The path length difference between waves effectively reflected by two adjacent planes is $2d \sin \theta$. (*d*) A different orientation of the incident x rays relative to the structure. A different family of parallel planes now effectively reflects the x rays.

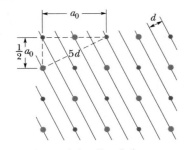

FIG. 36-31 A family of planes through the structure of Fig. 36-30a, and a way to relate the edge length a_0 of a unit cell to the interplanar spacing d.

FIG. 36-32 Color-shifting by a rubber sheet with embedded polystyrene spheres. (a) and (c) Unstretched, with the spheres forming planes of separation d. (b) and (d) Stretched, with the spheres forming planes of smaller separation d'. *(Courtesy Hiroshi Fudouzi)*

planes, with a different interplanar spacing d and different Bragg angle θ, in order to explain the x-ray diffraction via Bragg's law.

Figure 36-31 shows how the interplanar spacing d can be related to the unit cell dimension a_0. For the particular family of planes shown there, the Pythagorean theorem gives

$$5d = \sqrt{\tfrac{5}{4}a_0^2},$$

or

$$d = \frac{a_0}{\sqrt{20}} = 0.2236a_0. \tag{36-35}$$

Figure 36-31 suggests how the dimensions of the unit cell can be found once the interplanar spacing has been measured by means of x-ray diffraction.

X-ray diffraction is a powerful tool for studying both x-ray spectra and the arrangement of atoms in crystals. To study spectra, a particular set of crystal planes, having a known spacing d, is chosen. These planes effectively reflect different wavelengths at different angles. A detector that can discriminate one angle from another can then be used to determine the wavelength of radiation reaching it. The crystal itself can be studied with a monochromatic x-ray beam, to determine not only the spacing of various crystal planes but also the structure of the unit cell.

Structural Coloring by Diffraction

Figure 36-32a shows a transparent unstretched rubber sheet in which polystyrene spheres have been embedded fairly uniformly, forming planes with separation d (Fig. 36-32c). When visible light penetrates the sheet, it is diffracted by the spheres and a portion emerges toward us. However, because there is a path length difference between the light that reaches us from any two planes, the light undergoes interference. With the sheet unstretched, constructive interference occurs at wavelengths around 590 nm, giving us red light (Fig. 36-32a). When the sheet is stretched, it thins, which causes the separation between planes to narrow to d' (Fig. 36-32d) and thus the wavelengths for constructive interference to decrease. In Fig. 36-32b the stretched sheet gives us green light at wavelengths around 563 nm. This coloration is said to be an example of *structural coloring* because it is due to the interference of light caused by the structural arrangement within a material.

Similar structural coloring occurs in the facial skin of mandrill baboons (and the skin on some other animals) because of diffraction from parallel collagen fibers beneath the skin surface. Light penetrates the skin, diffracts from the fibers, and then emerges from the skin. The fiber separation results in constructive interference for wavelengths around 460 nm, so the skin is blue. It is not iridescent (like the wings on some types of butterflies) because the parallel organization of collagen fibers is uniform only in microscopic regions, an arrangement said to be *quasi-ordered.*

REVIEW & SUMMARY

Diffraction When waves encounter an edge, an obstacle, or an aperture the size of which is comparable to the wavelength of the waves, those waves spread out as they travel and, as a result, undergo interference. This is called **diffraction.**

Single-Slit Diffraction Waves passing through a long narrow slit of width a produce, on a viewing screen, a **single-slit diffraction pattern** that includes a central maximum and other maxima, separated by minima located at angles θ to the central axis that satisfy

$$a \sin \theta = m\lambda, \quad \text{for } m = 1, 2, 3, \ldots \quad \text{(minima).} \tag{36-3}$$

The intensity of the diffraction pattern at any given angle θ is

$$I(\theta) = I_m \left(\frac{\sin \alpha}{\alpha} \right)^2, \quad \text{where } \alpha = \frac{\pi a}{\lambda} \sin \theta \tag{36-5, 36-6}$$

and I_m is the intensity at the center of the pattern.

Circular-Aperture Diffraction Diffraction by a circular aperture or a lens with diameter d produces a central maximum and concentric maxima and minima, with the first minimum at an angle θ given by

$$\sin \theta = 1.22 \frac{\lambda}{d} \quad \text{(first minimum—circular aperture).} \tag{36-12}$$

Rayleigh's Criterion *Rayleigh's criterion* suggests that two objects are on the verge of resolvability if the central diffraction maximum of one is at the first minimum of the other. Their angular separation must then be at least

$$\theta_R = 1.22 \frac{\lambda}{d} \qquad \text{(Rayleigh's criterion)}, \qquad (36\text{-}14)$$

in which d is the diameter of the aperture through which the light passes.

Double-Slit Diffraction Waves passing through two slits, each of width a, whose centers are a distance d apart, display diffraction patterns whose intensity I at angle θ is

$$I(\theta) = I_m(\cos^2 \beta) \left(\frac{\sin \alpha}{\alpha} \right)^2 \qquad \text{(double slit)}, \quad (36\text{-}19)$$

with $\beta = (\pi d / \lambda) \sin \theta$ and α as for single-slit diffraction.

Diffraction Gratings A *diffraction grating* is a series of "slits" used to separate an incident wave into its component wavelengths by separating and displaying their diffraction maxima. Diffraction by N (multiple) slits results in maxima (lines) at angles θ such that

$$d \sin \theta = m\lambda, \qquad \text{for } m = 0, 1, 2, \dots \qquad \text{(maxima)}, \quad (36\text{-}25)$$

with the **half-widths** of the lines given by

$$\Delta\theta_{hw} = \frac{\lambda}{Nd \cos \theta} \qquad \text{(half-widths)}. \quad (36\text{-}28)$$

The dispersion D and resolving power R are given by

$$D = \frac{\Delta\theta}{\Delta\lambda} = \frac{m}{d \cos \theta} \qquad (36\text{-}29, 36\text{-}30)$$

$$R = \frac{\lambda_{avg}}{\Delta\lambda} = Nm. \qquad (36\text{-}31, 36\text{-}32)$$

X-Ray Diffraction The regular array of atoms in a crystal is a three-dimensional diffraction grating for short-wavelength waves such as x rays. For analysis purposes, the atoms can be visualized as being arranged in planes with characteristic interplanar spacing d. Diffraction maxima (due to constructive interference) occur if the incident direction of the wave, measured from the surfaces of these planes, and the wavelength λ of the radiation satisfy **Bragg's law:**

$$2d \sin \theta = m\lambda, \quad \text{for } m = 1, 2, 3, \dots \quad \text{(Bragg's law)}. \quad (36\text{-}34)$$

QUESTIONS

1 For three experiments, Fig. 36-33 gives the parameter β of Eq. 36-20 versus angle θ for two-slit interference using light of wavelength 500 nm. The slit separations in the three experiments differ. Rank the experiments according to (a) the slit separations and (b) the total number of two-slit interference maxima in the pattern, greatest first.

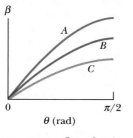

FIG. 36-33   Question 1.

2 For three experiments, Fig. 36-34 gives α versus angle θ in one-slit diffraction using light of wavelength 500 nm. Rank the experiments according to (a) the slit widths and (b) the total number of diffraction minima in the pattern, greatest first.

FIG. 36-34   Question 2.

3 Figure 36-35 shows four choices for the rectangular opening of a source of either sound waves or light waves. The sides have lengths of either L or $2L$, with L being 3.0 times the wavelength of the waves. Rank the openings according to the extent of (a) left–right spreading and (b) up–down spreading of the waves due to diffraction, greatest first.

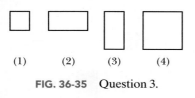

FIG. 36-35   Question 3.

4 Light of frequency f illuminating a long narrow slit produces a diffraction pattern. (a) If we switch to light of frequency 1.3f, does the pattern expand away from the center or contract toward the center? (b) Does the pattern expand or contract if, instead, we submerge the equipment in clear corn syrup?

5 You are conducting a single-slit diffraction experiment with light of wavelength λ. What appears, on a distant viewing screen, at a point at which the top and bottom rays through the slit have a path length difference equal to (a) 5λ and (b) 4.5λ?

6 In a single-slit diffraction experiment, the top and bottom rays through the slit arrive at a certain point on the viewing screen with a path length difference of 4.0 wavelengths. In a phasor representation like those in Fig 36-7, how many overlapping circles does the chain of phasors make?

7 Figure 36-36 shows a red line and a green line of the same order in the pattern produced by a diffraction grating. If we increased the number of rulings in the grating—say, by removing tape that had covered the outer half of the rulings— would (a) the half-widths of the lines and (b) the separation of the lines increase, decrease, or remain the same? (c) Would the lines shift to the right, shift to the left, or remain in place?

FIG. 36-36   Questions 7 and 8.

8 For the situation of Question 7 and Fig. 36-36, if instead we increased the grating spacing, would (a) the half-widths of the lines and (b) the separation of the lines increase, decrease, or remain the same? (c) Would the lines shift to the right, shift to the left, or remain in place?

9 At night many people see rings (called *entoptic halos*) surrounding bright outdoor lamps in otherwise dark surroundings. The rings are the first of the side maxima in diffraction patterns produced by structures that are thought to be within the cornea (or possibly the lens) of the observer's eye. (The central maxima of such patterns overlap the lamp.) (a) Would a particular ring become smaller or larger if the lamp were switched from blue to red light? (b) If a lamp emits white light, is blue or red on the outside edge of the ring?

10 (a) For a given diffraction grating, does the smallest difference $\Delta\lambda$ in two wavelengths that can be resolved increase, decrease, or remain the same as the wavelength increases? (b) For a given wavelength region (say, around 500 nm), is $\Delta\lambda$ greater in the first order or in the third order?

11 (a) Figure 36-37a shows the lines produced by diffraction gratings A and B using light of the same wavelength; the lines

(a)   (b)

FIG. 36-37 Question 11.

are of the same order and appear at the same angles θ. Which grating has the greater number of rulings? (b) Figure 36-37b shows lines of two orders produced by a single diffraction grating using light of two wavelengths, both in the red region of the spectrum. Which lines, the left pair or right pair, are in the order with greater m? Is the center of the diffraction pattern located to the left or to the right in (c) Fig. 36-37a and (d) Fig. 36-37b?

12 Figure 36-38 shows the bright fringes that lie within the central diffraction envelope in two double-slit diffraction experiments using the same wavelength of light. Are (a) the slit width a, (b) the slit separation d, and (c) the ratio d/a in experiment B greater than, less than, or the same as those quantities in experiment A?

FIG. 36-38 Question 12.

PROBLEMS

sec. 36-3 Diffraction by a Single Slit: Locating the Minima

•1 A single slit is illuminated by light of wavelengths λ_a and λ_b, chosen so that the first diffraction minimum of the λ_a component coincides with the second minimum of the λ_b component. (a) If $\lambda_b = 350$ nm, what is λ_a? For what order number m_b (if any) does a minimum of the λ_b component coincide with the minimum of the λ_a component in the order number (b) $m_a = 2$ and (c) $m_a = 3$?

•2 Monochromatic light of wavelength 441 nm is incident on a narrow slit. On a screen 2.00 m away, the distance between the second diffraction minimum and the central maximum is 1.50 cm. (a) Calculate the angle of diffraction θ of the second minimum. (b) Find the width of the slit.

•3 Light of wavelength 633 nm is incident on a narrow slit. The angle between the first diffraction minimum on one side of the central maximum and the first minimum on the other side is 1.20°. What is the width of the slit?

•4 What must be the ratio of the slit width to the wavelength for a single slit to have the first diffraction minimum at $\theta = 45.0°$?

•5 A plane wave of wavelength 590 nm is incident on a slit with a width of $a = 0.40$ mm. A thin converging lens of focal length +70 cm is placed between the slit and a viewing screen and focuses the light on the screen. (a) How far is the screen from the lens? (b) What is the distance on the screen from the center of the diffraction pattern to the first minimum?

•6 In conventional television, signals are broadcast from towers to home receivers. Even when a receiver is not in direct view of a tower because of a hill or building, it can still intercept a signal if the signal diffracts enough around the obstacle, into the obstacle's "shadow region." Previously, television signals had a wavelength of about 50 cm, but digital

television signals that are transmitted from towers have a wavelength of about 10 mm. (a) Did this change in wavelength increase or decrease the diffraction of the signals into the shadow regions of obstacles? Assume that a signal passes through an opening of 5.0 m width between two adjacent buildings. What is the angular spread of the central diffraction maximum (out to the first minima) for wavelengths of (b) 50 cm and (c) 10 mm?

•7 The distance between the first and fifth minima of a single-slit diffraction pattern is 0.35 mm with the screen 40 cm away from the slit, when light of wavelength 550 nm is used. (a) Find the slit width. (b) Calculate the angle θ of the first diffraction minimum. **GO**

••8 Manufacturers of wire (and other objects of small dimension) sometimes use a laser to continually monitor the thickness of the product. The wire intercepts the laser beam, producing a diffraction pattern like that of a single slit of the same width as the wire diameter (Fig. 36-39). Suppose a helium–neon laser, of wavelength 632.8 nm, illuminates a

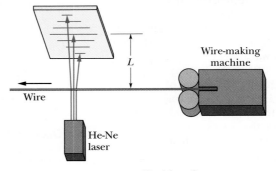

FIG. 36-39 Problem 8.

wire, and the diffraction pattern appears on a screen at distance $L = 2.60$ m. If the desired wire diameter is 1.37 mm, what is the observed distance between the two tenth-order minima (one on each side of the central maximum)?

••9 A slit 1.00 mm wide is illuminated by light of wavelength 589 nm. We see a diffraction pattern on a screen 3.00 m away. What is the distance between the first two diffraction minima on the same side of the central diffraction maximum? **SSM ILW**

••10 Sound waves with frequency 3000 Hz and speed 343 m/s diffract through the rectangular opening of a speaker cabinet and into a large auditorium of length $d =$ 100 m. The opening, which has a horizontal width of 30.0 cm, faces a wall 100 m away (Fig. 36-40). Along that wall, how far from the central axis will a listener be at the first diffraction minimum and thus have difficulty hearing the sound? (Neglect reflections.)

FIG. 36-40 Problem 10.

sec. 36-5 Intensity in Single-Slit Diffraction, Quantitatively

•11 Monochromatic light with wavelength 538 nm is incident on a slit with width 0.025 mm. The distance from the slit to a screen is 3.5 m. Consider a point on the screen 1.1 cm from the central maximum. Calculate (a) θ for that point, (b) α, and (c) the ratio of the intensity at that point to the intensity at the central maximum.

•12 In the single-slit diffraction experiment of Fig. 36-4, let the wavelength of the light be 500 nm, the slit width be 6.00 μm, and the viewing screen be at distance $D = 3.00$ m. Let a y axis extend upward along the viewing screen, with its origin at the center of the diffraction pattern. Also let I_P represent the intensity of the diffracted light at point P at $y = 15.0$ cm. (a) What is the ratio of I_P to the intensity I_m at the center of the pattern? (b) Determine where point P is in the diffraction pattern by giving the maximum and minimum between which it lies, or the two minima between which it lies.

•13 A 0.10-mm-wide slit is illuminated by light of wavelength 589 nm. Consider a point P on a viewing screen on which the diffraction pattern of the slit is viewed; the point is at 30° from the central axis of the slit. What is the phase difference between the Huygens wavelets arriving at point P from the top and midpoint of the slit? (*Hint:* See Eq. 36-4.)

•14 Figure 36-41 gives α versus the sine of the angle θ in a single-slit diffraction experiment using light of wavelength 610 nm. The vertical axis scale is set by $\alpha_s = 12$ rad. What are (a) the slit width, (b) the total number of diffraction minima in the pattern (count them on both sides of the center of the diffraction pattern), (c) the least angle for a minimum, and (d) the greatest angle for a minimum?

FIG. 36-41 Problem 14.

••15 (a) Show that the values of α at which intensity maxima for single-slit diffraction occur can be found exactly by differentiating Eq. 36-5 with respect to α and equating the result to zero, obtaining the condition $\tan \alpha = \alpha$. To find values

of α satisfying this relation, plot the curve $y = \tan \alpha$ and the straight line $y = \alpha$ and then find their intersections, or use a calculator to find an appropriate value of α by trial and error. Next, from $\alpha = (m + \frac{1}{2})\pi$, determine the values of m associated with the maxima in the single-slit pattern. (These m values are *not* integers because secondary maxima do not lie exactly halfway between minima.) What are the (b) smallest α and (c) associated m, the (d) second smallest α and (e) associated m, and the (f) third smallest α and (g) associated m?

••16 *Babinet's principle.* A monochromatic beam of parallel light is incident on a "collimating" hole of diameter $x \gg \lambda$. Point P lies in the geometrical shadow region on a *distant* screen (Fig. 36-42a). Two diffracting objects, shown in Fig. 36-42b, are placed in turn over the collimating hole. Object A is an opaque circle with a hole in it, and B is the "photographic negative" of A. Using superposition concepts, show that the intensity at P is identical for the two diffracting objects A and B.

FIG. 36-42 Problem 16.

••17 The full width at half-maximum (FWHM) of a central diffraction maximum is defined as the angle between the two points in the pattern where the intensity is one-half that at the center of the pattern. (See Fig. 36-8b.) (a) Show that the intensity drops to one-half the maximum value when $\sin^2 \alpha = \alpha^2/2$. (b) Verify that $\alpha = 1.39$ rad (about 80°) is a solution to the transcendental equation of (a). (c) Show that the FWHM is $\Delta\theta = 2 \sin^{-1}(0.443\lambda/a)$, where a is the slit width. Calculate the FWHM of the central maximum for slit width (d) 1.00λ, (e) 5.00λ, and (f) 10.0λ. **SSM WWW**

sec. 36-6 Diffraction by a Circular Aperture

•18 The telescopes on some commercial surveillance satellites can resolve objects on the ground as small as 85 cm across (see Google Earth), and the telescopes on military surveillance satellites reportedly can resolve objects as small as 10 cm across. Assume first that object resolution is determined entirely by Rayleigh's criterion and is not degraded by turbulence in the atmosphere. Also assume that the satellites are at a typical altitude of 400 km and that the wavelength of visible light is 550 nm. What would be the required diameter of the telescope aperture for (a) 85 cm resolution and (b) 10 cm resolution? (c) Now, considering that turbulence is certain to degrade resolution and that the aperture diameter of the Hubble Space Telescope is 2.4 m, what can you say about the answer to (b) and about how the military surveillance resolutions are accomplished?

•19 If Superman really had x-ray vision at 0.10 nm wavelength and a 4.0 mm pupil diameter, at what maximum altitude could he distinguish villains from heroes, assuming that he needs to resolve points separated by 5.0 cm to do this?

•20 Assume that Rayleigh's criterion gives the limit of resolution of an astronaut's eye looking down on Earth's surface from a typical space shuttle altitude of 400 km. (a) Under that idealized assumption, estimate the smallest linear width on Earth's surface that the astronaut can resolve. Take the astronaut's pupil diameter to be 5 mm and the wavelength of visible light to be 550 nm. (b) Can the astronaut resolve the Great Wall of China (Fig. 36-43), which is more than 3000 km long, 5 to 10 m thick at its base, 4 m thick at its top, and 8 m in height? (c) Would the astronaut be able to resolve any unmistakable sign of intelligent life on Earth's surface? ✈

FIG. 36-43 Problem 20. The Great Wall of China. *(AP/Wide World Photos)*

•21 The two headlights of an approaching automobile are 1.4 m apart. At what (a) angular separation and (b) maximum distance will the eye resolve them? Assume that the pupil diameter is 5.0 mm, and use a wavelength of 550 nm for the light. Also assume that diffraction effects alone limit the resolution so that Rayleigh's criterion can be applied. SSM

•22 *Entoptic halos.* If someone looks at a bright outdoor lamp in otherwise dark surroundings, the lamp appears to be surrounded by bright and dark rings (hence *halos*) that are actually a circular diffraction pattern as in Fig. 36-9, with the central maximum overlapping the direct light from the lamp. The diffraction is produced by structures within the cornea or lens of the eye (hence *entoptic*). If the lamp is monochromatic at wavelength 550 nm and the first dark ring subtends angular diameter 2.5° in the observer's view, what is the (linear) diameter of the structure producing the diffraction? ✈

•23 Find the separation of two points on the Moon's surface that can just be resolved by the 200 in. (= 5.1 m) telescope at Mount Palomar, assuming that this separation is determined by diffraction effects. The distance from Earth to the Moon is 3.8×10^5 km. Assume a wavelength of 550 nm for the light. ILW

•24 The radar system of a navy cruiser transmits at a wavelength of 1.6 cm, from a circular antenna with a diameter of 2.3 m. At a range of 6.2 km, what is the smallest distance that two speedboats can be from each other and still be resolved as two separate objects by the radar system?

•25 Estimate the linear separation of two objects on Mars that can just be resolved under ideal conditions by an observer on Earth (a) using the naked eye and (b) using the 200 in. (= 5.1 m) Mount Palomar telescope. Use the following data: distance to Mars = 8.0×10^7 km, diameter of pupil = 5.0 mm, wavelength of light = 550 nm. SSM WWW

•26 The wall of a large room is covered with acoustic tile in which small holes are drilled 5.0 mm from center to center. How far can a person be from such a tile and still distinguish the individual holes, assuming ideal conditions, the pupil diameter of the observer's eye to be 4.0 mm, and the wavelength of the room light to be 550 nm?

•27 (a) How far from grains of red sand must you be to position yourself just at the limit of resolving the grains if your pupil diameter is 1.5 mm, the grains are spherical with radius 50 μm, and the light from the grains has wavelength 650 nm? (b) If the grains were blue and the light from them had wavelength 400 nm, would the answer to (a) be larger or smaller?

••28 *Floaters.* The floaters you see when viewing a bright, featureless background are diffraction patterns of defects in the vitreous humor that fills most of your eye. Sighting through a pinhole sharpens the diffraction pattern. If you also view a small circular dot, you can approximate the defect's size. Assume that the defect diffracts light as a circular aperture does. Adjust the dot's distance L from your eye (or eye lens) until the dot and the circle of the first minimum in the diffraction pattern appear to have the same size in your view. That is, until they have the same diameter D' on the retina at distance $L' = 2.0$ cm from the front of the eye, as suggested in Fig. 36-44a, where the angles on the two sides of the eye lens are equal. Assume that the wavelength of visible light is $\lambda = 550$ nm. If the dot has diameter $D = 2.0$ mm and is distance $L = 45.0$ cm from the eye and the defect is $x = 6.0$ mm in front of the retina (Fig. 36-44b), what is the diameter of the defect? ✈

FIG. 36-44 Problem 28.

••29 Millimeter-wave radar generates a narrower beam than conventional microwave radar, making it less vulnerable to antiradar missiles than conventional radar. (a) Calculate the angular width 2θ of the central maximum, from first minimum to first minimum, produced by a 220 GHz radar beam emitted by a 55.0-cm-diameter circular antenna. (The frequency is chosen to coincide with a low-absorption atmospheric "window.") (b) What is 2θ for a more conventional circular antenna that has a diameter of 2.3 m and emits at wavelength 1.6 cm? SSM

••30 (a) A circular diaphragm 60 cm in diameter oscillates at a frequency of 25 kHz as an underwater source of sound used for submarine detection. Far from the source, the sound intensity is distributed as the diffraction pattern of a circular hole whose diameter equals that of the diaphragm. Take the speed of sound in water to be 1450 m/s and find the angle between the normal to the diaphragm and a line from the diaphragm to the first minimum. (b) Is there such a minimum for a source having an (audible) frequency of 1.0 kHz?

••**31** Nuclear-pumped x-ray lasers are seen as a possible weapon to destroy ICBM booster rockets at ranges up to 2000 km. One limitation on such a device is the spreading of the beam due to diffraction, with resulting dilution of beam intensity. Consider such a laser operating at a wavelength of 1.40 nm. The element that emits light is the end of a wire with diameter 0.200 mm. (a) Calculate the diameter of the central beam at a target 2000 km away from the beam source. (b) By what factor is the beam intensity reduced in transit to the target? (The laser is fired from space, so that atmospheric absorption can be ignored.) **GO**

••**32** The wings of tiger beetles (Fig. 36-45) are colored by interference due to thin cuticle-like layers. In addition, these layers are arranged in patches that are 60 μm across and produce different colors. The color you see is a pointillistic mixture of thin-film interference colors that varies with perspective. Approximately what viewing distance from a wing puts you at the limit of resolving the different colored patches according to Rayleigh's criterion? Use 550 nm as the wavelength of light and 3.00 mm as the diameter of your pupil.

FIG. 36-45 Problem 32. Tiger beetles are colored by pointillistic mixtures of thin-film interference colors. *(Kjell B. Sandved/Bruce Coleman, Inc.)*

••**33** (a) What is the angular separation of two stars if their images are barely resolved by the Thaw refracting telescope at the Allegheny Observatory in Pittsburgh? The lens diameter is 76 cm and its focal length is 14 m. Assume λ = 550 nm. (b) Find the distance between these barely resolved stars if each of them is 10 light-years distant from Earth. (c) For the image of a single star in this telescope, find the diameter of the first dark ring in the diffraction pattern, as measured on a photographic plate placed at the focal plane of the telescope lens. Assume that the structure of the image is associated entirely with diffraction at the lens aperture and not with lens "errors."

•••**34** A circular obstacle produces the same diffraction pattern as a circular hole of the same diameter (except very near θ = 0). Airborne water drops are examples of such obstacles. When you see the Moon through suspended water drops, such as in a fog, you intercept the diffraction pattern from many drops. The composite of the central diffraction maxima

of those drops forms a white region that surrounds the Moon and may obscure it. Figure 36-46 is a photograph in which the Moon is obscured. There are two faint, colored rings around the Moon (the larger one may be too faint to be seen in your copy of the photograph). The smaller ring is on the outer edge of the central maxima from the drops; the somewhat larger ring is on the outer edge of the smallest of the secondary maxima from the drops (see Fig. 36-10). The color is visible because the rings are adjacent to the diffraction minima (dark rings) in the patterns. (Colors in other parts of the pattern overlap too much to be visible.)

(a) What is the color of these rings on the outer edges of the diffraction maxima? (b) The colored ring around the central maxima in Fig. 36-46 has an angular diameter that is 1.35 times the angular diameter of the Moon, which is 0.50°. Assume that the drops all have about the same diameter. Approximately what is that diameter? **GO**

FIG. 36-46 Problem 34. The corona around the Moon is a composite of the diffraction patterns of airborne water drops. *(Pekka Parvianen/Photo Researchers)*

sec. 36-7 Diffraction by a Double Slit

•**35** In a double-slit experiment, the slit separation d is 2.00 times the slit width w. How many bright interference fringes are in the central diffraction envelope?

•**36** A beam of light of a single wavelength is incident perpendicularly on a double-slit arrangement, as in Fig. 35-10. The slit widths are each 46 μm and the slit separation is 0.30 mm. How many complete bright fringes appear between the two first-order minima of the diffraction pattern?

•**37** Suppose that the central diffraction envelope of a double-slit diffraction pattern contains 11 bright fringes and the first diffraction minima eliminate (are coincident with) bright fringes. How many bright fringes lie between the first and second minima of the diffraction envelope?

••**38** Two slits of width a and separation d are illuminated by a coherent beam of light of wavelength λ. What is the linear separation of the bright interference fringes observed on a screen that is at a distance D away?

••**39** (a) How many bright fringes appear between the

first diffraction-envelope minima to either side of the central maximum in a double-slit pattern if $\lambda = 550$ nm, $d = 0.150$ mm, and $a = 30.0$ μm? (b) What is the ratio of the intensity of the third bright fringe to the intensity of the central fringe? **SSM WWW**

••40 (a) In a double-slit experiment, what ratio of d to a causes diffraction to eliminate the fourth bright side fringe? (b) What other bright fringes are also eliminated?

••41 Light of wavelength 440 nm passes through a double slit, yielding a diffraction pattern whose graph of intensity I versus angular position θ is shown in Fig. 36-47. Calculate (a) the slit width and (b) the slit separation. (c) Verify the displayed intensities of the $m = 1$ and $m = 2$ interference fringes.

FIG. 36-47 Problem 41.

••42 Figure 36-48 gives the parameter β of Eq. 36-20 versus the sine of the angle θ in a two-slit interference experiment using light of wavelength 435 nm. The vertical axis scale is set by $\beta_s = 80.0$ rad. What are (a) the slit separation, (b) the total number of interference maxima (count them on both sides of the pattern's center), (c) the smallest angle for a maxima, and (d) the greatest angle for a minimum? Assume that none of the interference maxima are completely eliminated by a diffraction minimum.

FIG. 36-48 Problem 42.

••43 In the two-slit interference experiment of Fig. 35-10, the slit widths are each 12.0 μm, their separation is 24.0 μm, the wavelength is 600 nm, and the viewing screen is at a distance of 4.00 m. Let I_P represent the intensity at point P on the screen, at height $y = 70.0$ cm. (a) What is the ratio of I_P to the intensity I_m at the center of the pattern? (b) Determine where P is in the two-slit interference pattern by giving the maximum or minimum on which it lies or the maximum and minimum between which it lies. (c) Next, for the diffraction that occurs, determine where point P is in the diffraction pattern by giving the minimum on which it lies or the two minima between which it lies.

sec. 36-8 Diffraction Gratings

•44 Visible light is incident perpendicularly on a grating with 315 rulings/mm. What is the longest wavelength that can be seen in the fifth-order diffraction?

•45 A grating has 400 lines/mm. How many orders of the entire visible spectrum (400–700 nm) can it produce in a diffraction experiment, in addition to the $m = 0$ order? **SSM ILW**

•46 Perhaps to confuse a predator, some tropical gyrinid beetles (whirligig beetles) are colored by optical interference that is due to scales whose alignment forms a diffraction grating (which scatters light instead of transmiting it). When the incident light rays are perpendicular to the grating, the angle between the first-order maxima (on opposite sides of the zeroth-order maximum) is about 26° in light with a wavelength of 550 nm. What is the grating spacing of the beetle?

•47 A diffraction grating 20.0 mm wide has 6000 rulings. Light of wavelength 589 nm is incident perpendicularly on the grating. What are the (a) largest, (b) second largest, and (c) third largest values of θ at which maxima appear on a distant viewing screen?

•48 In a certain two-slit interference pattern, 10 bright fringes lie within the second side peak of the diffraction envelope and diffraction minima coincide with two-slit interference maxima. What is the ratio of the slit separation to the slit width?

••49 A diffraction grating having 180 lines/mm is illuminated with a light signal containing only two wavelengths, $\lambda_1 = 400$ nm and $\lambda_2 = 500$ nm. The signal is incident perpendicularly on the grating. (a) What is the angular separation between the second-order maxima of these two wavelengths? (b) What is the smallest angle at which two of the resulting maxima are superimposed? (c) What is the highest order for which maxima for both wavelengths are present in the diffraction pattern?

••50 A diffraction grating is made up of slits of width 300 nm with separation 900 nm. The grating is illuminated by monochromatic plane waves of wavelength $\lambda = 600$ nm at normal incidence. (a) How many maxima are there in the full diffraction pattern? (b) What is the angular width of a spectral line observed in the first order if the grating has 1000 slits?

••51 Light of wavelength 600 nm is incident normally on a diffraction grating. Two adjacent maxima occur at angles given by $\sin \theta = 0.2$ and $\sin \theta = 0.3$. The fourth-order maxima are missing. (a) What is the separation between adjacent slits? (b) What is the smallest slit width this grating can have? For that slit width, what are the (c) largest, (d) second largest, and (e) third largest values of the order number m of the maxima produced by the grating? **SSM WWW**

••52 With light from a gaseous discharge tube incident normally on a grating with slit separation 1.73 μm, sharp maxima of green light are experimentally found at angles $\theta = \pm17.6°$, $37.3°$, $-37.1°$, $65.2°$, and $-65.0°$. Compute the wavelength of the green light that best fits these data.

••53 Assume that the limits of the visible spectrum are arbitrarily chosen as 430 and 680 nm. Calculate the number of rulings per millimeter of a grating that will spread the first-order spectrum through an angle of 20.0°.

••54 A beam of light consisting of wavelengths from 460.0 nm to 640.0 nm is directed perpendicularly onto a diffraction grating with 160 lines/mm. (a) What is the lowest order that is overlapped by another order? (b) What is the highest order for which the complete wavelength range of the beam is present? In that highest order, at what angle does the light at wavelength (c) 460.0 nm and (d) 640.0 nm appear? (e) What is the greatest angle at which the light at wavelength 460.0 nm appears? **GO**

••55 A grating has 350 rulings/mm and is illuminated at normal incidence by white light. A spectrum is formed on a screen 30.0 cm from the grating. If a hole 10.0 mm square is cut in the screen, its inner edge being 50.0 mm from the central maximum and parallel to it, what are the (a) shortest and (b) longest wavelengths of the light that passes through the hole?

•••56 Derive this expression for the intensity pattern for a three-slit "grating":

$$I = \tfrac{1}{9} I_m (1 + 4 \cos \phi + 4 \cos^2 \phi),$$

where $\phi = (2\pi d \sin \theta)/\lambda$ and $a \ll \lambda$.

sec. 36-9 Gratings: Dispersion and Resolving Power

•57 A diffraction grating with a width of 2.0 cm contains 1000 lines/cm across that width. For an incident wavelength of 600 nm, what is the smallest wavelength difference this grating can resolve in the second order?

•58 The D line in the spectrum of sodium is a doublet with wavelengths 589.0 and 589.6 nm. Calculate the minimum number of lines needed in a grating that will resolve this doublet in the second-order spectrum. See Sample Problem 36-6.

•59 Light at wavelength 589 nm from a sodium lamp is incident perpendicularly on a grating with 40 000 rulings over width 76 nm. What are the first-order (a) dispersion D and (b) resolving power R, the second-order (c) D and (d) R, and the third-order (e) D and (f) R?

•60 (a) How many rulings must a 4.00-cm-wide diffraction grating have to resolve the wavelengths 415.496 and 415.487 nm in the second order? (b) At what angle are the second-order maxima found?

•61 A source containing a mixture of hydrogen and deuterium atoms emits red light at two wavelengths whose mean is 656.3 nm and whose separation is 0.180 nm. Find the minimum number of lines needed in a diffraction grating that can resolve these lines in the first order. **SSM ILW**

•62 A grating has 600 rulings/mm and is 5.0 mm wide. (a) What is the smallest wavelength interval it can resolve in the third order at $\lambda = 500$ nm? (b) How many higher orders of maxima can be seen?

•63 With a particular grating the sodium doublet (see Sample Problem 36-6) is viewed in the third order at 10° to the normal and is barely resolved. Find (a) the grating spacing and (b) the total width of the rulings.

••64 A diffraction grating illuminated by monochromatic light normal to the grating produces a certain line at angle θ. (a) What is the product of that line's half-width and the grating's resolving power? (b) Evaluate that product for the first order of a grating of slit separation 900 nm in light of wavelength 600 nm.

sec. 36-10 Diffraction by Organized Layers

•65 Figure 36-49 is a graph of intensity versus angular position θ for the diffraction of an x-ray beam by a crystal. The horizontal scale is set by $\theta_s = 2.00°$. The beam consists of two

wavelengths, and the spacing between the reflecting planes is 0.94 nm. What are the (a) shorter and (b) longer wavelengths in the beam?

FIG. 36-49 Problem 65.

•66 If first-order reflection occurs in a crystal at Bragg angle 3.4°, at what Bragg angle does second-order reflection occur from the same family of reflecting planes?

•67 X rays of wavelength 0.12 nm are found to undergo second-order reflection at a Bragg angle of 28° from a lithium fluoride crystal. What is the interplanar spacing of the reflecting planes in the crystal?

•68 What is the smallest Bragg angle for x rays of wavelength 30 pm to reflect from reflecting planes spaced 0.30 nm apart in a calcite crystal?

•69 An x-ray beam of wavelength A undergoes first-order reflection from a crystal when its angle of incidence to a crystal face is 23°, and an x-ray beam of wavelength 97 pm undergoes third-order reflection when its angle of incidence to that face is 60°. Assuming that the two beams reflect from the same family of reflecting planes, find (a) the interplanar spacing and (b) the wavelength A.

•70 An x-ray beam of a certain wavelength is incident on a NaCl crystal, at 30.0° to a certain family of reflecting planes of spacing 39.8 pm. If the reflection from those planes is of the first order, what is the wavelength of the x rays?

••71 Consider a two-dimensional square crystal structure, such as one side of the structure shown in Fig. 36-30a. The largest interplanar spacing of reflecting planes is the unit cell size a_0. Calculate and sketch the (a) second largest, (b) third largest, (c) fourth largest, (d) fifth largest, and (e) sixth largest interplanar spacing. (f) Show that your results in (a) through (e) are consistent with the general formula

$$d = \frac{a_0}{\sqrt{h^2 + k^2}},$$

where h and k are relatively prime integers (they have no common factor other than unity).

••72 In Fig. 36-50, first-order reflection from the reflection planes shown occurs when an x-ray beam of wavelength 0.260 nm makes an angle $\theta = 63.8°$ with the top face of the crystal. What is the unit cell size a_0? **GO**

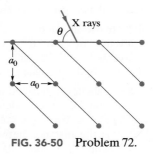

FIG. 36-50 Problem 72.

••73 In Fig. 36-51, let a beam of x rays of wavelength 0.125 nm be incident on an NaCl crystal at angle $\theta = 45.0°$ to the top face of the crystal and a family of reflecting planes. Let the reflecting planes have separation $d = 0.252$ nm. The crystal is turned through angle ϕ around an axis perpendicular to the plane of the page until these reflecting planes give diffraction maxima. What are the (a) smaller and (b) larger value of ϕ if the crystal is turned clockwise and the (c) smaller and (d) larger value of ϕ if it is turned counterclockwise? **SSM**

FIG. 36-51 Problems 73 and 74.

••74 In Fig. 36-51, an x-ray beam of wavelengths from 95.0 to 140 pm is incident at $\theta = 45.0°$ to a family of reflecting planes with spacing $d = 275$ pm. What are the (a) longest wavelength λ and (b) associated order number m and the (c) shortest λ and (d) associated m of the intensity maxima in the diffraction of the beam?

Additional Problems

75 In June 1985, a laser beam was sent out from the Air Force Optical Station on Maui, Hawaii, and reflected back from the shuttle *Discovery* as it sped by 354 km overhead. The diameter of the central maximum of the beam at the shuttle position was said to be 9.1 m, and the beam wavelength was 500 nm. What is the effective diameter of the laser aperture at the Maui ground station? (*Hint:* A laser beam spreads only because of diffraction; assume a circular exit aperture.)

76 An astronaut in a space shuttle claims she can just barely resolve two point sources on Earth's surface, 160 km below. Calculate their (a) angular and (b) linear separation, assuming ideal conditions. Take $\lambda = 540$ nm and the pupil diameter of the astronaut's eye to be 5.0 mm.

77 Visible light is incident perpendicularly on a diffraction grating of 200 rulings/mm. What are the (a) longest, (b) second longest, and (c) third longest wavelengths that can be associated with an intensity maximum at $\theta = 30.0°$? **SSM**

78 When monochromatic light is incident on a slit 22.0 μm wide, the first diffraction minimum lies at 1.80° from the direction of the incident light. What is the wavelength?

79 In a single-slit diffraction experiment, there is a minimum of intensity for orange light ($\lambda = 600$ nm) and a minimum of intensity for blue-green light ($\lambda = 500$ nm) at the same angle of 1.00 mrad. For what minimum slit width is this possible? **SSM**

80 In a two-slit interference pattern, what is the ratio of slit separation to slit width if there are 17 bright fringes within the central diffraction envelope and the diffraction minima coincide with two-slit interference maxima?

81 In two-slit interference, if the slit separation is 14 μm and the slit widths are each 2.0 μm, (a) how many two-slit maxima are in the central peak of the diffraction envelope and (b) how many are in either of the first side peak of the diffraction envelope? **SSM**

82 A single-slit diffraction experiment is set up with light of wavelength 420 nm, incident perpendicularly on a slit of width 5.10 μm. The viewing screen is 3.20 m distant. On the screen, what is the distance between the center of the diffraction pattern and the second diffraction minimum?

83 A beam of light with a narrow wavelength range centered on 450 nm is incident perpendicularly on a diffraction grating with a width of 1.80 cm and a line density of 1400 lines/cm across that width. For this light, what is the smallest wavelength difference this grating can resolve in the third order?

84 If you look at something 40 m from you, what is the smallest length (perpendicular to your line of sight) that you can resolve, according to Rayleigh's criterion? Assume the pupil of your eye has a diameter of 4.00 mm, and use 500 nm as the wavelength of the light reaching you.

85 Two yellow flowers are separated by 60 cm along a line perpendicular to your line of sight to the flowers. How far are you from the flowers when they are at the limit of resolution according to the Rayleigh criterion? Assume the light from the flowers has a single wavelength of 550 nm and that your pupil has a diameter of 5.5 mm.

86 In a single-slit diffraction experiment, what must be the ratio of the slit width to the wavelength if the second diffraction minima are to occur at an angle of 37.0° from the center of the diffraction pattern on a viewing screen?

87 A diffraction grating 3.00 cm wide produces the second order at 33.0° with light of wavelength 600 nm. What is the total number of lines on the grating?

88 A beam of light consists of two wavelengths, 590.159 nm and 590.220 nm, that are to be resolved with a diffraction grating. If the grating has lines across a width of 3.80 cm, what is the minimum number of lines required for the two wavelengths to be resolved in the second order?

89 A spy satellite orbiting at 160 km above Earth's surface has a lens with a focal length of 3.6 m and can resolve objects on the ground as small as 30 cm. For example, it can easily measure the size of an aircraft's air intake port. What is the effective diameter of the lens as determined by diffraction consideration alone? Assume $\lambda = 550$ nm.

90 The pupil of a person's eye has a diameter of 5.00 mm. According to Rayleigh's criterion, what distance apart must two small objects be if their images are just barely resolved when they are 250 mm from the eye? Assume they are illuminated with light of wavelength 500 nm.

91 Light is incident on a grating at an angle ψ as shown in Fig. 36-52. Show that bright fringes occur at angles θ that satisfy the equation

FIG. 36-52 Problem 91.

$$d(\sin \psi + \sin \theta) = m\lambda, \quad \text{for } m = 0, 1, 2, \dots.$$

(Compare this equation with Eq. 36-25.) Only the special case $\psi = 0$ has been treated in this chapter.

92 A grating with $d = 1.50$ μm is illuminated at various angles of incidence by light of wavelength 600 nm. Plot, as a function of the angle of incidence (0 to 90°), the angular deviation of the first-order maximum from the incident direction. (See Problem 91.)

93 If you double the width of a single slit, the intensity of the central maximum of the diffraction pattern increases by a factor of 4, even though the energy passing through the slit only doubles. Explain this quantitatively. **SSM**

94 In an experiment to monitor the Moon's surface with a light beam, pulsed radiation from a ruby laser ($\lambda = 0.69 \ \mu$m) was directed to the Moon through a reflecting telescope with a mirror radius of 1.3 m. A reflector on the Moon behaved like a circular flat mirror with radius 10 cm, reflecting the light directly back toward the telescope on Earth. The reflected light was then detected after being brought to a focus by this telescope. Approximately what fraction of the original light energy was picked up by the detector? Assume that for each direction of travel all the energy is in the central diffraction peak.

95 A diffraction grating has resolving power $R = \lambda_{avg}/\Delta\lambda = Nm$. (a) Show that the corresponding frequency range Δf that can just be resolved is given by $\Delta f = c/Nm\lambda$. (b) From Fig. 36-22, show that the times required for light to travel along the ray at the bottom of the figure and the ray at the top differ by $\Delta t = (Nd/c) \sin\theta$. (c) Show that $(\Delta f)(\Delta t) = 1$, this relation being independent of the various grating parameters. Assume $N \gg 1$. **SSM**

96 A double-slit system with individual slit widths of 0.030 mm and a slit separation of 0.18 mm is illuminated with 500 nm light directed perpendicular to the plane of the slits. What is the total number of complete bright fringes appearing between the two first-order minima of the diffraction pattern? (Do not count the fringes that coincide with the minima of the diffraction pattern.) **GO**

97 A diffraction grating has 8900 slits across 1.20 cm. If light with a wavelength of 500 nm is sent through it, how many orders (maxima) lie to one side of the central maximum?

98 A diffraction grating 1.00 cm wide has 10 000 parallel slits. Monochromatic light that is incident normally is diffracted through 30° in the first order. What is the wavelength of the light?

99 A diffraction grating has 200 lines/mm. Light consisting of a continuous range of wavelengths between 550 nm and 700 nm is incident perpendicularly on the grating. (a) What is the lowest order that is overlapped by another order? (b) What is the highest order for which the complete spectrum is present?

100 Suppose that two points are separated by 2.0 cm. If they are viewed by an eye with a pupil opening of 5.0 mm, what distance from the viewer puts them at the Rayleigh limit of resolution? Assume a light wavelength of 500 nm.

101 Show that the dispersion of a grating is $D = (\tan \theta)/\lambda$. **SSM**

102 Monochromatic light (wavelength = 450 nm) is incident perpendicularly on a single slit (width = 0.40 mm). A screen is placed parallel to the slit plane, and on it the distance between the two minima on either side of the central maximum is 1.8 mm. (a) What is the distance from the slit to the screen? (*Hint:* The angle to either minimum is small enough that $\sin \theta \approx \tan \theta$.) (b) What is the distance on the screen between the first minimum and the third minimum on the same side of the central maximum?

103 Light containing a mixture of two wavelengths, 500 and 600 nm, is incident normally on a diffraction grating. It is desired (1) that the first and second maxima for each wavelength appear at $\theta \leq 30°$, (2) that the dispersion be as high as possible, and (3) that the third order for the 600 nm light be a missing order. (a) What should be the slit separation?

(b) What is the smallest individual slit width that can be used? (c) For the values calculated in (a) and (b) and the light of wavelength 600 nm, what is the largest order of maxima produced by the grating?

104 A beam of x rays with wavelengths ranging from 0.120 nm to 0.0700 nm scatters from a family of reflecting planes in a crystal. The plane separation is 0.250 nm. It is observed that scattered beams are produced for 0.100 nm and 0.0750 nm. What is the angle between the incident and scattered beams?

105 Show that a grating made up of alternately transparent and opaque strips of equal width eliminates all the even orders of maxima (except $m = 0$).

106 Light of wavelength 500 nm diffracts through a slit of width 2.00 μm and onto a screen that is 2.00 m away. On the screen, what is the distance between the center of the diffraction pattern and the third diffraction minimum?

107 If, in a two-slit interference pattern, there are 8 bright fringes within the first side peak of the diffraction envelope and diffraction minima coincide with two-slit interference maxima, then what is the ratio of slit separation to slit width?

108 White light (consisting of wavelengths from 400 nm to 700 nm) is normally incident on a grating. Show that, no matter what the value of the grating spacing d, the second order and third order overlap.

109 If we make $d = a$ in Fig. 36-53, the two slits coalesce into a single slit of width $2a$. Show that Eq. 36-19 reduces to give the diffraction pattern for such a slit.

FIG. 36-53 Problem 109.

110 Derive Eq. 36-28, the expression for the half-width of lines in a grating's diffraction pattern.

111 Prove that it is not possible to determine both wavelength of incident radiation and spacing of reflecting planes in a crystal by measuring the Bragg angles for several orders.

112 How many orders of the entire visible spectrum (400–700 nm) can be produced by a grating of 500 lines/mm?

113 An acoustic double-slit system (of slit separation d and slit width a) is driven by two loudspeakers as shown in Fig. 36-54. By use of a variable delay line, the phase of one of the speakers may be varied relative to the other speaker. Describe in detail what changes occur in the double-slit diffraction pattern at large distances as the phase difference between the speakers is varied from zero to 2π. Take both interference and diffraction effects into account.

FIG. 36-54 Problem 113.

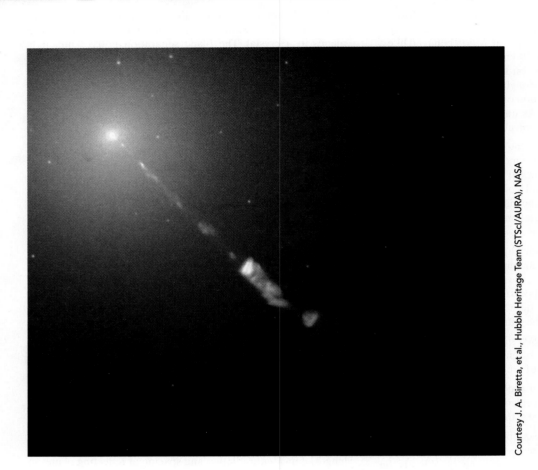

Courtesy J. A. Biretta, et al., Hubble Heritage Team (STScI/AURA), NASA

A jet 5000 light-years long streams from the center of galaxy M87 (the bright spot at the upper left end of the jet), which is 5×10^7 light-years from us. The jet is formed by electrons moving at nearly the speed of light. Something very strange must be occurring at the center of M87 in order for the electrons to be ejected from it at such high speed. Is there some type of monster there? Unfortunately, M87 is too distant for us to see any object at its center.

How can we tell what monster lies at the center of M87?

The answer is in this chapter.

37-1 WHAT IS PHYSICS?

One principal subject of physics is **relativity,** the field of study that measures events (things that happen): where and when they happen, and by how much any two events are separated in space and in time. In addition, relativity has to do with transforming such measurements between reference frames that move relative to each other. (Hence the name *relativity.*)

Transformations and moving reference frames, such as those we discussed in Sections 4-8 and 4-9, were well understood and quite routine to physicists in 1905. Then Albert Einstein (Fig. 37-1) published his **special theory of relativity.** The adjective *special* means that the theory deals only with **inertial reference frames,** which are frames in which Newton's laws are valid. (Einstein's *general theory of relativity* treats the more challenging situation in which reference frames can undergo gravitational acceleration; in this chapter the term *relativity* implies only inertial reference frames.)

Starting with two deceivingly simple postulates, Einstein stunned the scientific world by showing that the old ideas about relativity were wrong, even though everyone was so accustomed to them that they seemed to be unquestionable common sense. This supposed common sense, however, was derived only from experience with things that move rather slowly. Einstein's relativity, which turns out to be correct for all possible speeds, predicted many effects that were, at first study, bizarre because no one had ever experienced them.

In particular, Einstein demonstrated that space and time are entangled; that is, the time between two events depends on how far apart they occur, and vice versa. Also, the entanglement is different for observers who move relative to each other. One result is that time does not pass at a fixed rate, as if it were ticked off with mechanical regularity on some master grandfather clock that controls the universe. Rather, that rate is adjustable: Relative motion can change the rate at which time passes. Prior to 1905, no one but a few daydreamers would have thought that. Now, engineers and scientists take it for granted because their experience with special relativity has reshaped their common sense. For example, any engineer involved with the Global Positioning System of the NAVSTAR satellites must routinely use relativity to determine the rate at which time passes on the satellites because that rate differs from the rate on Earth's surface.

Special relativity has the reputation of being difficult. It is not difficult mathematically, at least not here. However, it is difficult in that we must be very careful about *who* measures *what* about an event and just *how* that measurement is made — and it can be difficult because it can contradict routine experience.

FIG. 37-1 Einstein posing for a photograph as fame began to accumulate. *(Corbis Images)*

37-2 | The Postulates

We now examine the two postulates of relativity, on which Einstein's theory is based:

➤ **1. The Relativity Postulate:** The laws of physics are the same for observers in all inertial reference frames. No one frame is preferred over any other.

Galileo assumed that the laws of *mechanics* were the same in all inertial reference frames. Einstein extended that idea to include *all* the laws of physics, especially those of electromagnetism and optics. This postulate does *not* say that the measured values of all physical quantities are the same for all inertial observers; most are not the same. It is the *laws of physics,* which relate these measurements to one another, that are the same.

➤ **2. The Speed of Light Postulate:** The speed of light in vacuum has the same value *c* in all directions and in all inertial reference frames.

FIG. 37-2 The dots show measured values of the kinetic energy of an electron plotted against its measured speed. No matter how much energy is given to an electron (or to any other particle having mass), its speed can never equal or exceed the ultimate limiting speed c. (The plotted curve through the dots shows the predictions of Einstein's special theory of relativity.)

We can also phrase this postulate to say that there is in nature an *ultimate speed c,* the same in all directions and in all inertial reference frames. Light happens to travel at this ultimate speed. However, no entity that carries energy or information can exceed this limit. Moreover, no particle that has mass can actually reach speed c, no matter how much or for how long that particle is accelerated. (Alas, the faster-than-light warp drive used in many science fiction stories appears to be impossible.)

Both postulates have been exhaustively tested, and no exceptions have ever been found.

The Ultimate Speed

The existence of a limit to the speed of accelerated electrons was shown in a 1964 experiment by W. Bertozzi, who accelerated electrons to various measured speeds and—by an independent method—measured their kinetic energies. He found that as the force on a very fast electron is increased, the electron's measured kinetic energy increases toward very large values but its speed does not increase appreciably (Fig. 37-2). Electrons have been accelerated in laboratories to at least 0.999 999 999 95 times the speed of light but—close though it may be—that speed is still less than the ultimate speed c.

This ultimate speed has been defined to be exactly

$$c = 299\ 792\ 458 \text{ m/s.} \tag{37-1}$$

Caution: So far in this book we have (appropriately) approximated c as 3.0×10^8 m/s, but in this chapter we shall often use the exact value. You might want to store the exact value in your calculator's memory (if it is not there already), to be called up when needed.

Testing the Speed of Light Postulate

If the speed of light is the same in all inertial reference frames, then the speed of light emitted by a source relative to, say, a laboratory should be the same as the speed of light that is emitted by a source at rest in the laboratory. This claim has been tested directly, in an experiment of high precision. The "light source" was the *neutral pion* (symbol π^0), an unstable, short-lived particle that can be produced by collisions in a particle accelerator. It decays (transforms) into two gamma rays by the process

$$\pi^0 \rightarrow \gamma + \gamma. \tag{37-2}$$

Gamma rays are part of the electromagnetic spectrum (at very high frequencies) and so obey the speed of light postulate, just as visible light does. (In this chapter we shall use the term light for any type of electromagnetic wave, visible or not.)

In 1964, physicists at CERN, the European particle-physics laboratory near Geneva, generated a beam of pions moving at a speed of 0.999 75c with respect to the laboratory. The experimenters then measured the speed of the gamma rays emitted from these very rapidly moving sources. They found that the speed of the light emitted by the pions was the same as it would be if the pions were at rest in the laboratory, namely c.

37-3 | Measuring an Event

An **event** is something that happens, and every event can be assigned three space coordinates and one time coordinate. Among many possible events are (1) the turning on or off of a tiny lightbulb, (2) the collision of two particles, (3) the passage of a pulse of light through a specified point, (4) an explosion, and (5) the sweeping of the hand of a clock past a marker on the rim of the clock. A certain observer, fixed in a certain inertial reference frame, might, for example, assign to an event A the coordinates given in Table 37-1. Because space and time are entangled with each other in relativity, we can describe these coordinates collectively as *spacetime* coordinates. The coordinate system itself is part of the reference frame of the observer.

A given event may be recorded by any number of observers, each in a different inertial reference frame. In general, different observers will assign different spacetime coordinates to the same event. Note that an event does not "belong" to any particular inertial reference frame. An event is just something that happens, and anyone in any reference frame may detect it and assign spacetime coordinates to it.

Making such an assignment can be complicated by a practical problem. For example, suppose a balloon bursts 1 km to your right while a firecracker pops 2 km to your left, both at 9:00 A.M. However, you do not detect either event precisely at 9:00 A.M. because at that instant light from the events has not yet reached you. Because light from the firecracker pop has farther to go, it arrives at your eyes later than does light from the balloon burst, and thus the pop will seem to have occurred later than the burst. To sort out the actual times and to assign 9:00 A.M. as the happening time for both events, you must calculate the travel times of the light and then subtract these times from the arrival times.

This procedure can be very messy in more challenging situations, and we need an easier procedure that automatically eliminates any concern about the travel time from an event to an observer. To set up such a procedure, we shall construct an imaginary array of measuring rods and clocks throughout the observer's inertial frame (the array moves rigidly with the observer). This construction may seem contrived, but it spares us much confusion and calculation and allows us to find the space coordinates, the time coordinate, and the spacetime coordinates, as follows.

1. **The Space Coordinates.** We imagine the observer's coordinate system fitted with a close-packed, three-dimensional array of measuring rods, one set of rods parallel to each of the three coordinate axes. These rods provide a way to determine coordinates along the axes. Thus, if the event is, say, the turning on of a small lightbulb, the observer, in order to locate the position of the event, need only read the three space coordinates at the bulb's location.

2. **The Time Coordinate.** For the time coordinate, we imagine that every point of intersection in the array of measuring rods includes a tiny clock, which the observer can read because the clock is illuminated by the light generated by the event. Figure 37-3 suggests one plane in the "jungle gym" of clocks and measuring rods we have described.

The array of clocks must be synchronized properly. It is not enough to assemble a set of identical clocks, set them all to the same time, and then move them to their assigned positions. We do not know, for example, whether moving the clocks will change their rates. (Actually, it will.) We must put the clocks in place and *then* synchronize them.

If we had a method of transmitting signals at infinite speed, synchronization would be a simple matter. However, no known signal has this property. We therefore choose light (any part of the electromagnetic spectrum) to send out our synchronizing signals because, in vacuum, light travels at the greatest possible speed, the limiting speed c.

Here is one of many ways in which an observer might synchronize an array of clocks using light signals: The observer enlists the help of a great number of temporary helpers, one for each clock. The observer then stands at a point selected as the origin and sends out a pulse of light when the origin clock reads $t = 0$. When the light pulse reaches the location of a helper, that helper sets the clock there to read $t = r/c$, where r is the distance between the helper and the origin. The clocks are then synchronized.

3. **The Spacetime Coordinates.** The observer can now assign spacetime coordinates to an event by simply recording the time on the clock nearest the event

TABLE 37-1	
Record of Event _A_	
Coordinate	Value
x	3.58 m
y	1.29 m
z	0 m
t	34.5 s

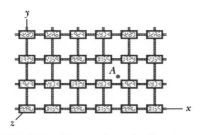

FIG. 37-3 One section of a three-dimensional array of clocks and measuring rods by which an observer can assign spacetime coordinates to an event, such as a flash of light at point A. The event's space coordinates are approximately $x = 3.6$ rod lengths, $y = 1.3$ rod lengths, and $z = 0$. The time coordinate is whatever time appears on the clock closest to A at the instant of the flash.

and the position as measured on the nearest measuring rods. If there are two events, the observer computes their separation in time as the difference in the times on clocks near each and their separation in space from the differences in coordinates on rods near each. We thus avoid the practical problem of calculating the travel times of the signals to the observer from the events.

37-4 | The Relativity of Simultaneity

Suppose that one observer (Sam) notes that two independent events (event Red and event Blue) occur at the same time. Suppose also that another observer (Sally), who is moving at a constant velocity $\vec{v}$ with respect to Sam, also records these same two events. Will Sally also find that they occur at the same time?

The answer is that in general she will not:

> If two observers are in relative motion, they will not, in general, agree as to whether two events are simultaneous. If one observer finds them to be simultaneous, the other generally will not.

We cannot say that one observer is right and the other wrong. Their observations are equally valid, and there is no reason to favor one over the other.

The realization that two contradictory statements about the same natural event can be correct is a seemingly strange outcome of Einstein's theory. However, in Chapter 17 we saw another way in which motion can affect measurement without balking at the contradictory results: In the Doppler effect, the frequency an observer measures for a sound wave depends on the relative motion of observer and source. Thus, two observers moving relative to each other can measure different frequencies for the same wave, and both measurements are correct.

We conclude the following:

> Simultaneity is not an absolute concept but rather a relative one, depending on the motion of the observer.

If the relative speed of the observers is very much less than the speed of light, then measured departures from simultaneity are so small that they are not noticeable. Such is the case for all our experiences of daily living; that is why the relativity of simultaneity is unfamiliar.

A Closer Look at Simultaneity

Let us clarify the relativity of simultaneity with an example based on the postulates of relativity, no clocks or measuring rods being directly involved. Figure 37-4 shows two long spaceships (the SS *Sally* and the SS *Sam*), which can serve as inertial reference frames for observers Sally and Sam. The two observers are stationed at the midpoints of their ships. The ships are separating along a common *x* axis, the relative velocity of *Sally* with respect to *Sam* being $\vec{v}$. Figure 37-4a shows the ships with the two observer stations momentarily aligned opposite each other.

Two large meteorites strike the ships, one setting off a red flare (event Red) and the other a blue flare (event Blue), not necessarily simultaneously. Each event leaves a permanent mark on each ship, at positions RR' and BB'.

Let us suppose that the expanding wavefronts from the two events happen to reach Sam at the same time, as Fig. 37-4b shows. Let us further suppose that, after the episode, Sam finds, by measuring the marks on his spaceship, that he was indeed stationed exactly halfway between the markers B and R on his ship when the two events occurred. He will say:

Sam   Light from event Red and light from event Blue reached me at the same time. From the marks on my spaceship, I find that I was standing halfway between the two sources. Therefore, event Red and event Blue were simultaneous events.

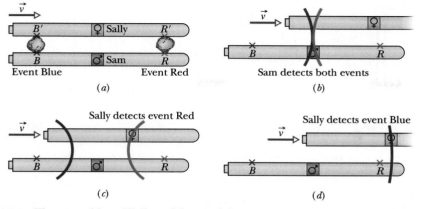

FIG. 37-4   The spaceships of Sally and Sam and the occurrences of events from Sam's view. Sally's ship moves rightward with velocity $\vec{v}$. (a) Event Red occurs at positions RR' and event Blue occurs at positions BB'; each event sends out a wave of light. (b) Sam simultaneously detects the waves from event Red and event Blue. (c) Sally detects the wave from event Red. (d) Sally detects the wave from event Blue.

As study of Fig. 37-4 shows, Sally and the expanding wavefront from event Red are moving *toward* each other, while she and the expanding wavefront from event Blue are moving in the *same direction*. Thus, the wavefront from event Red will reach Sally *before* the wavefront from event Blue does. She will say:

Sally   Light from event Red reached me before light from event Blue did. From the marks on my spaceship, I found that I too was standing halfway between the two sources. Therefore, the events were not simultaneous; event Red occurred first, followed by event Blue.

These reports do not agree. Nevertheless, *both* observers are correct.

Note carefully that there is only one wavefront expanding from the site of each event and that *this wavefront travels with the same speed c in both reference frames,* exactly as the speed of light postulate requires.

It *might* have happened that the meteorites struck the ships in such a way that the two hits appeared to Sally to be simultaneous. If that had been the case, then Sam would have declared them not to be simultaneous.

37-5 | The Relativity of Time

If observers who move relative to each other measure the time interval (or *temporal separation*) between two events, they generally will find different results. Why? Because the spatial separation of the events can affect the time intervals measured by the observers.

> The time interval between two events depends on how far apart they occur in both space and time; that is, their spatial and temporal separations are entangled.

In this section we discuss this entanglement by means of an example; however, the example is restricted in a crucial way: *To one of two observers, the two events occur at the same location.* We shall not get to more general examples until Section 37-7.

Figure 37-5a shows the basics of an experiment Sally conducts while she and her equipment — a light source, a mirror, and a clock — ride in a train moving with constant velocity $\vec{v}$ relative to a station. A pulse of light leaves the light source B (event 1), travels vertically upward, is reflected vertically downward by the mirror, and then is detected back at the source (event 2). Sally measures a certain time interval Δt_0 between the two events, related to the distance D from source to mirror by

$$\Delta t_0 = \frac{2D}{c} \qquad \text{(Sally).} \qquad (37\text{-}3)$$

FIG. 37-5 (*a*) Sally, on the train, measures the time interval Δt_0 between events 1 and 2 using a single clock C on the train. That clock is shown twice: first for event 1 and then for event 2. (*b*) Sam, watching from the station as the events occur, requires two synchronized clocks, C_1 at event 1 and C_2 at event 2, to measure the time interval between the two events; his measured time interval is Δt.

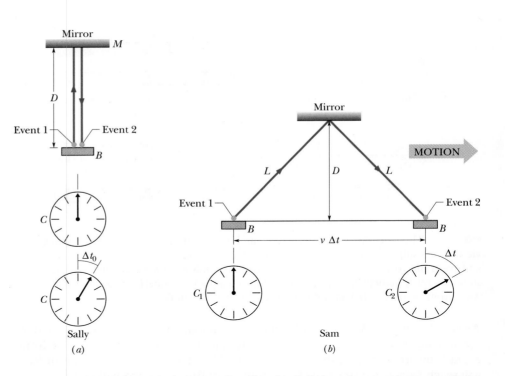

The two events occur at the same location in Sally's reference frame, and she needs only one clock C at that location to measure the time interval. Clock C is shown twice in Fig. 37-5*a*, at the beginning and end of the interval.

Consider now how these same two events are measured by Sam, who is standing on the station platform as the train passes. Because the equipment moves with the train during the travel time of the light, Sam sees the path of the light as shown in Fig. 37-5*b*. For him, the two events occur at different places in his reference frame, and so to measure the time interval between events, Sam must use *two* synchronized clocks, C_1 and C_2, one at each event. According to Einstein's speed of light postulate, the light travels at the same speed c for Sam as for Sally. Now, however, the light travels distance $2L$ between events 1 and 2. The time interval measured by Sam between the two events is

$$\Delta t = \frac{2L}{c} \qquad \text{(Sam)}, \qquad (37\text{-}4)$$

in which

$$L = \sqrt{(\tfrac{1}{2}v\,\Delta t)^2 + D^2}. \qquad (37\text{-}5)$$

From Eq. 37-3, we can write this as

$$L = \sqrt{(\tfrac{1}{2}v\,\Delta t)^2 + (\tfrac{1}{2}c\,\Delta t_0)^2}. \qquad (37\text{-}6)$$

If we eliminate L between Eqs. 37-4 and 37-6 and solve for Δt, we find

$$\Delta t = \frac{\Delta t_0}{\sqrt{1 - (v/c)^2}}. \qquad (37\text{-}7)$$

Equation 37-7 tells us how Sam's measured interval Δt between the events compares with Sally's interval Δt_0. Because v must be less than c, the denominator in Eq. 37-7 must be less than unity. Thus, Δt must be greater than Δt_0: Sam measures a *greater* time interval between the two events than does Sally. Sam and Sally have measured the time interval between the *same* two events, but the relative motion between Sam and Sally made their measurements *different*. We conclude that relative motion can change the *rate* at which time passes between two events; the key to this effect is the fact that the speed of light is the same for both observers.

We distinguish between the measurements of Sam and Sally with the following terminology:

> When two events occur at the same location in an inertial reference frame, the time interval between them, measured in that frame, is called the **proper time interval** or the **proper time.** Measurements of the same time interval from any other inertial reference frame are always greater.

Thus, Sally measures a proper time interval, and Sam measures a greater time interval. (The term *proper* is unfortunate in that it implies that any other measurement is improper or nonreal. That is just not so.) The amount by which a measured time interval is greater than the corresponding proper time interval is called **time dilation.** (To dilate is to expand or stretch; here the time interval is expanded or stretched.)

Often the dimensionless ratio v/c in Eq. 37-7 is replaced with β, called the **speed parameter,** and the dimensionless inverse square root in Eq. 37-7 is often replaced with γ, called the **Lorentz factor:**

$$\gamma = \frac{1}{\sqrt{1-\beta^2}} = \frac{1}{\sqrt{1-(v/c)^2}}. \qquad (37\text{-}8)$$

With these replacements, we can rewrite Eq. 37-7 as

$$\Delta t = \gamma \, \Delta t_0 \qquad \text{(time dilation).} \qquad (37\text{-}9)$$

The speed parameter β is always less than unity, and, provided v is not zero, γ is always greater than unity. However, the difference between γ and 1 is not significant unless $v > 0.1c$. Thus, in general, "old relativity" works well enough for $v < 0.1c$, but we must use special relativity for greater values of v. As shown in Fig. 37-6, γ increases rapidly in magnitude as β approaches 1 (as v approaches c). Therefore, the greater the relative speed between Sally and Sam is, the greater will be the time interval measured by Sam, until at a great enough speed, the interval takes "forever."

You might wonder what Sally says about Sam's having measured a greater time interval than she did. His measurement comes as no surprise to her, because to her, he failed to synchronize his clocks C_1 and C_2 in spite of his insistence that he did. Recall that observers in relative motion generally do not agree about simultaneity. Here, Sam insists that his two clocks simultaneously read the same time when event 1 occurred. To Sally, however, Sam's clock C_2 was erroneously set ahead during the synchronization process. Thus, when Sam read the time of event 2 on it, to Sally he was reading off a time that was too large, and that is why the time interval he measured between the two events was greater than the interval she measured.

Two Tests of Time Dilation

1. **Microscopic Clocks.** Subatomic particles called *muons* are unstable; that is, when a muon is produced, it lasts for only a short time before it *decays* (transforms into particles of other types). The *lifetime* of a muon is the time interval between its production (event 1) and its decay (event 2). When muons are stationary and their lifetimes are measured with stationary clocks (say, in a laboratory), their average lifetime is 2.200 μs. This is a proper time interval because, for each muon, events 1 and 2 occur at the same location in the reference frame of the muon—namely, at the muon itself. We can represent this proper time interval with Δt_0; moreover, we can call the reference frame in which it is measured the *rest frame* of the muon.

   If, instead, the muons are moving, say, through a laboratory, then measurements of their lifetimes made with the laboratory clocks should yield a greater average lifetime (a dilated average lifetime). To check this conclusion, measurements were made of the average lifetime of muons moving with a

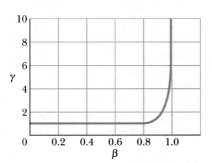

FIG. 37-6 A plot of the Lorentz factor γ as a function of the speed parameter $\beta \, (= v/c)$.

speed of 0.9994c relative to laboratory clocks. From Eq. 37-8, with $\beta = 0.9994$, the Lorentz factor for this speed is

$$\gamma = \frac{1}{\sqrt{1 - \beta^2}} = \frac{1}{\sqrt{1 - (0.9994)^2}} = 28.87.$$

Equation 37-9 then yields, for the average dilated lifetime,

$$\Delta t = \gamma \, \Delta t_0 = (28.87)(2.200 \; \mu s) = 63.51 \; \mu s.$$

The actual measured value matched this result within experimental error.

2. **Macroscopic Clocks.** In October 1977, Joseph Hafele and Richard Keating carried out what must have been a grueling experiment. They flew four portable atomic clocks twice around the world on commercial airlines, once in each direction. Their purpose was "to test Einstein's theory of relativity with macroscopic clocks." As we have just seen, the time dilation predictions of Einstein's theory have been confirmed on a microscopic scale, but there is great comfort in seeing a confirmation made with an actual clock. Such macroscopic measurements became possible only because of the very high precision of modern atomic clocks. Hafele and Keating verified the predictions of the theory to within 10%. (Einstein's *general* theory of relativity, which predicts that the rate at which time passes on a clock is influenced by the gravitational force on the clock, also plays a role in this experiment.)

A few years later, physicists at the University of Maryland carried out a similar experiment with improved precision. They flew an atomic clock round and round over Chesapeake Bay for flights lasting 15 h and succeeded in checking the time dilation prediction to better than 1%. Today, when atomic clocks are transported from one place to another for calibration or other purposes, the time dilation caused by their motion is always taken into account.

✓ **CHECKPOINT 1**   Standing beside railroad tracks, we are suddenly startled by a relativistic boxcar traveling past us as shown in the figure. Inside, a well-equipped hobo fires a laser pulse from the front of the boxcar to its rear. (a) Is our measurement of the speed of the pulse greater than, less than, or the same as that measured by the hobo? (b) Is his measurement of the flight time of the pulse a proper time? (c) Are his measurement and our measurement of the flight time related by Eq. 37-9?

Sample Problem 37-1

Your starship passes Earth with a relative speed of 0.9990c. After traveling 10.0 y (your time), you stop at lookout post LP13, turn, and then travel back to Earth with the same relative speed. The trip back takes another 10.0 y (your time). How long does the round trip take according to measurements made on Earth? (Neglect any effects due to the accelerations involved with stopping, turning, and getting back up to speed.)

KEY IDEAS   We begin by analyzing the outward trip:

1. This problem involves measurements made from two (inertial) reference frames, one attached to

Earth and the other (your reference frame) attached to your ship.

2. The outward trip involves two events: the start of the trip at Earth and the end of the trip at LP13.

3. Your measurement of 10.0 y for the outward trip is the proper time Δt_0 between those two events, because the events occur at the same location in your reference frame—namely, on your ship.

4. The Earth-frame measurement of the time interval Δt for the outward trip must be greater than Δt_0, according to Eq. 37-9 ($\Delta t = \gamma \, \Delta t_0$) for time dilation.

Calculations: Using Eq. 37-8 to substitute for γ in Eq.

37-9, we find

$$\Delta t = \frac{\Delta t_0}{\sqrt{1 - (v/c)^2}}$$

$$= \frac{10.0 \text{ y}}{\sqrt{1 - (0.9990c/c)^2}} = (22.37)(10.0 \text{ y}) = 224 \text{ y}.$$

On the return trip, we have the same situation and the same data. Thus, the round trip requires 20 y of your time but

$$\Delta t_{total} = (2)(224 \text{ y}) = 448 \text{ y} \qquad \text{(Answer)}$$

of Earth time. In other words, you have aged 20 y while the Earth has aged 448 y. Although you cannot travel into the past (as far as we know), you can travel into the future of, say, Earth, by using high-speed relative motion to adjust the rate at which time passes.

Sample Problem   37-2   Build your skill

The elementary particle known as the *positive kaon* (K^+) has, on average, a lifetime of 0.1237 μs when stationary—that is, when the lifetime is measured in the rest frame of the kaon. If a positive kaon has a speed of 0.990c relative to a laboratory reference frame when the kaon is produced, how far can it travel in that frame during its lifetime according to *classical physics* (which is a reasonable approximation for speeds much less than c) and according to special relativity (which is correct for all physically possible speeds)?

KEY IDEAS

1. This problem involves measurements made from two (inertial) reference frames, one attached to the kaon and the other attached to the laboratory.

2. This problem also involves two events: the start of the kaon's travel (when the kaon is produced) and the end of that travel (at the end of the kaon's lifetime).

3. The distance traveled by the kaon between those two events is related to its speed v and the time interval for the travel by

$$v = \frac{\text{distance}}{\text{time interval}}. \qquad (37\text{-}10)$$

With these ideas in mind, let us solve for the distance first with classical physics and then with special relativity.

Classical physics: In classical physics we would find the same distance and time interval (in Eq. 37-10) whether we measured them from the kaon frame or from the laboratory frame. Thus, we need not be careful about the frame in which the measurements are made. To find the kaon's travel distance d_{cp} according to classical physics, we first rewrite Eq. 37-10 as

$$d_{cp} = v \Delta t, \qquad (37\text{-}11)$$

where Δt is the time interval between the two events in either frame. Then, substituting 0.990c for v and 0.1237 μs for Δt in Eq. 37-11, we find

$$d_{cp} = (0.990c) \Delta t$$
$$= (0.990)(299\,792\,458 \text{ m/s})(0.1237 \times 10^{-6} \text{ s})$$
$$= 36.7 \text{ m}. \qquad \text{(Answer)}$$

This is how far the kaon would travel if classical physics were correct at speeds close to c.

Special relativity: In special relativity we must be very careful that both the distance and the time interval in Eq. 37-10 are measured in the *same* reference frame—especially when the speed is close to c, as here. Thus, to find the actual travel distance d_{sr} of the kaon *as measured from the laboratory frame* and according to special relativity, we rewrite Eq. 37-10 as

$$d_{sr} = v \Delta t, \qquad (37\text{-}12)$$

where Δt is the time interval between the two events *as measured from the laboratory frame.*

Before we can evaluate d_{sr} in Eq. 37-12, we must find Δt. The 0.1237 μs time interval is a proper time because the two events occur at the same location in the kaon frame—namely, at the kaon itself. Therefore, let Δt_0 represent this proper time interval. Then we can use Eq. 37-9 ($\Delta t = \gamma \Delta t_0$) for time dilation to find the time interval Δt as measured from the laboratory frame. Using Eq. 37-8 to substitute for γ in Eq. 37-9 leads to

$$\Delta t = \frac{\Delta t_0}{\sqrt{1 - (v/c)^2}} = \frac{0.1237 \times 10^{-6} \text{ s}}{\sqrt{1 - (0.990c/c)^2}} = 8.769 \times 10^{-7} \text{ s}.$$

This is about seven times longer than the kaon's proper lifetime. That is, the kaon's lifetime is about seven times longer in the laboratory frame than in its own frame—the kaon's lifetime is dilated. We can now evaluate Eq. 37-12 for the travel distance d_{sr} in the laboratory frame as

$$d_{sr} = v \Delta t = (0.990c) \Delta t$$
$$= (0.990)(299\,792\,458 \text{ m/s})(8.769 \times 10^{-7} \text{ s})$$
$$= 260 \text{ m}. \qquad \text{(Answer)}$$

This is about seven times d_{cp}. Experiments like the one outlined here, which verify special relativity, became routine in physics laboratories decades ago. The engineering design and the construction of any scientific or medical facility that employs high-speed particles must take relativity into account.

37-6 | The Relativity of Length

If you want to measure the length of a rod that is at rest with respect to you, you can—at your leisure—note the positions of its end points on a long stationary scale and subtract one reading from the other. If the rod is moving, however, you must note the positions of the end points *simultaneously* (in your reference frame) or your measurement cannot be called a length. Figure 37-7 suggests the difficulty of trying to measure the length of a moving penguin by locating its front and back at different times. Because simultaneity is relative and it enters into length measurements, length should also be a relative quantity. It is.

Let L_0 be the length of a rod that you measure when the rod is stationary (meaning you and it are in the same reference frame, the rod's rest frame). If, instead, there is relative motion at speed v between you and the rod *along the length of the rod*, then with simultaneous measurements you obtain a length L given by

$$L = L_0\sqrt{1 - \beta^2} = \frac{L_0}{\gamma} \qquad \text{(length contraction).} \qquad (37\text{-}13)$$

Because the Lorentz factor γ is always greater than unity if there is relative motion, L is less than L_0. The relative motion causes a *length contraction,* and L is called a *contracted length.* Because γ increases with speed v, the length contraction also increases with v.

> The length L_0 of an object measured in the rest frame of the object is its **proper length** or **rest length.** Measurements of the length from any reference frame that is in relative motion parallel to that length are always less than the proper length.

Be careful: Length contraction occurs only along the direction of relative motion. Also, the length that is measured does not have to be that of an object like a rod or a circle. Instead, it can be the length (or distance) between two objects in the same rest frame—for example, the Sun and a nearby star (which are, at least approximately, at rest relative to each other).

Does a moving object *really* shrink? Reality is based on observations and measurements; if the results are always consistent and if no error can be determined, then what is observed and measured is real. In that sense, the object really does shrink. However, a more precise statement is that the object *is really measured* to shrink—motion affects that measurement and thus reality.

When you measure a contracted length for, say, a rod, what does an observer moving with the rod say of your measurement? To that observer, you did not locate the two ends of the rod simultaneously. (Recall that observers in motion relative to each other do not agree about simultaneity.) To the observer, you first located the rod's front end and then, slightly later, its rear end, and that is why you measured a length that is less than the proper length.

Proof of Eq. 37-13

Length contraction is a direct consequence of time dilation. Consider once more our two observers. This time, both Sally, seated on a train moving through a station, and Sam, again on the station platform, want to measure the length of the platform. Sam, using a tape measure, finds the length to be L_0, a proper length because the platform is at rest with respect to him. Sam also notes that Sally, on the train, moves through this length in a time $\Delta t = L_0/v$, where v is the speed of the train; that is,

$$L_0 = v\,\Delta t \qquad \text{(Sam).} \qquad (37\text{-}14)$$

This time interval Δt is not a proper time interval because the two events that

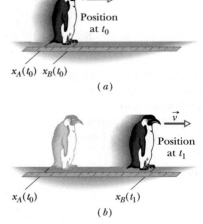

FIG. 37-7 If you want to measure the front-to-back length of a penguin while it is moving, you must mark the positions of its front and back simultaneously (in your reference frame), as in (*a*), rather than at different times, as in (*b*).

define it (Sally passes the back of the platform and Sally passes the front of the platform) occur at two different places, and therefore Sam must use two synchronized clocks to measure the time interval Δt.

For Sally, however, the platform is moving past her. She finds that the two events measured by Sam occur *at the same place* in her reference frame. She can time them with a single stationary clock, and so the interval Δt_0 that she measures is a proper time interval. To her, the length L of the platform is given by

$$L = v\,\Delta t_0 \qquad \text{(Sally)}. \qquad (37\text{-}15)$$

If we divide Eq. 37-15 by Eq. 37-14 and apply Eq. 37-9, the time dilation equation, we have

$$\frac{L}{L_0} = \frac{v\,\Delta t_0}{v\,\Delta t} = \frac{1}{\gamma},$$

or
$$L = \frac{L_0}{\gamma}, \qquad (37\text{-}16)$$

which is Eq. 37-13, the length contraction equation.

Sample Problem | **37-3**

In Fig. 37-8, Sally (at point A) and Sam's spaceship (of proper length $L_0 = 230$ m) pass each other with constant relative speed v. Sally measures a time interval of 3.57 μs for the ship to pass her (from the passage of point B to the passage of point C). In terms of c, what is the relative speed v between Sally and the ship?

FIG. 37-8 Sally, at point A, measures the time a spaceship takes to pass her.

KEY IDEAS Let's assume that speed v is near c. Then,

1. This problem involves measurements made from two (inertial) reference frames, one attached to Sally and the other attached to Sam and his spaceship.

2. This problem also involves two events: the first is the passage of point B past Sally and the second is the passage of point C past her.

3. From either reference frame, the other reference frame passes at speed v and moves a certain distance in the time interval between the two events:

$$v = \frac{\text{distance}}{\text{time interval}}. \qquad (37\text{-}17)$$

Because speed v is assumed to be near the speed of light, we must be careful that the distance and the time interval in Eq. 37-17 are measured in the *same* reference frame.

Calculations: We are free to use either frame for the measurements. Because we know that the time interval Δt between the two events measured from Sally's frame is 3.57 μs, let us also use the distance L between the two events measured from her frame. Equation 37-17 then becomes

$$v = \frac{L}{\Delta t}. \qquad (37\text{-}18)$$

We do not know L, but we can relate it to the given

L_0: The distance between the two events as measured from Sam's frame is the ship's proper length L_0. Thus, the distance L measured from Sally's frame must be less than L_0, as given by Eq. 37-13 ($L = L_0/\gamma$) for length contraction. Substituting L_0/γ for L in Eq. 37-18 and then substituting Eq. 37-8 for γ, we find

$$v = \frac{L_0/\gamma}{\Delta t} = \frac{L_0\sqrt{1 - (v/c)^2}}{\Delta t}.$$

Solving this equation for v (notice that it is on the left and also buried in the Lorentz factor) leads us to

$$v = \frac{L_0 c}{\sqrt{(c\,\Delta t)^2 + L_0^2}}$$

$$= \frac{(230\text{ m})c}{\sqrt{(299\ 792\ 458\text{ m/s})^2(3.57 \times 10^{-6}\text{ s})^2 + (230\text{ m})^2}}$$

$$= 0.210c. \qquad \text{(Answer)}$$

Thus, the relative speed between Sally and the ship is 21% of the speed of light. Note that only the relative motion of Sally and Sam matters here; whether either is stationary relative to, say, a space station is irrelevant. In Fig. 37-8 we took Sally to be stationary, but we could instead have taken the ship to be stationary, with Sally moving to the left past it. Nothing would have changed in our result.

Sample Problem ▌**37-4**

Caught by surprise near a supernova, you race away from the explosion in your spaceship, hoping to outrun the high-speed material ejected toward you. Your Lorentz factor γ relative to the inertial reference frame of the local stars is 22.4.

(a) To reach a safe distance, you figure you need to cover 9.00×10^{16} m as measured in the reference frame of the local stars. How long will the flight take, as measured in that frame?

KEY IDEAS Just as we did in Chapter 2, we can find the time required to travel a certain distance at constant speed with the definition of speed v:

$$\text{speed} = \frac{\text{distance}}{\text{time interval}}. \tag{37-19}$$

From Fig. 37-6, we see that because your Lorentz factor γ relative to the stars is 22.4 (large), your relative speed v is almost c—so close that we can approximate it as c. Then for speed $v \approx c$, we must be careful that the distance and the time interval in Eq. 37-19 are measured in the *same* reference frame.

Calculations: The given distance (9.00×10^{16} m) for the length of your travel path is measured in the reference frame of the stars, and the requested time interval Δt is to be measured in that same frame. Thus, we can write

$$\left(\begin{array}{c} \text{time interval} \\ \text{relative to stars} \end{array}\right) = \frac{\text{distance relative to stars}}{c}.$$

Then substituting the given distance, we find that

$$\left(\begin{array}{c} \text{time interval} \\ \text{relative to stars} \end{array}\right) = \frac{9.00 \times 10^{16} \text{ m}}{299\ 792\ 458 \text{ m/s}}$$

$$= 3.00 \times 10^8 \text{ s} = 9.51 \text{ y.} \quad \text{(Answer)}$$

(b) How long does that trip take according to you (in your reference frame)?

KEY IDEAS

1. We now want the time interval measured in a different reference frame—namely, yours. Thus, we need to transform the data given in the reference frame of the stars to your frame.

2. The given path length of 9.00×10^{16} m, measured in the reference frame of the stars, is a proper length L_0, because the two ends of the path are at rest in that frame. As observed from your reference frame, the stars' reference frame and those two ends of the path race past you at a relative speed of $v \approx c$.

3. You measure a contracted length L_0/γ for the path, not the proper length L_0.

Calculations: We can now rewrite Eq. 37-19 as

$$\left(\begin{array}{c} \text{time interval} \\ \text{relative to you} \end{array}\right) = \frac{\text{distance relative to you}}{c} = \frac{L_0/\gamma}{c}.$$

Substituting known data, we find

$$\left(\begin{array}{c} \text{time interval} \\ \text{relative to you} \end{array}\right) = \frac{(9.00 \times 10^{16} \text{ m})/22.4}{299\ 792\ 458 \text{ m/s}}$$

$$= 1.340 \times 10^7 \text{ s} = 0.425 \text{ y.} \quad \text{(Answer)}$$

In part (a) we found that the flight takes 9.51 y in the reference frame of the stars. However, here we find that it takes only 0.425 y in your frame, due to the relative motion and the resulting contracted length of the path.

37-7 | The Lorentz Transformation

Figure 37-9 shows inertial reference frame S' moving with speed v relative to frame S, in the common positive direction of their horizontal axes (marked x and x'). An observer in S reports spacetime coordinates x, y, z, t for an event, and an observer in S' reports x', y', z', t' for the same event. How are these sets of numbers related?

We claim at once (although it requires proof) that the y and z coordinates, which are perpendicular to the motion, are not affected by the motion; that is, $y = y'$ and $z = z'$. Our interest then reduces to the relation between x and x' and that between t and t'.

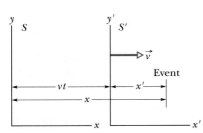

FIG. 37-9 Two inertial reference frames: frame S' has velocity $\vec{v}$ relative to frame S.

The Galilean Transformation Equations

Prior to Einstein's publication of his special theory of relativity, the four coordinates of interest were assumed to be related by the *Galilean transformation equations:*

$$x' = x - vt$$
$$t' = t$$

(Galilean transformation equations; approximately valid at low speeds). $\qquad$ (37-20)

(These equations are written with the assumption that $t = t' = 0$ when the origins of S and S' coincide.) You can verify the first equation with Fig. 37-9. The second equation effectively claims that time passes at the same rate for observers in both reference frames. That would have been so obviously true to a scientist prior to Einstein that it would not even have been mentioned. When speed v is small compared to c, Eqs. 37-20 generally work well.

The Lorentz Transformation Equations

We state without proof that the correct transformation equations, which remain valid for all speeds up to the speed of light, can be derived from the postulates of relativity. The results, called the **Lorentz transformation equations*** or sometimes (more loosely) just the Lorentz transformations, are

$$x' = \gamma(x - vt),$$
$$y' = y,$$
$$z' = z,$$
$$t' = \gamma(t - vx/c^2)$$

(Lorentz transformation equations; valid at all physically possible speeds). $\qquad$ (37-21)

(The equations are written with the assumption that $t = t' = 0$ when the origins of S and S' coincide.) Note that the spatial values x and the temporal values t are bound together in the first and last equations. This entanglement of space and time was a prime message of Einstein's theory, a message that was long rejected by many of his contemporaries.

It is a formal requirement of relativistic equations that they should reduce to familiar classical equations if we let c approach infinity. That is, if the speed of light were infinitely great, *all* finite speeds would be "low" and classical equations would never fail. If we let $c \rightarrow \infty$ in Eqs. 37-21, $\gamma \rightarrow 1$ and these equations reduce—as we expect—to the Galilean equations (Eqs. 37-20). You should check this.

Equations 37-21 are written in a form that is useful if we are given x and t and wish to find x' and t'. We may wish to go the other way, however. In that case we simply solve Eqs. 37-21 for x and t, obtaining

$$x = \gamma(x' + vt') \quad \text{and} \quad t = \gamma(t' + vx'/c^2). \qquad (37-22)$$

Comparison shows that, starting from either Eqs. 37-21 or Eqs. 37-22, you can find the other set by interchanging primed and unprimed quantities and reversing the sign of the relative velocity v.

Equations 37-21 and 37-22 relate the coordinates of a single event as seen by two observers. Sometimes we want to know not the coordinates of a single event but the differences between coordinates for a pair of events. That is, if we label our events 1 and 2, we may want to relate

$$\Delta x = x_2 - x_1 \quad \text{and} \quad \Delta t = t_2 - t_1,$$

as measured by an observer in S, and

$$\Delta x' = x'_2 - x'_1 \quad \text{and} \quad \Delta t' = t'_2 - t'_1,$$

as measured by an observer in S'.

---

* You may wonder why we do not call these the *Einstein transformation equations* (and why not the *Einstein factor* for γ). H. A. Lorentz actually derived these equations before Einstein did, but as the great Dutch physicist graciously conceded, he did not take the further bold step of interpreting these equations as describing the true nature of space and time. It is this interpretation, first made by Einstein, that is at the heart of relativity.

TABLE 37-2

The Lorentz Transformation Equations for Pairs of Events

1. $\Delta x = \gamma(\Delta x' + v\,\Delta t')$	1'. $\Delta x' = \gamma(\Delta x - v\,\Delta t)$
2. $\Delta t = \gamma(\Delta t' + v\,\Delta x'/c^2)$	2'. $\Delta t' = \gamma(\Delta t - v\,\Delta x/c^2)$

$$\gamma = \frac{1}{\sqrt{1-(v/c)^2}} = \frac{1}{\sqrt{1-\beta^2}}$$

Frame S' moves at velocity v relative to frame S.

Table 37-2 displays the Lorentz equations in difference form, suitable for analyzing pairs of events. The equations in the table were derived by simply substituting differences (such as Δx and $\Delta x'$) for the four variables in Eqs. 37-21 and 37-22.

Be careful: When substituting values for these differences, you must be consistent and not mix the values for the first event with those for the second event. Also, if, say, Δx is a negative quantity, you must be certain to include the minus sign in a substitution.

✓ **CHECKPOINT 2**   In Fig. 37-9, frame S' has velocity $0.90c$ relative to frame S. An observer in frame S' measures two events as occurring at the following spacetime coordinates: event Yellow at (5.0 m, 20 ns) and event Green at (-2.0 m, 45 ns). An observer in frame S wants to find the temporal separation $\Delta t_{GY} = t_G - t_Y$ between the events. (a) Which equation in Table 37-2 should be used? (b) Should $+0.90c$ or $-0.90c$ be substituted for v in the parentheses on the equation's right side and in the Lorentz factor γ? What value should be substituted into the (c) first and (d) second term in the parentheses?

37-8 | Some Consequences of the Lorentz Equations

Here we use the equations of Table 37-2 to affirm some of the conclusions that we reached earlier by arguments based directly on the postulates.

Simultaneity

Consider Eq. 2 of Table 37-2,

$$\Delta t = \gamma\left(\Delta t' + \frac{v\,\Delta x'}{c^2}\right). \qquad (37\text{-}23)$$

If two events occur at different places in reference frame S' of Fig. 37-9, then $\Delta x'$ in this equation is not zero. It follows that even if the events are simultaneous in S' (thus $\Delta t' = 0$), they will not be simultaneous in frame S. (This is in accord with our conclusion in Section 37-4.) The time interval between the events in S will be

$$\Delta t = \gamma\frac{v\,\Delta x'}{c^2} \qquad \text{(simultaneous events in } S'\text{).}$$

Time Dilation

Suppose now that two events occur at the same place in S' (thus $\Delta x' = 0$) but at different times (thus $\Delta t' \neq 0$). Equation 37-23 then reduces to

$$\Delta t = \gamma\,\Delta t' \qquad \text{(events in same place in } S'\text{).} \qquad (37\text{-}24)$$

This confirms time dilation between frames S and S'. Moreover, because the two events occur at the same place in S', the time interval $\Delta t'$ between them can be measured with a single clock, located at that place. Under these conditions, the measured interval is a proper time interval, and we can label it Δt_0 as we have previously labeled proper times. Thus, with that label Eq. 37-24 becomes

$$\Delta t = \gamma \, \Delta t_0 \quad \text{(time dilation)},$$

which is exactly Eq. 37-9, the time dilation equation.

Length Contraction

Consider Eq. 1' of Table 37-2,

$$\Delta x' = \gamma(\Delta x - v \, \Delta t). \tag{37-25}$$

If a rod lies parallel to the x and x' axes of Fig. 37-9 and is at rest in reference frame S', an observer in S' can measure its length at leisure. One way to do so is by subtracting the coordinates of the end points of the rod. The value of $\Delta x'$ that is obtained will be the proper length L_0 of the rod because the measurements are made in a frame where the rod is at rest.

Suppose the rod is moving in frame S. This means that Δx can be identified as the length L of the rod in frame S only if the coordinates of the rod's end points are measured *simultaneously*—that is, if $\Delta t = 0$. If we put $\Delta x' = L_0$, $\Delta x = L$, and $\Delta t = 0$ in Eq. 37-25, we find

$$L = \frac{L_0}{\gamma} \quad \text{(length contraction)}, \tag{37-26}$$

which is exactly Eq. 37-13, the length contraction equation.

Sample Problem | **37-5**

An Earth starship has been sent to check an Earth outpost on the planet P1407, whose moon houses a battle group of the often hostile Reptulians. As the ship follows a straight-line course first past the planet and then past the moon, it detects a high-energy microwave burst at the Reptulian moon base and then, 1.10 s later, an explosion at the Earth outpost, which is 4.00×10^8 m from the Reptulian base as measured from the ship's reference frame. The Reptulians have obviously attacked the Earth outpost, and so the starship begins to prepare for a confrontation with them.

(a) The speed of the ship relative to the planet and its moon is 0.980c. What are the distance and time interval between the burst and the explosion as measured in the planet–moon inertial frame (and thus according to the occupants of the stations)?

KEY IDEAS

1. This problem involves measurements made from two reference frames, the planet–moon frame and the starship frame.

2. This problem involves two events: the burst and the explosion.

3. We need to transform the given data about the time and distance between the two events as measured in the starship frame to the corresponding data as measured in the planet–moon frame.

FIG. 37-10 A planet and its moon in reference frame S' move rightward with speed v relative to a starship in reference frame S.

Starship frame: Before we get to the transformation, we need to carefully choose our notation. We begin with a sketch of the situation as shown in Fig. 37-10. There, we have chosen the ship's frame S to be stationary and the planet–moon frame S' to be moving with positive velocity (rightward). (This is an arbitrary choice; we could, instead, have chosen the planet–moon frame to be stationary. Then we would redraw $\vec{v}$ in Fig. 37-10 as being attached to the S frame and indicating leftward motion; v would then be a negative quantity. The results would be the same.) Let subscripts e and b represent the explosion and burst, respectively. Then the given data, all in the unprimed (starship) reference frame, are

$$\Delta x = x_e - x_b = +4.00 \times 10^8 \text{ m}$$

and
$$\Delta t = t_e - t_b = +1.10 \text{ s}.$$

Here, Δx is a positive quantity because in Fig. 37-10, the coordinate x_e for the explosion is greater than the coordinate x_b for the burst; Δt is also a positive quantity because the time t_e of the explosion is greater (later) than the time t_b of the burst.

Planet–moon frame: We seek $\Delta x'$ and $\Delta t'$, which we shall get by transforming the given S-frame data to the planet–moon frame S'. Because we are considering a pair of events, we choose transformation equations from Table 37-2—namely, Eqs. 1' and 2':

$$\Delta x' = \gamma(\Delta x - v\,\Delta t) \qquad (37\text{-}27)$$

and
$$\Delta t' = \gamma\left(\Delta t - \frac{v\,\Delta x}{c^2}\right). \qquad (37\text{-}28)$$

Here, $v = +0.980c$ and the Lorentz factor is

$$\gamma = \frac{1}{\sqrt{1-(v/c)^2}} = \frac{1}{\sqrt{1-(+0.980c/c)^2}} = 5.0252.$$

Equation 37-27 then becomes

$$\Delta x' = (5.0252)[4.00 \times 10^8\ \text{m} - (+0.980c)(1.10\ \text{s})]$$
$$= 3.86 \times 10^8\ \text{m}, \qquad \text{(Answer)}$$

and Eq. 37-28 becomes

$$\Delta t' = (5.0252)\left[(1.10\,\text{s}) - \frac{(+0.980c)(4.00 \times 10^8\,\text{m})}{c^2}\right]$$
$$= -1.04\ \text{s}. \qquad \text{(Answer)}$$

(b) What is the meaning of the minus sign in the value for $\Delta t'$?

Reasoning: We must be consistent with the notation we set up in part (a). Recall how we originally defined the time interval between burst and explosion: $\Delta t = t_e - t_b = +1.10$ s. To be consistent with that choice of notation, our definition of $\Delta t'$ must be $t'_e - t'_b$; thus, we have found that

$$\Delta t' = t'_e - t'_b = -1.04\ \text{s}.$$

The minus sign here tells us that $t'_b > t'_e$; that is, in the planet–moon reference frame, the burst occurred 1.04 s *after* the explosion, not 1.10 s *before* the explosion as detected in the ship frame.

(c) Did the burst cause the explosion, or vice versa?

KEY IDEA The sequence of events measured in the planet–moon reference frame is the reverse of that measured in the ship frame. In either situation, if there is a causal relationship between the two events, information must travel from the location of one event to the location of the other to cause it.

Checking the speed: Let us check the required speed of the information. In the ship frame, this speed is

$$v_{\text{info}} = \frac{\Delta x}{\Delta t} = \frac{4.00 \times 10^8\ \text{m}}{1.10\ \text{s}} = 3.64 \times 10^8\ \text{m/s},$$

but that speed is impossible because it exceeds c. In the planet–moon frame, the speed comes out to be 3.70×10^8 m/s, also impossible. Therefore, neither event could possibly have caused the other event; that is, they are *unrelated* events. Thus, the starship should not confront the Reptulians.

37-9 | The Relativity of Velocities

Here we wish to use the Lorentz transformation equations to compare the velocities that two observers in different inertial reference frames S and S' would measure for the same moving particle. Let S' move with velocity v relative to S.

Suppose that the particle, moving with constant velocity parallel to the x and x' axes in Fig. 37-11, sends out two signals as it moves. Each observer measures the space interval and the time interval between these two events. These four measurements are related by Eqs. 1 and 2 of Table 37-2,

$$\Delta x = \gamma(\Delta x' + v\,\Delta t')$$

and
$$\Delta t = \gamma\left(\Delta t' + \frac{v\,\Delta x'}{c^2}\right).$$

If we divide the first of these equations by the second, we find

$$\frac{\Delta x}{\Delta t} = \frac{\Delta x' + v\,\Delta t'}{\Delta t' + v\,\Delta x'/c^2}.$$

Dividing the numerator and denominator of the right side by $\Delta t'$, we find

$$\frac{\Delta x}{\Delta t} = \frac{\Delta x'/\Delta t' + v}{1 + v(\Delta x'/\Delta t')/c^2}.$$

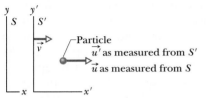

FIG. 37-11 Reference frame S' moves with velocity $\vec{v}$ relative to frame S. A particle has velocity $\vec{u}'$ relative to reference frame S' and velocity $\vec{u}$ relative to reference frame S.

However, in the differential limit, $\Delta x/\Delta t$ is u, the velocity of the particle as measured in S, and $\Delta x'/\Delta t'$ is u', the velocity of the particle as measured in S'. Then we have, finally,

$$u = \frac{u' + v}{1 + u'v/c^2} \qquad \text{(relativistic velocity transformation)} \qquad (37\text{-}29)$$

as the relativistic velocity transformation equation. This equation reduces to the classical, or Galilean, velocity transformation equation,

$$u = u' + v \qquad \text{(classical velocity transformation),} \qquad (37\text{-}30)$$

when we apply the formal test of letting $c \to \infty$. In other words, Eq. 37-29 is correct for all physically possible speeds, but Eq. 37-30 is approximately correct for speeds much less than c.

37-10 | Doppler Effect for Light

In Section 17-9 we discussed the Doppler effect (a shift in detected frequency) for sound waves traveling in air. For such waves, the Doppler effect depends on two velocities—namely, the velocities of the source and detector with respect to the air. Air is the medium that transmits the waves.

   That is not the situation with light waves, for they (and other electromagnetic waves) require no medium, being able to travel even through vacuum. The Doppler effect for light waves depends on only one velocity, the relative velocity $\vec{v}$ between source and detector, as measured from the reference frame of either. Let f_0 represent the **proper frequency** of the source—that is, the frequency that is measured by an observer in the rest frame of the source. Let f represent the frequency detected by an observer moving with velocity $\vec{v}$ relative to that rest frame. Then, when the direction of $\vec{v}$ is directly away from the source,

$$f = f_0 \sqrt{\frac{1 - \beta}{1 + \beta}} \qquad \text{(source and detector separating),} \qquad (37\text{-}31)$$

where $\beta = v/c$. When the direction of $\vec{v}$ is directly toward the source, we must change the signs in front of both β symbols in Eq. 37-31.

Low-Speed Doppler Effect

For low speeds ($\beta \ll 1$), Eq. 37-31 can be expanded in a power series in β and approximated as

$$f = f_0(1 - \beta + \tfrac{1}{2}\beta^2) \qquad \text{(source and detector separating, } \beta \ll 1\text{).} \qquad (37\text{-}32)$$

The corresponding low-speed equation for the Doppler effect with sound waves (or any waves except light waves) has the same first two terms but a different coefficient in the third term. Thus, the relativistic effect for low-speed light sources and detectors shows up only with the β^2 term.

   A police radar unit employs the Doppler effect with microwaves to measure the speed v of a car. A source in the radar unit emits a microwave beam at a certain (proper) frequency f_0 along the road. A car that is moving toward the unit intercepts that beam but at a frequency that is shifted upward by the Doppler effect due to the car's motion toward the radar unit. The car reflects the beam back toward the radar unit. Because the car is moving toward the radar unit, the detector in the unit intercepts a reflected beam that is further shifted up in frequency. The unit compares that detected frequency with f_0 and computes the speed v of the car.

Astronomical Doppler Effect

In astronomical observations of stars, galaxies, and other sources of light, we can determine how fast the sources are moving, either directly away from us or directly toward us, by measuring the *Doppler shift* of the light that reaches us. If a certain star were at rest relative to us, we would detect light from it with a certain proper frequency f_0. However, if the star is moving either directly away from us or directly toward us, the light we detect has a frequency f that is shifted from f_0 by the Doppler effect. This Doppler shift is due only to the *radial* motion of the star (its motion directly toward us or away from us), and the speed we can determine by measuring this Doppler shift is only the *radial speed v* of the star—that is, only the radial component of the star's velocity relative to us.

Suppose a star (or any other light source) moves away from us with a radial speed v that is low enough (β is small enough) for us to neglect the β^2 term in Eq. 37-32. Then we have

$$f = f_0(1 - \beta).$$ (37-33)

Because astronomical measurements involving light are usually done in wavelengths rather than frequencies, let us replace f with c/λ and f_0 with c/λ_0, where λ is the measured wavelength and λ_0 is the **proper wavelength** (the wavelength associated with f_0). We then have

$$\frac{c}{\lambda} = \frac{c}{\lambda_0}(1 - \beta),$$

or

$$\lambda = \lambda_0(1 - \beta)^{-1}.$$ (37-34)

Because we assume β is small, we can expand $(1 - \beta)^{-1}$ in a power series. Doing so and retaining only the first power of β, we have

$$\lambda = \lambda_0(1 + \beta),$$

or

$$\beta = \frac{\lambda - \lambda_0}{\lambda_0}.$$ (37-35)

Replacing β with v/c and $\lambda - \lambda_0$ with $|\Delta\lambda|$ leads to

$$v = \frac{|\Delta\lambda|}{\lambda_0}c \quad \text{(radial speed of light source, } v \ll c\text{)}.$$ (37-36)

The difference $\Delta\lambda$ is the *wavelength Doppler shift* of the light source. We enclose it with an absolute sign so that we always have a magnitude of the shift.

Equation 37-36 is an approximation that can be applied only when $v \ll c$. Under that condition, Eq. 37-36 can be applied whether the light source is moving toward or away from us. If it is moving away from us, then λ is longer than λ_0, $\Delta\lambda$ is positive, and the Doppler shift is called a *red shift*. (The term *red* does not mean the detected light is red or even visible. It merely serves as a memory device because red is at the *long* wavelength end of the visible spectrum. Thus λ is longer than λ_0.) If the light source is moving toward us, then λ is shorter than λ_0, $\Delta\lambda$ is negative, and the Doppler shift is called a *blue shift.*

✓ CHECKPOINT 3   The figure shows a source that emits light of proper frequency f_0 while moving directly toward the right with speed $c/4$ as measured from reference frame S. The figure also shows a light detector, which measures a frequency $f > f_0$ for the emitted light. (a) Is the detector moving toward the left or the right? (b) Is the speed of the detector as measured from reference frame S more than $c/4$, less than $c/4$, or equal to $c/4$?

Transverse Doppler Effect

So far, we have discussed the Doppler effect, here and in Chapter 17, only for situations in which the source and the detector move either directly toward or directly away from each other. Figure 37-12 shows a different arrangement, in which a source S moves past a detector D. When S reaches point P, the velocity of S is perpendicular to the line joining P and D, and at that instant S is moving neither toward nor away from D. If the source is emitting sound waves of frequency f_0, D detects that frequency (with no Doppler effect) when it intercepts the waves that were emitted at point P. However, if the source is emitting light waves, there is still a Doppler effect, called the **transverse Doppler effect**. In this situation, the detected frequency of the light emitted when the source is at point P is

$$f = f_0 \sqrt{1 - \beta^2} \quad \text{(transverse Doppler effect).} \quad (37\text{-}37)$$

For low speeds ($\beta \ll 1$), Eq. 37-37 can be expanded in a power series in β and approximated as

$$f = f_0(1 - \tfrac{1}{2}\beta^2) \quad \text{(low speeds).} \quad (37\text{-}38)$$

Here the first term is what we would expect for sound waves, and again the relativistic effect for low-speed light sources and detectors appears with the β^2 term.

In principle, a police radar unit can determine the speed of a car even when the path of the radar beam is perpendicular (transverse) to the path of the car. However, Eq. 37-38 tells us that because β is small even for a fast car, the relativistic term $\beta^2/2$ in the transverse Doppler effect is extremely small. Thus, $f \approx f_0$ and the radar unit computes a speed of zero. For this reason, police officers always try to direct the radar beam along the car's path to get a Doppler shift that gives the car's actual speed. Any deviation from that alignment works in favor of the motorist because it reduces the measured speed.

The transverse Doppler effect is really another test of time dilation. If we rewrite Eq. 37-37 in terms of the period T of oscillation of the emitted light wave instead of the frequency, we have, because $T = 1/f$,

$$T = \frac{T_0}{\sqrt{1 - \beta^2}} = \gamma T_0, \quad (37\text{-}39)$$

in which T_0 ($= 1/f_0$) is the **proper period** of the source. As comparison with Eq. 37-9 shows, Eq. 37-39 is simply the time dilation formula because a period is a time interval.

The NAVSTAR Navigation System

Each NAVSTAR satellite in the Global Positioning System (GPS) continuously signals its location by broadcasting radio signals at a set frequency controlled by precise atomic clocks. When the signal is sensed by a GPS detector on, say, a commercial aircraft, the frequency has been Doppler shifted because of the relative motion between the satellite and the detector. By detecting the signals from several NAVSTAR satellites simultaneously, the detector can determine the direction to any one of them and the direction of the satellite velocity. From the Doppler shift of the signal, the detector then determines the speed of the aircraft.

Although the relativistic contribution to the Doppler shift is extremely slight (roughly 4.5×10^{-10} of the shift is due to the relativity of this chapter), engineers must take the contribution into account to keep the NAVSTAR system accurate. They must also take into account the effects of Einstein's general theory of relativity because the gravitational acceleration at the altitude of a satellite is different from that at the altitude of a ground-based detector. Einstein's special and general theories of relativity were considered bizarre for a long time after they were published. Now they are indispensable in global positioning and long-range navigation.

FIG. 37-12 A light source S travels with velocity $\vec{v}$ past a detector at D. The special theory of relativity predicts a transverse Doppler effect as the source passes through point P, where the direction of travel is perpendicular to the line extending through D. Classical theory predicts no such effect.

Sample Problem **37-6**

Figure 37-13*a* shows curves of intensity versus wavelength for light reaching us from interstellar gas on two opposite sides of galaxy M87 (Fig. 37-13*b*). One curve peaks at 499.8 nm, the other at 501.6 nm. The gas orbits the core of the galaxy at a radius $r = 100$ light-years, moving toward us on one side of the core and moving away from us on the opposite side.

(a) Which curve corresponds to the gas moving toward us? What is the speed of the gas relative to us (and relative to the galaxy's core)?

KEY IDEAS

1. If the gas were not moving, the light from it would be detected at a certain wavelength.

2. The motion of the gas changes the detected wavelength via the Doppler effect, increasing the wavelength for the gas moving away from us and decreasing it for the gas moving toward us.

Thus, the curve peaking at 501.6 nm corresponds to motion away from us, and that peaking at 499.8 nm corresponds to motion toward us.

Calculations: Let us assume that the increase and the decrease in wavelength due to the motion of the gas are equal in magnitude. Then the unshifted wavelength, which we shall take as the proper wavelength λ_0, must be the average of the two shifted wavelengths:

$$\lambda_0 = \frac{501.6 \text{ nm} + 499.8 \text{ nm}}{2} = 500.7 \text{ nm}.$$

The Doppler shift $\Delta\lambda$ of the light from the gas moving away from us is then

$$\Delta\lambda = |\lambda - \lambda_0| = 501.6 \text{ nm} - 500.7 \text{ nm}$$
$$= 0.90 \text{ nm}.$$

Substituting this and $\lambda_0 = 500.7$ nm into Eq. 37-36, we find that the speed of the gas is

$$v = \frac{\Delta\lambda}{\lambda_0} c = \frac{0.90 \text{ nm}}{500.7 \text{ nm}} 299\ 792\ 458 \text{ m/s}$$
$$= 5.39 \times 10^5 \text{ m/s}. \qquad \text{(Answer)}$$

(b) The gas orbits the core of the galaxy because the gas experiences a gravitational force due to the mass M of the core. What is that mass in multiples of the Sun's mass M_S ($= 1.99 \times 10^{30}$ kg)?

KEY IDEAS

1. From Eq. 13-1, the magnitude F of the gravitational force on an orbiting gas element of mass m at orbital radius r is

$$F = \frac{GMm}{r^2}.$$

FIG. 37-13 (*a*) Plots of intensity versus wavelength for light emitted by gas on opposite sides of galaxy M87 and detected on Earth. (*b*) The central region of M87. The circles indicate the locations of the gas whose intensity is given in (*a*). The core of M87 is halfway between the circles. (*Courtesy NASA*)

2. If the gas element orbits the galaxy core in a circle, then the element must have a centripetal acceleration of magnitude $a = v^2/r$, directed toward the core.

3. Newton's second law, written for a radial axis extending from the core to the gas element, tells us that $F = ma$.

Calculations: Putting these three ideas together, we have

$$\frac{GMm}{r^2} = m \frac{v^2}{r}.$$

Solving this for M and substituting known data, we find

$$M = \frac{v^2 r}{G}$$
$$= \frac{(5.39 \times 10^5 \text{ m/s})^2(100 \text{ ly})(9.46 \times 10^{15} \text{ m/ly})}{6.67 \times 10^{-11} \text{ N} \cdot \text{m}^2/\text{kg}^2}$$
$$= 4.12 \times 10^{39} \text{ kg} = (2.1 \times 10^9)M_S. \qquad \text{(Answer)}$$

This result tells us that a mass equivalent to two billion suns has been compacted into the core of the galaxy, strongly suggesting that a *supermassive black hole* occupies the core.

The monster of M87: The M87 black hole is about 1000 times larger than the supermassive black hole lying at the center of our Milky Way galaxy (Sample Problem 13-7) and thus is indeed a monster. It rotates about an axis, and as material (stars and dust) fall into it, the material generates strong electromagnetic forces that shoot electrons away from the black hole, outward along the rotation axis and out of M87. The ejection is so severe that the electrons move at nearly the speed of light. The image that opens this chapter shows the jet that is angled somewhat in our direction; we cannot see the jet in the opposite direction.

37-11 | A New Look at Momentum

Suppose that a number of observers, each in a different inertial reference frame, watch an isolated collision between two particles. In classical mechanics, we have seen that—even though the observers measure different velocities for the colliding particles—they all find that the law of conservation of momentum holds. That is, they find that the total momentum of the system of particles after the collision is the same as it was before the collision.

How is this situation affected by relativity? We find that if we continue to define the momentum $\vec{p}$ of a particle as $m\vec{v}$, the product of its mass and its velocity, total momentum is *not* conserved for the observers in different inertial frames. We have two choices: (1) Give up the law of conservation of momentum or (2) see whether we can refine our definition of momentum in some new way so that the law of conservation of momentum still holds. The correct choice is the second one.

Consider a particle moving with constant speed v in the positive direction of an x axis. Classically, its momentum has magnitude

$$p = mv = m\frac{\Delta x}{\Delta t} \quad \text{(classical momentum)}, \tag{37-40}$$

in which Δx is the distance it travels in time Δt. To find a relativistic expression for momentum, we start with the new definition

$$p = m\frac{\Delta x}{\Delta t_0}.$$

Here, as before, Δx is the distance traveled by a moving particle as viewed by an observer watching that particle. However, Δt_0 is the time required to travel that distance, measured not by the observer watching the moving particle but by an observer moving with the particle. The particle is at rest with respect to this second observer; thus that measured time is a proper time.

Using the time dilation formula, $\Delta t = \gamma\,\Delta t_0$ (Eq. 37-9), we can then write

$$p = m\frac{\Delta x}{\Delta t_0} = m\frac{\Delta x}{\Delta t}\frac{\Delta t}{\Delta t_0} = m\frac{\Delta x}{\Delta t}\gamma.$$

However, since $\Delta x/\Delta t$ is just the particle velocity v, we have

$$p = \gamma m v \quad \text{(momentum)}. \tag{37-41}$$

Note that this differs from the classical definition of Eq. 37-40 only by the Lorentz factor γ. However, that difference is important: Unlike classical momentum, relativistic momentum approaches an infinite value as v approaches c.

We can generalize the definition of Eq. 37-41 to vector form as

$$\vec{p} = \gamma m \vec{v} \quad \text{(momentum)}. \tag{37-42}$$

This equation gives the correct definition of momentum for all physically possible speeds. For a speed much less than c, it reduces to the classical definition of momentum ($\vec{p} = m\vec{v}$).

37-12 | A New Look at Energy

Mass Energy

The science of chemistry was initially developed with the assumption that in chemical reactions, energy and mass are conserved separately. In 1905, Einstein showed that as a consequence of his theory of special relativity, mass can be considered to be another form of energy. Thus, the law of conservation of energy is really the law of conservation of mass–energy.

In a *chemical reaction* (a process in which atoms or molecules interact), the amount of mass that is transferred into other forms of energy (or vice versa) is such a tiny fraction of the total mass involved that there is no hope of measuring the mass change with even the best laboratory balances. Mass and energy truly *seem* to be separately conserved. However, in a *nuclear reaction* (in which nuclei or fundamental particles interact), the energy released is often about a million times greater than in a chemical reaction, and the change in mass can easily be measured. Taking mass–energy transfers into account in nuclear reactions became routine long ago.

An object's mass m and the equivalent energy E_0 are related by

$$E_0 = mc^2, \tag{37-43}$$

which, without the subscript 0, is the best-known science equation of all time. This energy that is associated with the mass of an object is called **mass energy** or **rest energy.** The second name suggests that E_0 is an energy that the object has even when it is at rest, simply because it has mass. (If you continue your study of physics beyond this book, you will see more refined discussions of the relation between mass and energy. You might even encounter disagreements about just what that relation is and means.)

Table 37-3 shows the (approximate) mass energy, or rest energy, of a few objects. The mass energy of, say, a U.S. penny is enormous; the equivalent amount of electrical energy would cost well over a million dollars. On the other hand, the entire annual U.S. electrical energy production corresponds to a mass of only a few hundred kilograms of matter (stones, burritos, or anything else).

In practice, SI units are rarely used with Eq. 37-43 because they are too large to be convenient. Masses are usually measured in atomic mass units, where

$$1 \text{ u} = 1.660\,538\,86 \times 10^{-27} \text{ kg}, \tag{37-44}$$

and energies are usually measured in electron-volts or multiples of it, where

$$1 \text{ eV} = 1.602\,176\,462 \times 10^{-19} \text{ J}. \tag{37-45}$$

In the units of Eqs. 37-44 and 37-45, the multiplying constant c^2 has the values

$$c^2 = 9.314\,940\,13 \times 10^8 \text{ eV/u} = 9.314\,940\,13 \times 10^5 \text{ keV/u}$$
$$= 931.494\,013 \text{ MeV/u}. \tag{37-46}$$

Total Energy

Equation 37-43 gives, for any object, the mass energy E_0 that is associated with the object's mass m, regardless of whether the object is at rest or moving. If the object is moving, it has additional energy in the form of kinetic energy K. If we assume that the object's potential energy is zero, then its total energy E is the sum of its mass energy and its kinetic energy:

$$E = E_0 + K = mc^2 + K. \tag{37-47}$$

TABLE 37-3

The Energy Equivalents of a Few Objects

Object	Mass (kg)	Energy Equivalent	
Electron	$\approx 9.11 \times 10^{-31}$	$\approx 8.19 \times 10^{-14}$ J	(≈ 511 keV)
Proton	$\approx 1.67 \times 10^{-27}$	$\approx 1.50 \times 10^{-10}$ J	(≈ 938 MeV)
Uranium atom	$\approx 3.95 \times 10^{-25}$	$\approx 3.55 \times 10^{-8}$ J	(≈ 225 GeV)
Dust particle	$\approx 1 \times 10^{-13}$	$\approx 1 \times 10^4$ J	(≈ 2 kcal)
U.S. penny	$\approx 3.1 \times 10^{-3}$	$\approx 2.8 \times 10^{14}$ J	(≈ 78 GW·h)

Although we shall not prove it, the total energy E can also be written as

$$E = \gamma mc^2, \tag{37-48}$$

where γ is the Lorentz factor for the object's motion.

Since Chapter 7, we have discussed many examples involving changes in the total energy of a particle or a system of particles. However, we did not include mass energy in the discussions because the changes in mass energy were either zero or small enough to be neglected. The law of conservation of total energy still applies when changes in mass energy are significant. Thus, regardless of what happens to the mass energy, the following statement from Section 8-8 is still true:

> The total energy E of an *isolated system* cannot change.

For example, if the total mass energy of two interacting particles in an isolated system decreases, some other type of energy in the system must increase because the total energy cannot change.

In a system undergoing a chemical or nuclear reaction, a change in the total mass energy of the system due to the reaction is often given as a Q value. The Q value for a reaction is obtained from the relation

$$\begin{pmatrix} \text{system's initial} \\ \text{total mass energy} \end{pmatrix} = \begin{pmatrix} \text{system's final} \\ \text{total mass energy} \end{pmatrix} + Q$$

or

$$E_{0i} = E_{0f} + Q. \tag{37-49}$$

Using Eq. 37-43 ($E_0 = mc^2$), we can rewrite this in terms of the initial *total* mass M_i and the final *total* mass M_f as

$$M_i c^2 = M_f c^2 + Q$$

or

$$Q = M_i c^2 - M_f c^2 = -\Delta M c^2, \tag{37-50}$$

where the change in mass due to the reaction is $\Delta M = M_f - M_i$.

If a reaction results in the transfer of energy from mass energy to, say, kinetic energy of the reaction products, the system's total mass energy E_0 (and total mass M) decreases and Q is positive. If, instead, a reaction requires that energy be transferred to mass energy, the system's total mass energy E_0 (and its total mass M) increases and Q is negative.

For example, suppose two hydrogen nuclei undergo a *fusion reaction* in which they join together to form a single nucleus and release two particles in the process. The total mass energy (and total mass) of the resultant single nucleus and two released particles is less than the total mass energy (and total mass) of the initial hydrogen nuclei. Thus, the Q of the fusion reaction is positive, and energy is said to be *released* (transferred from mass energy) by the reaction. This release is important to you because the fusion of hydrogen nuclei in the Sun is one part of the process that results in sunshine on Earth and makes life here possible.

Kinetic Energy

In Chapter 7 we defined the kinetic energy K of an object of mass m moving at speed v well below c to be

$$K = \tfrac{1}{2}mv^2. \tag{37-51}$$

However, this classical equation is only an approximation that is good enough when the speed is well below the speed of light.

Let us now find an expression for kinetic energy that is correct for *all* physically possible speeds, including speeds close to c. Solving Eq. 37-47 for K and then

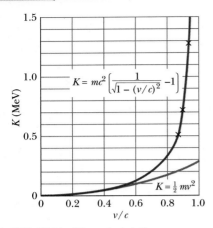

FIG. 37-14 The relativistic (Eq. 37-52) and classical (Eq. 37-51) equations for the kinetic energy of an electron, plotted as a function of v/c, where v is the speed of the electron and c is the speed of light. Note that the two curves blend together at low speeds and diverge widely at high speeds. Experimental data (at the × marks) show that at high speeds the relativistic curve agrees with experiment but the classical curve does not.

substituting for E from Eq. 37-48 lead to

$$K = E - mc^2 = \gamma mc^2 - mc^2$$
$$= mc^2(\gamma - 1) \quad \text{(kinetic energy)}, \quad (37\text{-}52)$$

where $\gamma \, (= 1/\sqrt{1 - (v/c)^2})$ is the Lorentz factor for the object's motion.

Figure 37-14 shows plots of the kinetic energy of an electron as calculated with the correct definition (Eq. 37-52) and the classical approximation (Eq. 37-51), both as functions of v/c. Note that on the left side of the graph the two plots coincide; this is the part of the graph—at lower speeds—where we have calculated kinetic energies so far in this book. This part of the graph tells us that we have been justified in calculating kinetic energy with the classical expression of Eq. 37-51. However, on the right side of the graph—at speeds near c—the two plots differ significantly. As v/c approaches 1.0, the plot for the classical definition of kinetic energy increases only moderately while the plot for the correct definition of kinetic energy increases dramatically, approaching an infinite value as v/c approaches 1.0. Thus, when an object's speed v is near c, we *must* use Eq. 37-52 to calculate its kinetic energy.

Figure 37-14 also tells us something about the work we must do on an object to increase its speed by, say, 1%. The required work W is equal to the resulting change ΔK in the object's kinetic energy. If the change is to occur on the low-speed, left side of Fig. 37-14, the required work might be modest. However, if the change is to occur on the high-speed, right side of Fig. 37-14, the required work could be enormous because the kinetic energy K increases so rapidly there with an increase in speed v. To increase an object's speed to c would require, in principle, an infinite amount of energy; thus, doing so is impossible.

The kinetic energies of electrons, protons, and other particles are often stated with the unit electron-volt or one of its multiples used as an adjective. For example, an electron with a kinetic energy of 20 MeV may be described as a 20 MeV electron.

Momentum and Kinetic Energy

In classical mechanics, the momentum p of a particle is mv and its kinetic energy K is $\frac{1}{2}mv^2$. If we eliminate v between these two expressions, we find a direct relation between momentum and kinetic energy:

$$p^2 = 2Km \quad \text{(classical)}. \quad (37\text{-}53)$$

We can find a similar connection in relativity by eliminating v between the relativistic definition of momentum (Eq. 37-41) and the relativistic definition of kinetic energy (Eq. 37-52). Doing so leads, after some algebra, to

$$(pc)^2 = K^2 + 2Kmc^2. \quad (37\text{-}54)$$

With the aid of Eq. 37-47, we can transform Eq. 37-54 into a relation between the momentum p and the total energy E of a particle:

$$E^2 = (pc)^2 + (mc^2)^2. \quad (37\text{-}55)$$

The right triangle of Fig. 37-15 can help you keep these useful relations in mind. You can also show that, in that triangle,

$$\sin \theta = \beta \quad \text{and} \quad \cos \theta = 1/\gamma. \quad (37\text{-}56)$$

With Eq. 37-55 we can see that the product pc must have the same unit as energy E; thus, we can express the unit of momentum p as an energy unit divided by c, usually as MeV/c or GeV/c in fundamental particle physics.

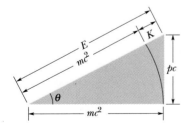

FIG. 37-15 A useful memory diagram for the relativistic relations among the total energy E, the rest energy or mass energy mc^2, the kinetic energy K, and the momentum magnitude p.

✓ **CHECKPOINT 4** Are (a) the kinetic energy and (b) the total energy of a 1 GeV electron more than, less than, or equal to those of a 1 GeV proton?

Sample Problem | **37-7**

(a) What is the total energy E of a 2.53 MeV electron?

KEY IDEA From Eq. 37-47, the total energy E is the sum of the electron's mass energy (or rest energy) mc^2 and its kinetic energy:

$$E = mc^2 + K. \qquad (37-57)$$

Calculations: The adjective "2.53 MeV" in the problem statement means that the electron's kinetic energy is 2.53 MeV. To evaluate the electron's mass energy mc^2, we substitute the electron's mass m from Appendix B, obtaining

$$mc^2 = (9.109 \times 10^{-31}\ \text{kg})(299\ 792\ 458\ \text{m/s})^2$$
$$= 8.187 \times 10^{-14}\ \text{J}.$$

Then dividing this result by 1.602×10^{-13} J/MeV gives us 0.511 MeV as the electron's mass energy (confirming the value in Table 37-3). Equation 37-57 then yields

$$E = 0.511\ \text{MeV} + 2.53\ \text{MeV} = 3.04\ \text{MeV}. \quad \text{(Answer)}$$

(b) What is the magnitude p of the electron's momentum, in the unit MeV/c?

KEY IDEA We can find p from the total energy E and the mass energy mc^2 via Eq. 37-55,

$$E^2 = (pc)^2 + (mc^2)^2.$$

Calculations: Solving for pc gives us

$$pc = \sqrt{E^2 - (mc^2)^2}$$
$$= \sqrt{(3.04\ \text{MeV})^2 - (0.511\ \text{MeV})^2} = 3.00\ \text{MeV}.$$

Finally, dividing both sides by c we find

$$p = 3.00\ \text{MeV}/c. \qquad \text{(Answer)}$$

Sample Problem | **37-8**

The most energetic proton ever detected in the cosmic rays coming to Earth from space had an astounding kinetic energy of 3.0×10^{20} eV (enough energy to warm a teaspoon of water by a few degrees).

(a) What were the proton's Lorentz factor γ and speed v (both relative to the ground-based detector)?

KEY IDEAS (1) The proton's Lorentz factor γ relates its total energy E to its mass energy mc^2 via Eq. 37-48 ($E = \gamma mc^2$). (2) The proton's total energy is the sum of its mass energy mc^2 and its (given) kinetic energy K.

Calculations: Putting these ideas together we have

$$\gamma = \frac{E}{mc^2} = \frac{mc^2 + K}{mc^2} = 1 + \frac{K}{mc^2}. \quad (37-58)$$

We can calculate the proton's mass energy mc^2 from its mass given in Appendix B, as we did for the electron in Sample Problem 37-7a. We find that mc^2 is 938 MeV (as listed in Table 37-3). Substituting this and the given kinetic energy into Eq. 37-58, we obtain

$$\gamma = 1 + \frac{3.0 \times 10^{20}\ \text{eV}}{938 \times 10^6\ \text{eV}}$$
$$= 3.198 \times 10^{11} \approx 3.2 \times 10^{11}. \quad \text{(Answer)}$$

This computed value for γ is so large that we cannot use the definition of γ (Eq. 37-8) to find v. Try it; your calculator will tell you that β is effectively equal to 1 and thus that v is effectively equal to c. Actually, v is almost c, but we want a more accurate answer, which we can obtain by first solving Eq. 37-8 for $1 - \beta$. To begin we write

$$\gamma = \frac{1}{\sqrt{1 - \beta^2}} = \frac{1}{\sqrt{(1 - \beta)(1 + \beta)}} \approx \frac{1}{\sqrt{2(1 - \beta)}},$$

where we have used the fact that β is so close to unity that $1 + \beta$ is very close to 2. The velocity we seek is contained in the $1 - \beta$ term. Solving for $1 - \beta$ then yields

$$1 - \beta = \frac{1}{2\gamma^2} = \frac{1}{(2)(3.198 \times 10^{11})^2}$$
$$= 4.9 \times 10^{-24} \approx 5 \times 10^{-24}.$$

Thus, $\qquad \beta = 1 - 5 \times 10^{-24}$

and, since $v = \beta c$,

$$v \approx 0.999\ 999\ 999\ 999\ 999\ 999\ 999\ 995c. \quad \text{(Answer)}$$

(b) Suppose that the proton travels along a diameter (9.8×10^4 ly) of the Milky Way galaxy. Approximately how long does the proton take to travel that diameter as measured from the common reference frame of Earth and the Galaxy?

Reasoning: We just saw that this *ultrarelativistic* proton is traveling at a speed barely less than c. By the definition of light-year, light takes 1 y to travel a distance of 1 ly, and so light should take 9.8×10^4 y to travel 9.8×10^4 ly, and this proton should take almost the same time. Thus, from our Earth–Milky Way reference frame, the proton's trip takes

$$\Delta t = 9.8 \times 10^4\ \text{y}. \qquad \text{(Answer)}$$

(c) How long does the trip take as measured in the reference frame of the proton?

1. This problem involves measurements made from two (inertial) reference frames: one is the Earth–Milky Way frame and the other is attached to the proton.

2. This problem also involves two events: the first is when the proton passes one end of the diameter along the Galaxy, and the second is when it passes the opposite end.

3. The time interval between those two events as measured in the proton's reference frame is the proper time interval Δt_0 because the events occur at the same location in that frame — namely, at the proton itself.

4. We can find the proper time interval Δt_0 from the time interval Δt measured in the Earth–Milky Way frame by using Eq. 37-9 ($\Delta t = \gamma \Delta t_0$) for time dilation.

Calculation: Solving Eq. 37-9 for Δt_0 and substituting γ from (a) and Δt from (b), we find

$$\Delta t_0 = \frac{\Delta t}{\gamma} = \frac{9.8 \times 10^4 \text{ y}}{3.198 \times 10^{11}}$$
$$= 3.06 \times 10^{-7} \text{ y} = 9.7 \text{ s.} \qquad \text{(Answer)}$$

In our frame, the trip takes 98 000 y. In the proton's frame, it takes 9.7 s! As promised at the start of this chapter, relative motion can alter the rate at which time passes, and we have here an extreme example.

REVIEW & SUMMARY

The Postulates Einstein's **special theory of relativity** is based on two postulates:

1. The laws of physics are the same for observers in all inertial reference frames. No one frame is preferred over any other.

2. The speed of light in vacuum has the same value c in all directions and in all inertial reference frames.

The speed of light c in vacuum is an ultimate speed that cannot be exceeded by any entity carrying energy or information.

Coordinates of an Event Three space coordinates and one time coordinate specify an **event**. One task of special relativity is to relate these coordinates as assigned by two observers who are in uniform motion with respect to each other.

Simultaneous Events If two observers are in relative motion, they will not, in general, agree as to whether two events are simultaneous. If one of the observers finds two events at different locations to be simultaneous, the other will not, and conversely. Simultaneity is *not* an absolute concept but a relative one, depending on the motion of the observer. The relativity of simultaneity is a direct consequence of the finite ultimate speed c.

Time Dilation If two successive events occur at the same place in an inertial reference frame, the time interval Δt_0 between them, measured on a single clock where they occur, is the **proper time** between the events. *Observers in frames moving relative to that frame will measure a larger value for this interval.* For an observer moving with relative speed v, the measured time interval is

$$\Delta t = \frac{\Delta t_0}{\sqrt{1 - (v/c)^2}} = \frac{\Delta t_0}{\sqrt{1 - \beta^2}}$$
$$= \gamma \Delta t_0 \qquad \text{(time dilation).} \qquad \text{(37-7 to 37-9)}$$

Here $\beta = v/c$ is the **speed parameter** and $\gamma = 1/\sqrt{1 - \beta^2}$ is the **Lorentz factor.** An important result of time dilation is that moving clocks run slow as measured by an observer at rest.

Length Contraction The length L_0 of an object measured by an observer in an inertial reference frame in which the object is at rest is called its **proper length.** *Observers in frames moving relative to that frame and parallel to that length will measure a shorter length.* For an observer moving with relative speed v, the measured length is

$$L = L_0\sqrt{1 - \beta^2} = \frac{L_0}{\gamma} \qquad \text{(length contraction).} \qquad \text{(37-13)}$$

The Lorentz Transformation The *Lorentz transformation* equations relate the spacetime coordinates of a single event as seen by observers in two inertial frames, S and S', where S' is moving relative to S with velocity v in the positive x and x' direction. The four coordinates are related by

$$
\begin{aligned}
x' &= \gamma(x - vt), \\
y' &= y, \\
z' &= z, \\
t' &= \gamma(t - vx/c^2)
\end{aligned}
\qquad (37\text{-}21)
$$

Relativity of Velocities When a particle is moving with speed u' in the positive x' direction in an inertial reference frame S' that itself is moving with speed v parallel to the x direction of a second inertial frame S, the speed u of the particle as measured in S is

$$u = \frac{u' + v}{1 + u'v/c^2} \qquad \text{(relativistic velocity).} \qquad (37\text{-}29)$$

Relativistic Doppler Effect If a source emitting light waves of frequency f_0 moves directly away from a detector with relative radial speed v (and speed parameter $\beta = v/c$), the frequency f measured by the detector is

$$f = f_0\sqrt{\frac{1 - \beta}{1 + \beta}}. \qquad (37\text{-}31)$$

If the source moves directly toward the detector, the signs in Eq. 37-31 are reversed.

For astronomical observations, the Doppler effect is measured in wavelengths. For speeds much less than c, Eq. 37-31 leads to

$$v = \frac{|\Delta\lambda|}{\lambda_0} c, \qquad (37\text{-}36)$$

where $\Delta\lambda$ ($= \lambda - \lambda_0$) is the *Doppler shift* in wavelength due to the motion.

Transverse Doppler Effect If the relative motion of the light source is perpendicular to a line joining the source and detector, the Doppler frequency formula is

$$f = f_0\sqrt{1 - \beta^2}. \qquad (37\text{-}37)$$

This **transverse Doppler effect** is due to time dilation.

Momentum and Energy The following definitions of linear momentum $\vec{p}$, kinetic energy K, and total energy E for a particle of mass m are valid at any physically possible speed:

$$\vec{p} = \gamma m\vec{v} \qquad \text{(momentum)}, \qquad (37\text{-}42)$$
$$E = mc^2 + K = \gamma mc^2 \quad \text{(total energy)}, \qquad (37\text{-}47, 37\text{-}48)$$
$$K = mc^2(\gamma - 1) \qquad \text{(kinetic energy)}. \qquad (37\text{-}52)$$

Here γ is the Lorentz factor for the particle's motion, and mc^2 is the *mass energy*, or *rest energy*, associated with the mass of the particle. These equations lead to the relationships

$$(pc)^2 = K^2 + 2Kmc^2, \qquad (37\text{-}54)$$

and

$$E^2 = (pc)^2 + (mc^2)^2. \qquad (37\text{-}55)$$

When a system of particles undergoes a chemical or nuclear reaction, the Q of the reaction is the negative of the change in the system's total mass energy:

$$Q = M_i c^2 - M_f c^2 = -\Delta M\, c^2, \qquad (37\text{-}50)$$

where M_i is the system's total mass before the reaction and M_f is its total mass after the reaction.

QUESTIONS

1 Figure 37-16 shows two clocks in stationary frame S (they are synchronized in that frame) and one clock in moving frame S'. Clocks C_1 and C_1' read zero when they pass each other. When clocks C_1' and C_2 pass each other, (a) which clock has the smaller reading and (b) which clock measures a proper time?

FIG. 37-16
Question 1.

2 Figure 37-17 shows two clocks in stationary frame S' (they are synchronized in that frame) and one clock in moving frame S. Clocks C_1 and C_1' read zero when they pass each other. When clocks C_1 and C_2' pass each other, (a) which clock has the smaller reading and (b) which clock measures a proper time?

FIG. 37-17   Question 2.

3 The plane of clocks and measuring rods in Fig. 37-18 is like that in Fig. 37-3. The clocks along the x axis are separated (center to center) by 1 light-second, as are the clocks along the y axis, and all the clocks are synchronized via the procedure described in Section 37-3. When the initial synchronizing signal of $t = 0$ from the origin reaches (a) clock A, (b) clock B,

FIG. 37-18   Question 3.

and (c) clock C, what initial time is then set on those clocks? An event occurs at clock A when it reads 10 s. (d) How long does the signal of that event take to travel to an observer stationed at the origin? (e) What time does that observer assign to the event?

4 Sam leaves Venus in a spaceship headed to Mars and passes Sally, who is on Earth, with a relative speed of $0.5c$. (a) Each measures the Venus–Mars voyage time. Who measures a proper time: Sam, Sally, or neither? (b) On the way, Sam sends a pulse of light to Mars. Each measures the travel time of the pulse. Who measures a proper time?

5 A rod is to move at constant speed v along the x axis of reference frame S, with the rod's length parallel to that axis. An observer in frame S is to measure the length L of the rod. Which of the curves in Fig. 37-19 best gives length L (vertical axis of the graph) versus speed parameter β?

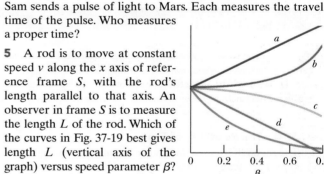

FIG. 37-19
Questions 5 and 7.

6 Figure 37-20 shows a ship (attached to reference frame S') passing us (standing in reference frame S). A proton is fired at nearly the speed of light along the length of the ship, from the front to the rear. (a) Is the spatial separation $\Delta x'$ between the point at which the proton is fired and the point at which it hits the ship's rear wall a positive or negative quantity? (b) Is the temporal separation $\Delta t'$ between those events a positive or negative quantity?

FIG. 37-20   Question 6 and Problem 64.

7 Reference frame S' is to pass reference frame S at speed v along the common direction of the x' and x axes, as in Fig. 37-9. An observer who rides along with frame S' is to count off 25 s on his wristwatch. The corresponding time interval Δt is to be measured by an observer in frame S. Which of the curves in Fig. 37-19 best gives Δt (vertical axis of the graph) versus speed parameter β?

8 While on board a starship, you intercept signals from four shuttle craft that are moving either directly toward or directly away from you. The signals have the same proper frequency f_0. The speed and direction (both relative to you) of the shuttle craft are (a) $0.3c$ toward, (b) $0.6c$ toward, (c) $0.3c$ away, and (d) $0.6c$ away. Rank the shuttle craft according to the frequency you receive, greatest first.

9 Figure 37-21 shows one of four star cruisers that are in a race. As each cruiser passes the starting line, a shuttle craft leaves the cruiser and races toward the finish line. You, judging the race, are stationary relative to the starting and finish lines. The speeds v_c of the cruisers relative to you and the speeds v_s of the shuttle craft relative to their respective starships are, in that order, (1) $0.70c$, $0.40c$; (2) $0.40c$, $0.70c$;

FIG. 37-21 Question 9.

(3) $0.20c$, $0.90c$; (4) $0.50c$, $0.60c$. (a) Without written calculation, rank the shuttle craft according to their speeds relative to you, greatest first. (b) Still without written calculation, rank the shuttle craft according to the distances their pilots measure from the starting line to the finish line, greatest first. (c) Each starship sends a signal to its shuttle craft at a certain frequency f_0 as measured on board the starship. Again without written calculation, rank the shuttle craft according to the frequencies they detect, greatest first.

10 The rest energy and total energy, respectively, of three particles, expressed in terms of a basic amount A are (1) A, $2A$; (2) A, $3A$; (3) $3A$, $4A$. Without written calculation, rank the particles according to their (a) mass, (b) kinetic energy, (c) Lorentz factor, and (d) speed, greatest first.

11 Figure 37-22 shows the triangle of Fig. 37-15 for six particles; the slanted lines 2 and 4 have the same length. Rank the particles according to (a) mass, (b) momentum magnitude, and (c) Lorentz factor, greatest first. (d) Identify which two particles have the same total energy. (e) Rank the three lowest-mass particles according to kinetic energy, greatest first.

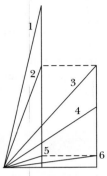

FIG. 37-22 Question 11.

PROBLEMS

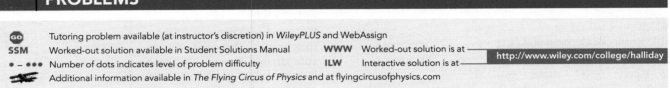

GO	Tutoring problem available (at instructor's discretion) in *WileyPLUS* and WebAssign
SSM	Worked-out solution available in Student Solutions Manual
• – •••	Number of dots indicates level of problem difficulty
	Additional information available in *The Flying Circus of Physics* and at flyingcircusofphysics.com

WWW Worked-out solution is at — http://www.wiley.com/college/halliday
ILW Interactive solution is at —

sec. 37-5 The Relativity of Time

•1 The mean lifetime of stationary muons is measured to be 2.2000 μs. The mean lifetime of high-speed muons in a burst of cosmic rays observed from Earth is measured to be 16.000 μs. To five significant figures, what is the speed parameter β of these cosmic-ray muons relative to Earth?

•2 To eight significant figures, what is the speed parameter β if the Lorentz factor γ is (a) 1.010 000 0, (b) 10.000 000, (c) 100.000 00, and (d) 1000.000 0?

••3 An unstable high-energy particle enters a detector and leaves a track of length 1.05 mm before it decays. Its speed relative to the detector was $0.992c$. What is its proper lifetime? That is, how long would the particle have lasted before decay had it been at rest with respect to the detector? **ILW**

••4 Reference frame S' is to pass reference frame S at speed v along the common direction of the x' and x axes, as in Fig. 37-9. An observer who rides along with frame S' is to count off a certain time interval on his wristwatch. The corresponding time interval Δt is to be measured by

FIG. 37-23 Problem 4.

an observer in frame S. Figure 37-23 gives Δt versus speed parameter β for a range of values for β. The vertical axis scale is set by $\Delta t_a = 14.0$ s. What is interval Δt if $v = 0.98c$?

••5 The premise of the *Planet of the Apes* movies and book is that hibernating astronauts travel far into Earth's future, to a time when human civilization has been replaced by an ape civilization. Considering only special relativity, determine how far into Earth's future the astronauts would travel if they slept for 120 y while traveling relative to Earth with a speed of $0.9990c$, first outward from Earth and then back again.

••6 *(Come) back to the future.* Suppose that a father is 20.00 y older than his daughter. He wants to travel outward from Earth for 2.000 y and then back to Earth for another 2.000 y (both intervals as he measures them) such that he is then 20.00 y *younger* than his daughter. What constant speed parameter β (relative to Earth) is required for the trip?

••7 You wish to make a round trip from Earth in a spaceship, traveling at constant speed in a straight line for exactly 6 months (as you measure the time interval) and then returning at the same constant speed. You wish further, on your return, to find Earth as it will be exactly 1000 years in the future. (a) To eight significant figures, at what speed parameter β must you travel? (b) Does it matter whether you travel in a straight line on your journey?

sec. 37-6   The Relativity of Length

•8   A meter stick in frame S' makes an angle of 30° with the x' axis. If that frame moves parallel to the x axis of frame S with speed 0.90c relative to frame S, what is the length of the stick as measured from S?

•9   A rod lies parallel to the x axis of reference frame S, moving along this axis at a speed of 0.630c. Its rest length is 1.70 m. What will be its measured length in frame S?

•10   An electron of $\beta = 0.999\ 987$ moves along the axis of an evacuated tube that has a length of 3.00 m as measured by a laboratory observer S at rest relative to the tube. An observer S' at rest relative to the electron, however, would see this tube moving with speed v (= βc). What length would observer S' measure for the tube?

•11   A spaceship of rest length 130 m races past a timing station at a speed of 0.740c. (a) What is the length of the spaceship as measured by the timing station? (b) What time interval will the station clock record between the passage of the front and back ends of the ship?   **SSM**

••12   A rod is to move at constant speed v along the x axis of reference frame S, with the rod's length parallel to that axis. An observer in frame S is to measure the length L of the rod. Figure 37-24 gives length L versus speed parameter β for a range of values for β. The vertical axis scale is set by L_a = 1.00 m. What is L if $v = 0.95c$?

FIG. 37-24   Problem 12.

••13   The center of our Milky Way galaxy is about 23 000 ly away. (a) To eight significant figures, at what constant speed parameter would you need to travel exactly 23 000 ly (measured in the Galaxy frame) in exactly 30 y (measured in your frame)? (b) Measured in your frame and in light-years, what length of the Galaxy would pass by you during the trip?   **GO**

••14   The length of a spaceship is measured to be exactly half its rest length. (a) To three significant figures, what is the speed parameter β of the spaceship relative to the observer's frame? (b) By what factor do the spaceship's clocks run slow relative to clocks in the observer's frame?

••15   A space traveler takes off from Earth and moves at speed 0.9900c toward the star Vega, which is 26.00 ly distant. How much time will have elapsed by Earth clocks (a) when the traveler reaches Vega and (b) when Earth observers receive word from the traveler that she has arrived? (c) How much older will Earth observers calculate the traveler to be (measured from her frame) when she reaches Vega than she was when she started the trip?   **GO**

sec. 37-8   Some Consequences of the Lorentz Equations

•16   Inertial frame S' moves at a speed of 0.60c with respect to frame S (Fig. 37-9). Further, $x = x' = 0$ at $t = t' = 0$. Two events are recorded. In frame S, event 1 occurs at the origin at $t = 0$ and event 2 occurs on the x axis at $x = 3.0$ km at $t = 4.0\ \mu$s. According to observer S', what is the time of (a) event 1 and (b) event 2? (c) Do the two observers see the two events in the same sequence or the reverse sequence?

•17   An experimenter arranges to trigger two flashbulbs simultaneously, producing a big flash located at the origin of his reference frame and a small flash at $x = 30.0$ km. An observer moving at a speed of 0.250c in the positive direction of x also views the flashes. (a) What is the time interval between them according to her? (b) Which flash does she say occurs first?

•18   Observer S reports that an event occurred on the x axis of his reference frame at $x = 3.00 \times 10^8$ m at time $t = 2.50$ s. Observer S' and her frame are moving in the positive direction of the x axis at a speed of 0.400c. Further, $x = x' = 0$ at $t = t' = 0$. What are the (a) spatial and (b) temporal coordinate of the event according to S'? If S' were, instead, moving in the *negative* direction of the x axis, what would be the (c) spatial and (d) temporal coordinate of the event according to S'?

•19   In Fig. 37-9, the origins of the two frames coincide at $t = t' = 0$ and the relative speed is 0.950c. Two micrometeorites collide at coordinates $x = 100$ km and $t = 200\ \mu$s according to an observer in frame S. What are the (a) spatial and (b) temporal coordinate of the collision according to an observer in frame S'?   **SSM   WWW**

••20   Bullwinkle in reference frame S' passes you in reference frame S along the common direction of the x' and x axes, as in Fig. 37-9. He carries three meter sticks: meter stick 1 is parallel to the x' axis, meter stick 2 is parallel to the y' axis, and meter stick 3 is parallel to the z' axis. On his wristwatch he counts off 15.0 s, which takes 30.0 s according to you. Two events occur during his passage. According to you, event 1 occurs at $x_1 = 33.0$ m and $t_1 = 22.0$ ns, and event 2 occurs at $x_2 = 53.0$ m and $t_2 = 62.0$ ns. According to your measurements, what is the length of (a) meter stick 1, (b) meter stick 2, and (c) meter stick 3? According to Bullwinkle, what are (d) the spatial separation and (e) the temporal separation between events 1 and 2, and (f) which event occurs first?

••21   A clock moves along an x axis at a speed of 0.600c and reads zero as it passes the origin. (a) Calculate the clock's Lorentz factor. (b) What time does the clock read as it passes $x = 180$ m?   **ILW**

••22   As in Fig. 37-9, reference frame S' passes reference frame S with a certain velocity. Events 1 and 2 are to have a certain temporal separation $\Delta t'$ according to the S' observer. However, their spatial separation $\Delta x'$ according to that observer has not been set yet. Figure 37-25 gives their temporal separation Δt according to the S observer as a function of

FIG. 37-25   Problem 22.

$\Delta x'$ for a range of $\Delta x'$ values. The vertical axis scale is set by $\Delta t_a = 6.00\ \mu$s. What is $\Delta t'$?

••23   In Fig. 37-9, observer S detects two flashes of light. A big flash occurs at $x_1 = 1200$ m and, 5.00 μs later, a small flash occurs at $x_2 = 480$ m. As detected by observer S', the two flashes occur at a single coordinate x'. (a) What is the speed parameter of S', and (b) is S' moving in the positive or negative direction of the x axis? To S', (c) which flash occurs first and (d) what is the time interval between the flashes?

••**24** In Fig. 37-9, observer S detects two flashes of light. A big flash occurs at $x_1 = 1200$ m and, slightly later, a small flash occurs at $x_2 = 480$ m. The time interval between the flashes is $\Delta t = t_2 - t_1$. What is the smallest value of Δt for which observer S' will determine that the two flashes occur at the same x' coordinate?

••**25** *Relativistic reversal of events.* Figures 37-26a and b show the (usual) situation in which a primed reference frame passes an unprimed reference frame, in the common positive direction of the x and x' axes, at a constant relative velocity of magnitude v. We are at rest in the unprimed frame; Bullwinkle, an astute student of relativity in spite of his cartoon upbringing, is at rest in the primed frame. The figures also indicate events A and B that occur at the following spacetime coordinates as measured in our unprimed frame and in Bullwinkle's primed frame:

Event	Unprimed	Primed
A	(x_A, t_A)	(x'_A, t'_A)
B	(x_B, t_B)	(x'_B, t'_B)

In our frame, event A occurs before event B, with temporal separation $\Delta t = t_B - t_A = 1.00 \ \mu s$ and spatial separation $\Delta x = x_B - x_A = 400$ m. Let $\Delta t'$ be the temporal separation of the events according to Bullwinkle. (a) Find an expression for $\Delta t'$ in terms of the speed parameter $\beta \ (= v/c)$ and the given data. Graph $\Delta t'$ versus β for the following two ranges of β:

(b) 0 to 0.01   (v is low, from 0 to 0.01c)
(c) 0.1 to 1   (v is high, from 0.1c to the limit c)

(d) At what value of β is $\Delta t' = 0$? For what range of β is the sequence of events A and B according to Bullwinkle (e) the same as ours and (f) the reverse of ours? (g) Can event A cause event B, or vice versa? Explain.

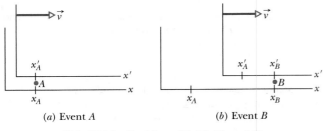

(a) Event A (b) Event B

FIG. 37-26 Problems 25, 26, 62, and 63.

••**26** For the passing reference frames in Fig. 37-26, events A and B occur at the following spacetime coordinates: according to the unprimed frame, (x_A, t_A) and (x_B, t_B); according to the primed frame, (x'_A, t'_A) and (x'_B, t'_B). In the unprimed frame, $\Delta t = t_B - t_A = 1.00 \ \mu s$ and $\Delta x = x_B - x_A = 400$ m. (a) Find an expression for $\Delta x'$ in terms of the speed parameter β and the given data. Graph $\Delta x'$ versus β for two ranges of β: (b) 0 to 0.01 and (c) 0.1 to 1. (d) At what value of β is $\Delta x'$ minimum, and (e) what is that minimum?

sec. 37-9 The Relativity of Velocities

•**27** Galaxy A is reported to be receding from us with a speed of 0.35c. Galaxy B, located in precisely the opposite direction, is also found to be receding from us at this same speed. What multiple of c gives the recessional speed an observer on Galaxy A would find for (a) our galaxy and (b) Galaxy B?

•**28** Stellar system Q_1 moves away from us at a speed of 0.800c. Stellar system Q_2, which lies in the same direction in space but is closer to us, moves away from us at speed 0.400c. What multiple of c gives the speed of Q_2 as measured by an observer in the reference frame of Q_1?

•**29** A particle moves along the x' axis of frame S' with velocity 0.40c. Frame S' moves with velocity 0.60c with respect to frame S. What is the velocity of the particle with respect to frame S? SSM

•**30** In Fig. 37-11, frame S' moves relative to frame S with velocity $0.62c\hat{i}$ while a particle moves parallel to the common x and x' axes. An observer attached to frame S' measures the particle's velocity to be $0.47c\hat{i}$. In terms of c, what is the particle's velocity as measured by an observer attached to frame S according to the (a) relativistic and (b) classical velocity transformation? Suppose, instead, that the S' measure of the particle's velocity is $-0.47c\hat{i}$. What velocity does the observer in S now measure according to the (c) relativistic and (d) classical velocity transformation?

••**31** An armada of spaceships that is 1.00 ly long (in its rest frame) moves with speed 0.800c relative to a ground station in frame S. A messenger travels from the rear of the armada to the front with a speed of 0.950c relative to S. How long does the trip take as measured (a) in the messenger's rest frame, (b) in the armada's rest frame, and (c) by an observer in frame S? GO

••**32** In Fig. 37-27a, particle P is to move parallel to the x and x' axes of reference frames S and S', at a certain velocity relative to frame S. Frame S' is to move parallel to the x axis of frame S at velocity v. Figure 37-27b gives the velocity u' of the particle relative to frame S' for a range of values for v. The vertical axis scale is set by $u'_a = 0.800c$. What value will u' have if (a) $v = 0.90c$ and (b) $v \to c$?

(a)

FIG. 37-27 Problem 32.

••**33** A spaceship whose rest length is 350 m has a speed of 0.82c with respect to a certain reference frame. A micrometeorite, also with a speed of 0.82c in this frame, passes the spaceship on an antiparallel track. How long does it take this object to pass the ship as measured on the ship? SSM ILW WWW

sec. 37-10 Doppler Effect for Light

•**34** Certain wavelengths in the light from a galaxy in the constellation Virgo are observed to be 0.4% longer than the corresponding light from Earth sources. (a) What is the radial speed of this galaxy with respect to Earth? (b) Is the galaxy approaching or receding from Earth?

•**35** Assuming that Eq. 37-36 holds, find how fast you would have to go through a red light to have it appear green. Take 620 nm as the wavelength of red light and 540 nm as the wavelength of green light.

•36   Figure 37-28 is a graph of intensity versus wavelength for light reaching Earth from galaxy NGC 7319, which is about 3×10^8 light-years away. The most intense light is emitted by the oxygen in NGC 7319. In a laboratory that emission is at wavelength $\lambda = 513$ nm, but in the light from NGC 7319 it has been shifted to 525 nm due to the Doppler effect (all the emissions from NGC 7319 have been shifted). (a) What is the radial speed of NGC 7319 relative to Earth? (b) Is the relative motion toward or away from our planet?

FIG. 37-28   Problem 36.

•37   A spaceship, moving away from Earth at a speed of 0.900c, reports back by transmitting at a frequency (measured in the spaceship frame) of 100 MHz. To what frequency must Earth receivers be tuned to receive the report?   SSM

•38   A sodium light source moves in a horizontal circle at a constant speed of 0.100c while emitting light at the proper wavelength of $\lambda_0 = 589.00$ nm. Wavelength λ is measured for that light by a detector fixed at the center of the circle. What is the wavelength shift $\lambda - \lambda_0$?

••39   A spaceship is moving away from Earth at speed 0.20c. A source on the rear of the ship emits light at wavelength 450 nm according to someone on the ship. What (a) wavelength and (b) color (blue, green, yellow, or red) are detected by someone on Earth watching the ship?   SSM

sec. 37-12   A New Look at Energy

•40   What is the minimum energy that is required to break a nucleus of ^{12}C (of mass 11.996 71 u) into three nuclei of ^{4}He (of mass 4.001 51 u each)?

•41   How much work must be done to increase the speed of an electron (a) from 0.18c to 0.19c and (b) from 0.98c to 0.99c? Note that the speed increase is 0.01c in both cases.

•42   In the reaction p + ^{19}F → α + ^{16}O, the masses are

$m(p) = $ 1.007825 u, $m(\alpha) = $ 4.002603 u,

$m(F) = $ 18.998405 u, $m(O) = $ 15.994915 u.

Calculate the Q of the reaction from these data.

•43   The mass of an electron is $9.109\ 381\ 88 \times 10^{-31}$ kg. To six significant figures, find (a) γ and (b) β for an electron with kinetic energy $K = 100.000$ MeV.   SSM WWW

•44   How much work must be done to increase the speed of an electron from rest to (a) 0.500c, (b) 0.990c, and (c) 0.9990c?

••45   What must be the momentum of a particle with mass

m so that the total energy of the particle is 3.00 times its rest energy?   ILW

••46   The mass of an electron is $9.109\ 381\ 88 \times 10^{-31}$ kg. To eight significant figures, find the following for the given electron kinetic energy: (a) γ and (b) β for $K = 1.000\ 000\ 0$ keV, (c) γ and (d) β for $K = 1.000\ 000\ 0$ MeV, and (e) γ and (f) β for $K = 1.000\ 000\ 0$ GeV.

••47   As you read this page (on paper or monitor screen), a cosmic ray proton passes along the left–right width of the page with relative speed v and a total energy of 14.24 nJ. According to your measurements, that left–right width is 21.0 cm. (a) What is the width according to the proton's reference frame? How much time did the passage take according to (b) your frame and (c) the proton's frame?   GO

••48   (a) The energy released in the explosion of 1.00 mol of TNT is 3.40 MJ. The molar mass of TNT is 0.227 kg/mol. What weight of TNT is needed for an explosive release of 1.80×10^{14} J? (b) Can you carry that weight in a backpack, or is a truck or train required? (c) Suppose that in an explosion of a fission bomb, 0.080% of the fissionable mass is converted to released energy. What weight of fissionable material is needed for an explosive release of 1.80×10^{14} J? (d) Can you carry that weight in a backpack, or is a truck or train required?

••49   A certain particle of mass m has momentum of magnitude mc. What are (a) β, (b) γ, and (c) the ratio K/E_0?

••50   What is β for a particle with (a) $K = 2.00E_0$ and (b) $E = 2.00E_0$?

••51   Quasars are thought to be the nuclei of active galaxies in the early stages of their formation. A typical quasar radiates energy at the rate of 10^{41} W. At what rate is the mass of this quasar being reduced to supply this energy? Express your answer in solar mass units per year, where one solar mass unit (1 smu = 2.0×10^{30} kg) is the mass of our Sun.

••52   (a) If m is a particle's mass, p is its momentum magnitude, and K is its kinetic energy, show that

$$m = \frac{(pc)^2 - K^2}{2Kc^2}.$$

(b) For low particle speeds, show that the right side of the equation reduces to m. (c) If a particle has $K = 55.0$ MeV when $p = 121$ MeV/c, what is the ratio m/m_e of its mass to the electron mass?

••53   A 5.00-grain aspirin tablet has a mass of 320 mg. For how many kilometers would the energy equivalent of this mass power an automobile? Assume 12.75 km/L and a heat of combustion of 3.65×10^7 J/L for the gasoline used in the automobile.   SSM

••54   To four significant figures, find the following when the kinetic energy is 10.00 MeV: (a) γ and (b) β for an electron ($E_0 = 0.510\ 998$ MeV), (c) γ and (d) β for a proton ($E_0 = 938.272$ MeV), and (e) γ and (f) β for an α particle ($E_0 = 3727.40$ MeV).

••55   In Section 28-6, we showed that a particle of charge q and mass m will move in a circle of radius $r = mv/|q|B$ when its velocity $\vec{v}$ is perpendicular to a uniform magnetic field $\vec{B}$. We also found that the period T of the motion is independent of speed v. These two results are approximately correct if $v \ll c$. For relativistic speeds, we must use the correct equation for the radius:

$$r = \frac{p}{|q|B} = \frac{\gamma mv}{|q|B}.$$

(a) Using this equation and the definition of period ($T = 2\pi r/v$), find the correct expression for the period. (b) Is T independent of v? If a 10.0 MeV electron moves in a circular path in a uniform magnetic field of magnitude 2.20 T, what are (c) the radius according to Chapter 28, (d) the correct radius, (e) the period according to Chapter 28, and (f) the correct period?

••**56** The mass of a muon is 207 times the electron mass; the average lifetime of muons at rest is 2.20 μs. In a certain experiment, muons moving through a laboratory are measured to have an average lifetime of 6.90 μs. For the moving muons, what are (a) β, (b) K, and (c) p (in MeV/c)? **GO**

••**57** In a high-energy collision between a cosmic-ray particle and a particle near the top of Earth's atmosphere, 120 km above sea level, a pion is created. The pion has a total energy E of 1.35×10^5 MeV and is traveling vertically downward. In the pion's rest frame, the pion decays 35.0 ns after its creation. At what altitude above sea level, as measured from Earth's reference frame, does the decay occur? The rest energy of a pion is 139.6 MeV.

••**58** Apply the binomial theorem (Appendix E) to the last part of Eq. 37-52 for the kinetic energy of a particle. (a) Retain the first two terms of the expansion to show the kinetic energy in the form

$$K = (\text{first term}) + (\text{second term}).$$

The first term is the classical expression for kinetic energy. The second term is the first-order correction to the classical expression. Assume the particle is an electron. If its speed v is $c/20$, what is the value of (b) the classical expression and (c) the first-order correction? If the electron's speed is $0.80c$, what is the value of (d) the classical expression and (e) the first-order correction? (f) At what speed parameter β does the first-order correction become 10% or greater of the classical expression?

•••**59** An alpha particle with kinetic energy 7.70 MeV collides with an ^{14}N nucleus at rest, and the two transform into an ^{17}O nucleus and a proton. The proton is emitted at 90° to the direction of the incident alpha particle and has a kinetic energy of 4.44 MeV. The masses of the various particles are alpha particle, 4.00260 u; ^{14}N, 14.00307 u; proton, 1.007825 u; and ^{17}O, 16.99914 u. In MeV, what are (a) the kinetic energy of the oxygen nucleus and (b) the Q of the reaction? (*Hint:* The speeds of the particles are much less than c.)

Additional Problems

60 In Fig. 37-29a, particle P is to move parallel to the x and x' axes of reference frames S and S', at a certain velocity relative to frame S. Frame S' is to move parallel to the x axis of frame S at velocity v. Figure 37-29b gives the velocity u' of the particle relative to frame S' for a range of values for v.

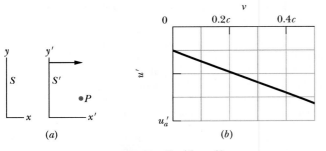

FIG. 37-29 Problem 60.

The vertical axis scale is set by $u'_a = -0.800c$. What value will u' have if (a) $v = 0.80c$ and (b) $v \rightarrow c$?

61 *Superluminal jets.* Figure 37-30a shows the path taken by a knot in a jet of ionized gas that has been expelled from a galaxy. The knot travels at constant velocity $\vec{v}$ at angle θ from the direction of Earth. The knot occasionally emits a burst of light, which is eventually detected on Earth. Two bursts are indicated in Fig. 37-30a, separated by time t as measured in a stationary frame near the bursts. The bursts are shown in Fig. 37-30b as if they were photographed on the same piece of film, first when light from burst 1 arrived on Earth and then later when light from burst 2 arrived. The apparent distance D_{app} traveled by the knot between the two bursts is the distance across an Earth-observer's view of the knot's path. The apparent time T_{app} between the bursts is the difference in the arrival times of the light from them. The apparent speed of the knot is then $V_{app} = D_{app}/T_{app}$. In terms of v, t, and θ, what are (a) D_{app} and (b) T_{app}? (c) Evaluate V_{app} for $v = 0.980c$ and $\theta = 30.0°$. When superluminal (faster than light) jets were first observed, they seemed to defy special relativity—at least until the correct geometry (Fig. 37-30a) was understood. **GO**

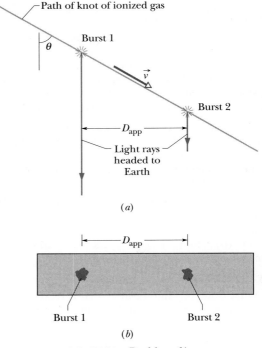

(a)

(b)

FIG. 37-30 Problem 61.

62 *Temporal separation between two events.* Events A and B occur with the following spacetime coordinates in the reference frames of Fig. 37-26: according to the unprimed frame, (x_A, t_A) and (x_B, t_B); according to the primed frame, (x'_A, t'_A) and (x'_B, t'_B). In the unprimed frame, $\Delta t = t_B - t_A = 1.00$ μs and $\Delta x = x_B - x_A = 240$ m. (a) Find an expression for $\Delta t'$ in terms of the speed parameter β and the given data. Graph $\Delta t'$ versus β for the following two ranges of β: (b) 0 to 0.01 and (c) 0.1 to 1. (d) At what value of β is $\Delta t'$ minimum and (e) what is that minimum? (f) Can one of these events cause the other? Explain.

63 *Spatial separation between two events.* For the passing reference frames of Fig. 37-26, events A and B occur with the following spacetime coordinates: according to the unprimed frame, (x_A, t_A) and (x_B, t_B); according to the primed frame, (x'_A, t'_A) and (x'_B, t'_B). In the unprimed frame, $\Delta t = t_B - t_A = 1.00$ μs and $\Delta x = x_B - x_A = 240$ m. (a) Find an expression for $\Delta x'$ in terms of the speed parameter β and the given data. Graph $\Delta x'$ versus β for two ranges of β: (b) 0 to 0.01 and (c) 0.1 to 1. (d) At what value of β is $\Delta x' = 0$?

64 Figure 37-20 shows a ship (attached to reference frame S') passing us (standing in reference frame S) with velocity $\vec{v} = 0.950c\hat{i}$. A proton is fired at speed $0.980c$ relative to the ship from the front of the ship to the rear. The proper length of the ship is 760 m. What is the temporal separation between the time the proton is fired and the time it hits the rear wall of the ship according to (a) a passenger in the ship and (b) us? Suppose that, instead, the proton is fired from the rear to the front. What then is the temporal separation between the time it is fired and the time it hits the front wall according to (c) the passenger and (d) us?

65 *The car-in-the-garage problem.* Carman has just purchased the world's longest stretch limo, which has a proper length of $L_c = 30.5$ m. In Fig. 37-31a, it is shown parked in front of a garage with a proper length of $L_g = 6.00$ m. The garage has a front door (shown open) and a back door (shown closed). The limo is obviously longer than the garage. Still, Garageman, who owns the garage and knows something about relativistic length contraction, makes a bet with Carman that the limo can fit in the garage with both doors closed. Carman, who dropped his physics course before reaching special relativity, says such a thing, even in principle, is impossible.

To analyze Garageman's scheme, an x_c axis is attached to the limo, with $x_c = 0$ at the rear bumper, and an x_g axis is attached to the garage, with $x_g = 0$ at the (now open) front door. Then Carman is to drive the limo directly toward the front door at a velocity of $0.9980c$ (which is, of course, both technically and financially impossible). Carman is stationary in the x_c reference frame; Garageman is stationary in the x_g reference frame.

There are two events to consider. *Event 1:* When the rear bumper clears the front door, the front door is closed. Let the time of this event be zero to both Carman and Garageman: $t_{g1} = t_{c1} = 0$. The event occurs at $x_c = x_g = 0$. Figure 37-31b shows event 1 according to the x_g reference frame. *Event 2:* When the front bumper reaches the back door, that door opens. Figure 37-31c shows event 2 according to the x_g reference frame.

According to Garageman, (a) what is the length of the limo, and what are the spacetime coordinates (b) x_{g2} and (c) t_{g2} of event 2? (d) For how long is the limo temporarily "trapped"

inside the garage with both doors shut? Now consider the situation from the x_c reference frame, in which the garage comes racing past the limo at a velocity of $-0.9980c$. According to Carman, (e) what is the length of the passing garage, what are the spacetime coordinates (f) x_{c2} and (g) t_{c2} of event 2, (h) is the limo ever in the garage with both doors shut, and (i) which event occurs first? (j) Sketch events 1 and 2 as seen by Carman. (k) Are the events causally related; that is, does one of them cause the other? (l) Finally, who wins the bet?

66 Reference frame S' passes reference frame S with a certain velocity as in Fig. 37-9. Events 1 and 2 are to have a certain spatial separation $\Delta x'$ according to the S' observer. However, their temporal separation $\Delta t'$ according to that observer has not been set yet. Figure 37-32 gives their spatial separation Δx according to the S observer as a function of $\Delta t'$ for a range of $\Delta t'$ values. The vertical axis scale is set by $\Delta x_a = 10.0$ m. What is $\Delta x'$?

FIG. 37-32 Problem 66.

67 *Another approach to velocity transformations.* In Fig. 37-33, reference frames B and C move past reference frame A in the common direction of their x axes. Represent the x components of the velocities of one frame relative to another with a two-letter subscript. For example, v_{AB} is the x component of the velocity of A relative to B.

FIG. 37-33 Problems 67, 68, and 69.

Similarly, represent the corresponding speed parameters with two-letter subscripts. For example, β_{AB} ($= v_{AB}/c$) is the speed parameter corresponding to v_{AB}. (a) Show that

$$\beta_{AC} = \frac{\beta_{AB} + \beta_{BC}}{1 + \beta_{AB}\beta_{BC}}.$$

Let M_{AB} represent the ratio $(1 - \beta_{AB})/(1 + \beta_{AB})$, and let M_{BC} and M_{AC} represent similar ratios. (b) Show that the relation

$$M_{AC} = M_{AB}M_{BC}$$

is true by deriving the equation of part (a) from it.

68 *Continuation of Problem 67.* Use the result of part (b) in Problem 67 for the motion along a single axis in the following situation. Frame A in Fig. 37-33 is attached to a particle that moves with velocity $+0.500c$ past frame B, which moves past frame C with a velocity of $+0.500c$. What are (a) M_{AC}, (b) β_{AC}, and (c) the velocity of the particle relative to frame C?

69 *Continuation of Problem 67.* Let reference frame C in Fig. 37-33 move past reference frame D (not shown). (a) Show that

$$M_{AD} = M_{AB}M_{BC}M_{CD}.$$

(b) Now put this general result to work: Three particles move parallel to a single axis on which an observer is stationed. Let plus and minus signs indicate the directions of motion along that axis. Particle A moves past particle B at $\beta_{AB} = +0.20$. Particle B moves past particle C at $\beta_{BC} = -0.40$. Particle C moves past observer D at $\beta_{CD} = +0.60$. What is the velocity of particle A relative to observer D? (The solution technique here is *much* faster than using Eq. 37-29.)

(a)

(b)   (c)

FIG. 37-31 Problem 65.

70 The total energy of a proton passing through a laboratory apparatus is 10.611 nJ. What is its speed parameter β? Use the proton mass given in Appendix B under "Best Value," not the commonly remembered rounded number.

71 If we intercept an electron having total energy 1533 MeV that came from Vega, which is 26 ly from us, how far in light-years was the trip in the rest frame of the electron? **SSM**

72 A pion is created in the higher reaches of Earth's atmosphere when an incoming high-energy cosmic-ray particle collides with an atomic nucleus. A pion so formed descends toward Earth with a speed of 0.99c. In a reference frame in which they are at rest, pions decay with an average life of 26 ns. As measured in a frame fixed with respect to Earth, how far (on the average) will such a pion move through the atmosphere before it decays?

73 What is the momentum in MeV/c of an electron with a kinetic energy of 2.00 MeV? **SSM**

74 Find the speed parameter of a particle that takes 2.0 y longer than light to travel a distance of 6.0 ly.

75 How much work is needed to accelerate a proton from a speed of 0.9850c to a speed of 0.9860c? **SSM**

76 An airplane whose rest length is 40.0 m is moving at uniform velocity with respect to Earth, at a speed of 630 m/s. (a) By what fraction of its rest length is it shortened to an observer on Earth? (b) How long would it take, according to Earth clocks, for the airplane's clock to fall behind by 1.00 μs? (Use special relativity in your calculations.)

77 To circle Earth in low orbit, a satellite must have a speed of about 2.7×10^4 km/h. Suppose that two such satellites orbit Earth in opposite directions. (a) What is their relative speed as they pass, according to the classical Galilean velocity transformation equation? (b) What fractional error do you make in (a) by not using the (correct) relativistic transformation equation? **SSM**

78 A radar transmitter T is fixed to a reference frame S' that is moving to the right with speed v relative to reference frame S (Fig. 37-34). A mechanical timer (essentially a clock) in frame S', having a period τ_0 (measured in S'), causes transmitter T to emit timed radar pulses, which travel at the speed of light and are received by R, a receiver fixed in frame S. (a) What is the period τ of the timer as detected by observer A, who is fixed in frame S? (b) Show that at receiver R the time interval between pulses arriving from T is not τ or τ_0, but

$$\tau_R = \tau_0 \sqrt{\frac{c + v}{c - v}}.$$

(c) Explain why receiver R and observer A, who are in the same reference frame, measure a different period for the

transmitter. (*Hint:* A clock and a radar pulse are not the same thing.)

79 A particle with mass m has speed $c/2$ relative to inertial frame S. The particle collides with an identical particle at rest relative to frame S. Relative to S, what is the speed of a frame S' in which the total momentum of these particles is zero? This frame is called the *center of momentum frame.*

80 An elementary particle produced in a laboratory experiment travels 0.230 mm through the lab at a relative speed of 0.960c before it decays (becomes another particle). (a) What is the proper lifetime of the particle? (b) What is the distance the particle travels as measured from its rest frame?

81 What are (a) K, (b) E, and (c) p (in GeV/c) for a proton moving at speed 0.990c? What are (d) K, (e) E, and (f) p (in MeV/c) for an electron moving at speed 0.990c?

82 In the red shift of radiation from a distant galaxy, a certain radiation, known to have a wavelength of 434 nm when observed in the laboratory, has a wavelength of 462 nm. (a) What is the radial speed of the galaxy relative to Earth? (b) Is the galaxy approaching or receding from Earth?

83 (a) What potential difference would accelerate an electron to speed c according to classical physics? (b) With this potential difference, what speed would the electron actually attain?

84 The radius of Earth is 6370 km, and its orbital speed about the Sun is 30 km/s. Suppose Earth moves past an observer at this speed. To the observer, by how much does Earth's diameter contract along the direction of motion?

85 A spaceship at rest in a certain reference frame S is given a speed increment of 0.50c. Relative to its new rest frame, it is then given a further 0.50c increment. This process is continued until its speed with respect to its original frame S exceeds 0.999c. How many increments does this process require?

86 A Foron cruiser moving directly toward a Reptulian scout ship fires a decoy toward the scout ship. Relative to the scout ship, the speed of the decoy is 0.980c and the speed of the Foron cruiser is 0.900c. What is the speed of the decoy relative to the cruiser?

87 One cosmic-ray particle approaches Earth along Earth's north–south axis with a speed of 0.80c toward the geographic north pole, and another approaches with a speed of 0.60c toward the geographic south pole (Fig. 37-35). What is the relative speed of approach of one particle with respect to the other?

88 (a) How much energy is released in the explosion of a fission bomb containing 3.0 kg of fissionable material? Assume that 0.10% of the mass is converted to released energy. (b) What mass of TNT would have to explode to provide the same energy release? Assume that each mole of TNT liberates 3.4 MJ of energy on exploding. The molecular mass of TNT is 0.227 kg/mol. (c) For the same mass of explosive, what is the ratio of the energy released in a nuclear explosion to that released in a TNT explosion?

FIG. 37-34 Problem 78.

0.80c

Geographic north pole

Geographic south pole

0.60c

FIG. 37-35
Problem 87.

The International System of Units (SI)*

TABLE 1

The SI Base Units

Quantity	Name	Symbol	Definition
length	meter	m	"...the length of the path traveled by light in vacuum in 1/299,792,458 of a second." (1983)
mass	kilogram	kg	"...this prototype [a certain platinum–iridium cylinder] shall henceforth be considered to be the unit of mass." (1889)
time	second	s	"...the duration of 9,192,631,770 periods of the radiation corresponding to the transition between the two hyperfine levels of the ground state of the cesium-133 atom." (1967)
electric current	ampere	A	"...that constant current which, if maintained in two straight parallel conductors of infinite length, of negligible circular cross section, and placed 1 meter apart in vacuum, would produce between these conductors a force equal to 2×10^{-7} newton per meter of length." (1946)
thermodynamic temperature	kelvin	K	"...the fraction 1/273.16 of the thermodynamic temperature of the triple point of water." (1967)
amount of substance	mole	mol	"...the amount of substance of a system which contains as many elementary entities as there are atoms in 0.012 kilogram of carbon-12." (1971)
luminous intensity	candela	cd	"...the luminous intensity, in a given direction, of a source that emits monochromatic radiation of frequency 540×10^{12} hertz and that has a radiant intensity in that direction of 1/683 watt per steradian." (1979)

*Adapted from "The International System of Units (SI)," National Bureau of Standards Special Publication 330, 1972 edition. The definitions above were adopted by the General Conference of Weights and Measures, an international body, on the dates shown. In this book we do not use the candela.

TABLE 2

Some SI Derived Units

Quantity	Name of Unit	Symbol	
area	square meter	m^2	
volume	cubic meter	m^3	
frequency	hertz	Hz	s^{-1}
mass density (density)	kilogram per cubic meter	kg/m^3	
speed, velocity	meter per second	m/s	
angular velocity	radian per second	rad/s	
acceleration	meter per second per second	m/s^2	
angular acceleration	radian per second per second	rad/s^2	
force	newton	N	$kg \cdot m/s^2$
pressure	pascal	Pa	N/m^2
work, energy, quantity of heat	joule	J	$N \cdot m$
power	watt	W	J/s
quantity of electric charge	coulomb	C	$A \cdot s$
potential difference, electromotive force	volt	V	W/A
electric field strength	volt per meter (or newton per coulomb)	V/m	N/C
electric resistance	ohm	Ω	V/A
capacitance	farad	F	$A \cdot s/V$
magnetic flux	weber	Wb	$V \cdot s$
inductance	henry	H	$V \cdot s/A$
magnetic flux density	tesla	T	Wb/m^2
magnetic field strength	ampere per meter	A/m	
entropy	joule per kelvin	J/K	
specific heat	joule per kilogram kelvin	$J/(kg \cdot K)$	
thermal conductivity	watt per meter kelvin	$W/(m \cdot K)$	
radiant intensity	watt per steradian	W/sr	

TABLE 3

The SI Supplementary Units

Quantity	Name of Unit	Symbol
plane angle	radian	rad
solid angle	steradian	sr

Some Fundamental B
Constants of Physics*

Constant	Symbol	Computational Value	Best (1998) Value Value[a]	Uncertainty[b]
Speed of light in a vacuum	c	3.00×10^8 m/s	2.997 924 58	exact
Elementary charge	e	1.60×10^{-19} C	1.602 176 462	0.039
Gravitational constant	G	6.67×10^{-11} m³/s²·kg	6.673	1500
Universal gas constant	R	8.31 J/mol·K	8.314 472	1.7
Avogadro constant	N_A	6.02×10^{23} mol⁻¹	6.022 141 99	0.079
Boltzmann constant	k	1.38×10^{-23} J/K	1.380 650 3	1.7
Stefan–Boltzmann constant	σ	5.67×10^{-8} W/m²·K⁴	5.670 400	7.0
Molar volume of ideal gas at STP[d]	V_m	2.27×10^{-2} m³/mol	2.271 098 1	1.7
Permittivity constant	ϵ_0	8.85×10^{-12} F/m	8.854 187 817 62	exact
Permeability constant	μ_0	1.26×10^{-6} H/m	1.256 637 061 43	exact
Planck constant	h	6.63×10^{-34} J·s	6.626 068 76	0.078
Electron mass[c]	m_e	9.11×10^{-31} kg	9.109 381 88	0.079
		5.49×10^{-4} u	5.485 799 110	0.0021
Proton mass[c]	m_p	1.67×10^{-27} kg	1.672 621 58	0.079
		1.0073 u	1.007 276 466 88	1.3×10^{-4}
Ratio of proton mass to electron mass	m_p/m_e	1840	1836.152 667 5	0.0021
Electron charge-to-mass ratio	e/m_e	1.76×10^{11} C/kg	1.758 820 174	0.040
Neutron mass[c]	m_n	1.68×10^{-27} kg	1.674 927 16	0.079
		1.0087 u	1.008 664 915 78	5.4×10^{-4}
Hydrogen atom mass[c]	m_{1_H}	1.0078 u	1.007 825 031 6	0.0005
Deuterium atom mass[c]	m_{2_H}	2.0141 u	2.014 101 777 9	0.0005
Helium atom mass[c]	$m_{4_{He}}$	4.0026 u	4.002 603 2	0.067
Muon mass	m_μ	1.88×10^{-28} kg	1.883 531 09	0.084
Electron magnetic moment	μ_e	9.28×10^{-24} J/T	9.284 763 62	0.040
Proton magnetic moment	μ_p	1.41×10^{-26} J/T	1.410 606 663	0.041
Bohr magneton	μ_B	9.27×10^{-24} J/T	9.274 008 99	0.040
Nuclear magneton	μ_N	5.05×10^{-27} J/T	5.050 783 17	0.040
Bohr radius	a	5.29×10^{-11} m	5.291 772 083	0.0037
Rydberg constant	R	1.10×10^7 m⁻¹	1.097 373 156 854 8	7.6×10^{-6}
Electron Compton wavelength	λ_C	2.43×10^{-12} m	2.426 310 215	0.0073

[a]Values given in this column should be given the same unit and power of 10 as the computational value.
[b]Parts per million.
[c]Masses given in u are in unified atomic mass units, where 1 u = 1.660 538 86 $\times 10^{-27}$ kg.
[d]STP means standard temperature and pressure: 0°C and 1.0 atm (0.1 MPa).

*The values in this table were selected from the 1998 CODATA recommended values (www.physics.nist.gov).

C | Some Astronomical Data

Some Distances from Earth

To the Moon*	3.82×10^8 m	To the center of our galaxy	2.2×10^{20} m
To the Sun*	1.50×10^{11} m	To the Andromeda Galaxy	2.1×10^{22} m
To the nearest star (Proxima Centauri)	4.04×10^{16} m	To the edge of the observable universe	$\sim 10^{26}$ m

*Mean distance.

The Sun, Earth, and the Moon

Property	Unit	Sun	Earth	Moon
Mass	kg	1.99×10^{30}	5.98×10^{24}	7.36×10^{22}
Mean radius	m	6.96×10^8	6.37×10^6	1.74×10^6
Mean density	kg/m^3	1410	5520	3340
Free-fall acceleration at the surface	m/s^2	274	9.81	1.67
Escape velocity	km/s	618	11.2	2.38
Period of rotation[a]	—	37 d at poles[b] 26 d at equator[b]	23 h 56 min	27.3 d
Radiation power[c]	W	3.90×10^{26}		

[a]Measured with respect to the distant stars.

[b]The Sun, a ball of gas, does not rotate as a rigid body.

[c]Just outside Earth's atmosphere solar energy is received, assuming normal incidence, at the rate of 1340 W/m^2.

Some Properties of the Planets

	Mercury	Venus	Earth	Mars	Jupiter	Saturn	Uranus	Neptune	Pluto
Mean distance from Sun, 10^6 km	57.9	108	150	228	778	1430	2870	4500	5900
Period of revolution, y	0.241	0.615	1.00	1.88	11.9	29.5	84.0	165	248
Period of rotation,[a] d	58.7	-243^b	0.997	1.03	0.409	0.426	-0.451^b	0.658	6.39
Orbital speed, km/s	47.9	35.0	29.8	24.1	13.1	9.64	6.81	5.43	4.74
Inclination of axis to orbit	<28°	≈3°	23.4°	25.0°	3.08°	26.7°	97.9°	29.6°	57.5°
Inclination of orbit to Earth's orbit	7.00°	3.39°		1.85°	1.30°	2.49°	0.77°	1.77°	17.2°
Eccentricity of orbit	0.206	0.0068	0.0167	0.0934	0.0485	0.0556	0.0472	0.0086	0.250
Equatorial diameter, km	4880	12 100	12 800	6790	143 000	120 000	51 800	49 500	2300
Mass (Earth = 1)	0.0558	0.815	1.000	0.107	318	95.1	14.5	17.2	0.002
Density (water = 1)	5.60	5.20	5.52	3.95	1.31	0.704	1.21	1.67	2.03
Surface value of g,[c] m/s^2	3.78	8.60	9.78	3.72	22.9	9.05	7.77	11.0	0.5
Escape velocity,[c] km/s	4.3	10.3	11.2	5.0	59.5	35.6	21.2	23.6	1.1
Known satellites	0	0	1	2	60 + ring	31 + rings	21 + rings	11 + rings	1

[a]Measured with respect to the distant stars.

[b]Venus and Uranus rotate opposite their orbital motion.

[c]Gravitational acceleration measured at the planet's equator.

Conversion Factors D

Conversion factors may be read directly from these tables. For example, 1 degree = 2.778×10^{-3} revolutions, so $16.7° = 16.7 \times 2.778 \times 10^{-3}$ rev. The SI units are fully capitalized. Adapted in part from G. Shortley and D. Williams, *Elements of Physics,* 1971, Prentice-Hall, Englewood Cliffs, NJ.

Plane Angle

	°	′	″	RADIAN	rev
1 degree =	1	60	3600	1.745×10^{-2}	2.778×10^{-3}
1 minute =	1.667×10^{-2}	1	60	2.909×10^{-4}	4.630×10^{-5}
1 second =	2.778×10^{-4}	1.667×10^{-2}	1	4.848×10^{-6}	7.716×10^{-7}
1 RADIAN =	57.30	3438	2.063×10^{5}	1	0.1592
1 revolution =	360	2.16×10^{4}	1.296×10^{6}	6.283	1

Solid Angle

1 sphere = 4π steradians = 12.57 steradians

Length

	cm	METER	km	in.	ft	mi
1 centimeter =	1	10^{-2}	10^{-5}	0.3937	3.281×10^{-2}	6.214×10^{-6}
1 METER =	100	1	10^{-3}	39.37	3.281	6.214×10^{-4}
1 kilometer =	10^{5}	1000	1	3.937×10^{4}	3281	0.6214
1 inch =	2.540	2.540×10^{-2}	2.540×10^{-5}	1	8.333×10^{-2}	1.578×10^{-5}
1 foot =	30.48	0.3048	3.048×10^{-4}	12	1	1.894×10^{-4}
1 mile =	1.609×10^{5}	1609	1.609	6.336×10^{4}	5280	1

1 angström = 10^{-10} m   1 fermi = 10^{-15} m   1 fathom = 6 ft   1 rod = 16.5 ft

1 nautical mile = 1852 m   1 light-year = 9.461×10^{12} km   1 Bohr radius = 5.292×10^{-11} m   1 mil = 10^{-3} in.

   = 1.151 miles = 6076 ft   1 parsec = 3.084×10^{13} km   1 yard = 3 ft   1 nm = 10^{-9} m

Area

	METER^2	cm^2	ft^2	in.^{2}
1 SQUARE METER =	1	10^{4}	10.76	1550
1 square centimeter =	10^{-4}	1	1.076×10^{-3}	0.1550
1 square foot =	9.290×10^{-2}	929.0	1	144
1 square inch =	6.452×10^{-4}	6.452	6.944×10^{-3}	1

1 square mile = 2.788×10^{7} ft^2 = 640 acres   1 acre = 43 560 ft^2

1 barn = 10^{-28} m^2 1 hectare = 10^{4} m^2 = 2.471 acres

Volume

	METER^3	cm^3	L	ft^3	in.^{3}
1 CUBIC METER = 1		10^6	1000	35.31	6.102×10^4
1 cubic centimeter = 10^{-6}		1	1.000×10^{-3}	3.531×10^{-5}	6.102×10^{-2}
1 liter = 1.000×10^{-3}		1000	1	3.531×10^{-2}	61.02
1 cubic foot = 2.832×10^{-2}		2.832×10^4	28.32	1	1728
1 cubic inch = 1.639×10^{-5}		16.39	1.639×10^{-2}	5.787×10^{-4}	1

1 U.S. fluid gallon = 4 U.S. fluid quarts = 8 U.S. pints = 128 U.S. fluid ounces = 231 in.^{3}
1 British imperial gallon = 277.4 in.^{3} = 1.201 U.S. fluid gallons

Mass

Quantities in the colored areas are not mass units but are often used as such. For example, when we write 1 kg "=" 2.205 lb, this means that a kilogram is a *mass* that *weighs* 2.205 pounds at a location where g has the standard value of 9.80665 m/s^2.

	g	KILOGRAM	slug	u	oz	lb	ton
1 gram = 1		0.001	6.852×10^{-5}	6.022×10^{23}	3.527×10^{-2}	2.205×10^{-3}	1.102×10^{-6}
1 KILOGRAM = 1000		1	6.852×10^{-2}	6.022×10^{26}	35.27	2.205	1.102×10^{-3}
1 slug = 1.459×10^4		14.59	1	8.786×10^{27}	514.8	32.17	1.609×10^{-2}
1 atomic mass unit = 1.661×10^{-24}		1.661×10^{-27}	1.138×10^{-28}	1	5.857×10^{-26}	3.662×10^{-27}	1.830×10^{-30}
1 ounce = 28.35		2.835×10^{-2}	1.943×10^{-3}	1.718×10^{25}	1	6.250×10^{-2}	3.125×10^{-5}
1 pound = 453.6		0.4536	3.108×10^{-2}	2.732×10^{26}	16	1	0.0005
1 ton = 9.072×10^5		907.2	62.16	5.463×10^{29}	3.2×10^4	2000	1

1 metric ton = 1000 kg

Density

Quantities in the colored areas are weight densities and, as such, are dimensionally different from mass densities. See the note for the mass table.

	slug/ft^3	KILOGRAM/ METER^3	g/cm^3	lb/ft^3	lb/in.^{3}
1 slug per foot^3 = 1		515.4	0.5154	32.17	1.862×10^{-2}
1 KILOGRAM per METER^3 = 1.940×10^{-3}		1	0.001	6.243×10^{-2}	3.613×10^{-5}
1 gram per centimeter^3 = 1.940		1000	1	62.43	3.613×10^{-2}
1 pound per foot^3 = 3.108×10^{-2}		16.02	16.02×10^{-2}	1	5.787×10^{-4}
1 pound per inch^3 = 53.71		2.768×10^4	27.68	1728	1

Time

	y	d	h	min	SECOND
1 year = 1		365.25	8.766×10^3	5.259×10^5	3.156×10^7
1 day = 2.738×10^{-3}		1	24	1440	8.640×10^4
1 hour = 1.141×10^{-4}		4.167×10^{-2}	1	60	3600
1 minute = 1.901×10^{-6}		6.944×10^{-4}	1.667×10^{-2}	1	60
1 SECOND = 3.169×10^{-8}		1.157×10^{-5}	2.778×10^{-4}	1.667×10^{-2}	1

Speed

	ft/s	km/h	METER/SECOND	mi/h	cm/s
1 foot per second = 1	1.097	0.3048	0.6818	30.48
1 kilometer per hour = 0.9113	1	0.2778	0.6214	27.78
1 METER per SECOND = 3.281	3.6	1	2.237	100
1 mile per hour = 1.467	1.609	0.4470	1	44.70
1 centimeter per second = 3.281×10^{-2}	3.6×10^{-2}	0.01	2.237×10^{-2}	1

1 knot = 1 nautical mi/h = 1.688 ft/s   1 mi/min = 88.00 ft/s = 60.00 mi/h

Force

Force units in the colored areas are now little used. To clarify: 1 gram-force (= 1 gf) is the force of gravity that would act on an object whose mass is 1 gram at a location where g has the standard value of 9.80665 m/s^2.

	dyne	NEWTON	lb	pdl	gf	kgf
1 dyne = 1	10^{-5}	2.248×10^{-6}	7.233×10^{-5}	1.020×10^{-3}	1.020×10^{-6}
1 NEWTON = 10^5	1	0.2248	7.233	102.0	0.1020
1 pound = 4.448×10^5	4.448	1	32.17	453.6	0.4536
1 poundal = 1.383×10^4	0.1383	3.108×10^{-2}	1	14.10	1.410×10^2
1 gram-force = 980.7	9.807×10^{-3}	2.205×10^{-3}	7.093×10^{-2}	1	0.001
1 kilogram-force = 9.807×10^5	9.807	2.205	70.93	1000	1

1 ton = 2000 lb

Pressure

	atm	dyne/cm^2	inch of water	cm Hg	PASCAL	lb/in.^{2}	lb/ft^2
1 atmosphere = 1	1.013×10^6	406.8	76	1.013×10^5	14.70	2116
1 dyne per centimeter^2 = 9.869×10^{-7}	1	4.015×10^{-4}	7.501×10^{-5}	0.1	1.405×10^{-5}	2.089×10^{-3}
1 inch of water^a at 4°C = 2.458×10^{-3}	2491	1	0.1868	249.1	3.613×10^{-2}	5.202
1 centimeter of mercury^a at 0°C = 1.316×10^{-2}	1.333×10^4	5.353	1	1333	0.1934	27.85
1 PASCAL = 9.869×10^{-6}	10	4.015×10^{-3}	7.501×10^{-4}	1	1.450×10^{-4}	2.089×10^{-2}
1 pound per inch^2 = 6.805×10^{-2}	6.895×10^4	27.68	5.171	6.895×10^3	1	144
1 pound per foot^2 = 4.725×10^{-4}	478.8	0.1922	3.591×10^{-2}	47.88	6.944×10^{-3}	1

^{a}Where the acceleration of gravity has the standard value of 9.80665 m/s^2.

1 bar = 10^6 dyne/cm^2 = 0.1 MPa   1 millibar = 10^3 dyne/cm^2 = 10^2 Pa   1 torr = 1 mm Hg

Energy, Work, Heat

Quantities in the colored areas are not energy units but are included for convenience. They arise from the relativistic mass–energy equivalence formula $E = mc^2$ and represent the energy released if a kilogram or unified atomic mass unit (u) is completely converted to energy (bottom two rows) or the mass that would be completely converted to one unit of energy (rightmost two columns).

	Btu	erg	ft·lb	hp·h	JOULE	cal	kW·h	eV	MeV	kg	u
1 British thermal unit =	1	1.055×10^{10}	777.9	3.929×10^{-4}	1055	252.0	2.930×10^{-4}	6.585×10^{21}	6.585×10^{15}	1.174×10^{-14}	7.070×10^{12}
1 erg =	9.481×10^{-11}	1	7.376×10^{-8}	3.725×10^{-14}	10^{-7}	2.389×10^{-8}	2.778×10^{-14}	6.242×10^{11}	6.242×10^{5}	1.113×10^{-24}	670.2
1 foot-pound =	1.285×10^{-3}	1.356×10^{7}	1	5.051×10^{-7}	1.356	0.3238	3.766×10^{-7}	8.464×10^{18}	8.464×10^{12}	1.509×10^{-17}	9.037×10^{9}
1 horsepower-hour =	2545	2.685×10^{13}	1.980×10^{6}	1	2.685×10^{6}	6.413×10^{5}	0.7457	1.676×10^{25}	1.676×10^{19}	2.988×10^{-11}	1.799×10^{16}
1 JOULE =	9.481×10^{-4}	10^{7}	0.7376	3.725×10^{-7}	1	0.2389	2.778×10^{-7}	6.242×10^{18}	6.242×10^{12}	1.113×10^{-17}	6.702×10^{9}
1 calorie =	3.968×10^{-3}	4.1868×10^{7}	3.088	1.560×10^{-6}	4.1868	1	1.163×10^{-6}	2.613×10^{19}	2.613×10^{13}	4.660×10^{-17}	2.806×10^{10}
1 kilowatt-hour =	3413	3.600×10^{13}	2.655×10^{6}	1.341	3.600×10^{6}	8.600×10^{5}	1	2.247×10^{25}	2.247×10^{19}	4.007×10^{-11}	2.413×10^{16}
1 electron-volt =	1.519×10^{-22}	1.602×10^{-12}	1.182×10^{-19}	5.967×10^{-26}	1.602×10^{-19}	3.827×10^{-20}	4.450×10^{-26}	1	10^{-6}	1.783×10^{-36}	1.074×10^{-9}
1 million electron-volts =	1.519×10^{-16}	1.602×10^{-6}	1.182×10^{-13}	5.967×10^{-20}	1.602×10^{-13}	3.827×10^{-14}	4.450×10^{-20}	10^{-6}	1	1.783×10^{-30}	1.074×10^{-3}
1 kilogram =	8.521×10^{13}	8.987×10^{23}	6.629×10^{16}	3.348×10^{10}	8.987×10^{16}	2.146×10^{16}	2.497×10^{10}	5.610×10^{35}	5.610×10^{29}	1	6.022×10^{26}
1 unified atomic mass unit =	1.415×10^{-13}	1.492×10^{-3}	1.101×10^{-10}	5.559×10^{-17}	1.492×10^{-10}	3.564×10^{-11}	4.146×10^{-17}	9.320×10^{8}	932.0	1.661×10^{-27}	1

Power

	Btu/h	ft·lb/s	hp	cal/s	kW	WATT
1 British thermal unit per hour =	1	0.2161	3.929×10^{-4}	6.998×10^{-2}	2.930×10^{-4}	0.2930
1 foot-pound per second =	4.628	1	1.818×10^{-3}	0.3239	1.356×10^{-3}	1.356
1 horsepower =	2545	550	1	178.1	0.7457	745.7
1 calorie per second =	14.29	3.088	5.615×10^{-3}	1	4.186×10^{-3}	4.186
1 kilowatt =	3413	737.6	1.341	238.9	1	1000
1 WATT =	3.413	0.7376	1.341×10^{-3}	0.2389	0.001	1

Magnetic Field

	gauss	TESLA	milligauss
1 gauss =	1	10^{-4}	1000
1 TESLA =	10^{4}	1	10^{7}
1 milligauss =	0.001	10^{-7}	1

1 tesla = 1 weber/meter^2

Magnetic Flux

	maxwell	WEBER
1 maxwell =	1	10^{-8}
1 WEBER =	10^{8}	1

Mathematical Formulas E

Geometry

Circle of radius r: circumference $= 2\pi r$; area $= \pi r^2$.

Sphere of radius r: area $= 4\pi r^2$; volume $= \frac{4}{3}\pi r^3$.

Right circular cylinder of radius r and height h:
area $= 2\pi r^2 + 2\pi rh$; volume $= \pi r^2 h$.

Triangle of base a and altitude h: area $= \frac{1}{2}ah$.

Quadratic Formula

If $ax^2 + bx + c = 0$, then $x = \dfrac{-b \pm \sqrt{b^2 - 4ac}}{2a}$.

Trigonometric Functions of Angle θ

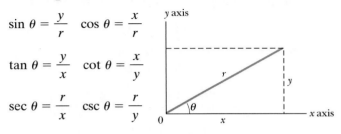

$$\sin\theta = \frac{y}{r} \qquad \cos\theta = \frac{x}{r}$$

$$\tan\theta = \frac{y}{x} \qquad \cot\theta = \frac{x}{y}$$

$$\sec\theta = \frac{r}{x} \qquad \csc\theta = \frac{r}{y}$$

Pythagorean Theorem

In this right triangle,
$$a^2 + b^2 = c^2$$

Triangles

Angles are A, B, C

Opposite sides are a, b, c

Angles $A + B + C = 180°$

$$\frac{\sin A}{a} = \frac{\sin B}{b} = \frac{\sin C}{c}$$

$c^2 = a^2 + b^2 - 2ab\cos C$

Exterior angle $D = A + C$

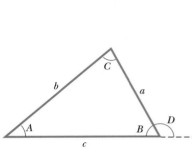

Mathematical Signs and Symbols

$=$ equals

$\approx$ equals approximately

$\sim$ is the order of magnitude of

$\neq$ is not equal to

$\equiv$ is identical to, is defined as

$>$ is greater than ($\gg$ is much greater than)

$<$ is less than ($\ll$ is much less than)

$\geq$ is greater than or equal to (or, is no less than)

$\leq$ is less than or equal to (or, is no more than)

$\pm$ plus or minus

$\propto$ is proportional to

Σ the sum of

x_{avg} the average value of x

Trigonometric Identities

$\sin(90° - \theta) = \cos\theta$

$\cos(90° - \theta) = \sin\theta$

$\sin\theta/\cos\theta = \tan\theta$

$\sin^2\theta + \cos^2\theta = 1$

$\sec^2\theta - \tan^2\theta = 1$

$\csc^2\theta - \cot^2\theta = 1$

$\sin 2\theta = 2\sin\theta\cos\theta$

$\cos 2\theta = \cos^2\theta - \sin^2\theta = 2\cos^2\theta - 1 = 1 - 2\sin^2\theta$

$\sin(\alpha \pm \beta) = \sin\alpha\cos\beta \pm \cos\alpha\sin\beta$

$\cos(\alpha \pm \beta) = \cos\alpha\cos\beta \mp \sin\alpha\sin\beta$

$$\tan(\alpha \pm \beta) = \frac{\tan\alpha \pm \tan\beta}{1 \mp \tan\alpha\tan\beta}$$

$\sin\alpha \pm \sin\beta = 2\sin\frac{1}{2}(\alpha \pm \beta)\cos\frac{1}{2}(\alpha \mp \beta)$

$\cos\alpha + \cos\beta = 2\cos\frac{1}{2}(\alpha + \beta)\cos\frac{1}{2}(\alpha - \beta)$

$\cos\alpha - \cos\beta = -2\sin\frac{1}{2}(\alpha + \beta)\sin\frac{1}{2}(\alpha - \beta)$

Binomial Theorem

$$(1 + x)^n = 1 + \frac{nx}{1!} + \frac{n(n-1)x^2}{2!} + \cdots \qquad (x^2 < 1)$$

Exponential Expansion

$$e^x = 1 + x + \frac{x^2}{2!} + \frac{x^3}{3!} + \cdots$$

Logarithmic Expansion

$$\ln(1 + x) = x - \tfrac{1}{2}x^2 + \tfrac{1}{3}x^3 - \cdots \qquad (|x| < 1)$$

Trigonometric Expansions (θ in radians)

$$\sin\theta = \theta - \frac{\theta^3}{3!} + \frac{\theta^5}{5!} - \cdots$$

$$\cos\theta = 1 - \frac{\theta^2}{2!} + \frac{\theta^4}{4!} - \cdots$$

$$\tan\theta = \theta + \frac{\theta^3}{3} + \frac{2\theta^5}{15} + \cdots$$

Cramer's Rule

Two simultaneous equations in unknowns x and y,

$$a_1 x + b_1 y = c_1 \quad \text{and} \quad a_2 x + b_2 y = c_2,$$

have the solutions

$$x = \frac{\begin{vmatrix} c_1 & b_1 \\ c_2 & b_2 \end{vmatrix}}{\begin{vmatrix} a_1 & b_1 \\ a_2 & b_2 \end{vmatrix}} = \frac{c_1 b_2 - c_2 b_1}{a_1 b_2 - a_2 b_1}$$

and

$$y = \frac{\begin{vmatrix} a_1 & c_1 \\ a_2 & c_2 \end{vmatrix}}{\begin{vmatrix} a_1 & b_1 \\ a_2 & b_2 \end{vmatrix}} = \frac{a_1 c_2 - a_2 c_1}{a_1 b_2 - a_2 b_1}.$$

Products of Vectors

Let $\hat{i}$, $\hat{j}$, and $\hat{k}$ be unit vectors in the x, y, and z directions. Then

$$\hat{i}\cdot\hat{i} = \hat{j}\cdot\hat{j} = \hat{k}\cdot\hat{k} = 1, \quad \hat{i}\cdot\hat{j} = \hat{j}\cdot\hat{k} = \hat{k}\cdot\hat{i} = 0,$$

$$\hat{i}\times\hat{i} = \hat{j}\times\hat{j} = \hat{k}\times\hat{k} = 0,$$

$$\hat{i}\times\hat{j} = \hat{k}, \quad \hat{j}\times\hat{k} = \hat{i}, \quad \hat{k}\times\hat{i} = \hat{j}$$

Any vector $\vec{a}$ with components a_x, a_y, and a_z along the x, y, and z axes can be written as

$$\vec{a} = a_x\hat{i} + a_y\hat{j} + a_z\hat{k}.$$

Let $\vec{a}$, $\vec{b}$, and $\vec{c}$ be arbitrary vectors with magnitudes a, b, and c. Then

$$\vec{a} \times (\vec{b} + \vec{c}) = (\vec{a} \times \vec{b}) + (\vec{a} \times \vec{c})$$

$$(s\vec{a}) \times \vec{b} = \vec{a} \times (s\vec{b}) = s(\vec{a} \times \vec{b}) \qquad (s = \text{a scalar}).$$

Let θ be the smaller of the two angles between $\vec{a}$ and $\vec{b}$. Then

$$\vec{a}\cdot\vec{b} = \vec{b}\cdot\vec{a} = a_x b_x + a_y b_y + a_z b_z = ab\cos\theta$$

$$\vec{a} \times \vec{b} = -\vec{b} \times \vec{a} = \begin{vmatrix} \hat{i} & \hat{j} & \hat{k} \\ a_x & a_y & a_z \\ b_x & b_y & b_z \end{vmatrix}$$

$$= \hat{i}\begin{vmatrix} a_y & a_z \\ b_y & b_z \end{vmatrix} - \hat{j}\begin{vmatrix} a_x & a_z \\ b_x & b_z \end{vmatrix} + \hat{k}\begin{vmatrix} a_x & a_y \\ b_x & b_y \end{vmatrix}$$

$$= (a_y b_z - b_y a_z)\hat{i} + (a_z b_x - b_z a_x)\hat{j}$$
$$+ (a_x b_y - b_x a_y)\hat{k}$$

$$|\vec{a} \times \vec{b}| = ab\sin\theta$$

$$\vec{a}\cdot(\vec{b} \times \vec{c}) = \vec{b}\cdot(\vec{c} \times \vec{a}) = \vec{c}\cdot(\vec{a} \times \vec{b})$$

$$\vec{a} \times (\vec{b} \times \vec{c}) = (\vec{a}\cdot\vec{c})\vec{b} - (\vec{a}\cdot\vec{b})\vec{c}$$

Derivatives and Integrals

In what follows, the letters u and v stand for any functions of x, and a and m are constants. To each of the indefinite integrals should be added an arbitrary constant of integration. The *Handbook of Chemistry and Physics* (CRC Press Inc.) gives a more extensive tabulation.

1. $\dfrac{dx}{dx} = 1$

2. $\dfrac{d}{dx}(au) = a\dfrac{du}{dx}$

3. $\dfrac{d}{dx}(u + v) = \dfrac{du}{dx} + \dfrac{dv}{dx}$

4. $\dfrac{d}{dx}x^m = mx^{m-1}$

5. $\dfrac{d}{dx}\ln x = \dfrac{1}{x}$

6. $\dfrac{d}{dx}(uv) = u\dfrac{dv}{dx} + v\dfrac{du}{dx}$

7. $\dfrac{d}{dx}e^x = e^x$

8. $\dfrac{d}{dx}\sin x = \cos x$

9. $\dfrac{d}{dx}\cos x = -\sin x$

10. $\dfrac{d}{dx}\tan x = \sec^2 x$

11. $\dfrac{d}{dx}\cot x = -\csc^2 x$

12. $\dfrac{d}{dx}\sec x = \tan x \sec x$

13. $\dfrac{d}{dx}\csc x = -\cot x \csc x$

14. $\dfrac{d}{dx}e^u = e^u\dfrac{du}{dx}$

15. $\dfrac{d}{dx}\sin u = \cos u\dfrac{du}{dx}$

16. $\dfrac{d}{dx}\cos u = -\sin u\dfrac{du}{dx}$

1. $\displaystyle\int dx = x$

2. $\displaystyle\int au\,dx = a\int u\,dx$

3. $\displaystyle\int (u + v)\,dx = \int u\,dx + \int v\,dx$

4. $\displaystyle\int x^m\,dx = \dfrac{x^{m+1}}{m+1} \quad (m \neq -1)$

5. $\displaystyle\int \dfrac{dx}{x} = \ln |x|$

6. $\displaystyle\int u\dfrac{dv}{dx}\,dx = uv - \int v\dfrac{du}{dx}\,dx$

7. $\displaystyle\int e^x\,dx = e^x$

8. $\displaystyle\int \sin x\,dx = -\cos x$

9. $\displaystyle\int \cos x\,dx = \sin x$

10. $\displaystyle\int \tan x\,dx = \ln |\sec x|$

11. $\displaystyle\int \sin^2 x\,dx = \tfrac{1}{2}x - \tfrac{1}{4}\sin 2x$

12. $\displaystyle\int e^{-ax}\,dx = -\dfrac{1}{a}e^{-ax}$

13. $\displaystyle\int xe^{-ax}\,dx = -\dfrac{1}{a^2}(ax + 1)e^{-ax}$

14. $\displaystyle\int x^2 e^{-ax}\,dx = -\dfrac{1}{a^3}(a^2x^2 + 2ax + 2)e^{-ax}$

15. $\displaystyle\int_0^\infty x^n e^{-ax}\,dx = \dfrac{n!}{a^{n+1}}$

16. $\displaystyle\int_0^\infty x^{2n}e^{-ax^2}\,dx = \dfrac{1\cdot3\cdot5\,\cdots\,(2n-1)}{2^{n+1}a^n}\sqrt{\dfrac{\pi}{a}}$

17. $\displaystyle\int \dfrac{dx}{\sqrt{x^2 + a^2}} = \ln(x + \sqrt{x^2 + a^2})$

18. $\displaystyle\int \dfrac{x\,dx}{(x^2 + a^2)^{3/2}} = -\dfrac{1}{(x^2 + a^2)^{1/2}}$

19. $\displaystyle\int \dfrac{dx}{(x^2 + a^2)^{3/2}} = \dfrac{x}{a^2(x^2 + a^2)^{1/2}}$

20. $\displaystyle\int_0^\infty x^{2n+1}e^{-ax^2}\,dx = \dfrac{n!}{2a^{n+1}} \quad (a > 0)$

21. $\displaystyle\int \dfrac{x\,dx}{x + d} = x - d\ln(x + d)$

Properties of the Elements

All physical properties are for a pressure of 1 atm unless otherwise specified.

Element	Symbol	Atomic Number Z	Molar Mass, g/mol	Density, g/cm³ at 20°C	Melting Point, °C	Boiling Point, °C	Specific Heat, J/(g·°C) at 25°C
Actinium	Ac	89	(227)	10.06	1323	(3473)	0.092
Aluminum	Al	13	26.9815	2.699	660	2450	0.900
Americium	Am	95	(243)	13.67	1541	—	—
Antimony	Sb	51	121.75	6.691	630.5	1380	0.205
Argon	Ar	18	39.948	1.6626×10^{-3}	−189.4	−185.8	0.523
Arsenic	As	33	74.9216	5.78	817 (28 atm)	613	0.331
Astatine	At	85	(210)	—	(302)	—	—
Barium	Ba	56	137.34	3.594	729	1640	0.205
Berkelium	Bk	97	(247)	14.79	—	—	—
Beryllium	Be	4	9.0122	1.848	1287	2770	1.83
Bismuth	Bi	83	208.980	9.747	271.37	1560	0.122
Bohrium	Bh	107	262.12	—	—	—	—
Boron	B	5	10.811	2.34	2030	—	1.11
Bromine	Br	35	79.909	3.12 (liquid)	−7.2	58	0.293
Cadmium	Cd	48	112.40	8.65	321.03	765	0.226
Calcium	Ca	20	40.08	1.55	838	1440	0.624
Californium	Cf	98	(251)	—	—	—	—
Carbon	C	6	12.01115	2.26	3727	4830	0.691
Cerium	Ce	58	140.12	6.768	804	3470	0.188
Cesium	Cs	55	132.905	1.873	28.40	690	0.243
Chlorine	Cl	17	35.453	3.214×10^{-3} (0°C)	−101	−34.7	0.486
Chromium	Cr	24	51.996	7.19	1857	2665	0.448
Cobalt	Co	27	58.9332	8.85	1495	2900	0.423
Copper	Cu	29	63.54	8.96	1083.40	2595	0.385
Curium	Cm	96	(247)	13.3	—	—	—
Darmstadtium	Ds	110	(271)	—	—	—	—
Dubnium	Db	105	262.114	—	—	—	—
Dysprosium	Dy	66	162.50	8.55	1409	2330	0.172
Einsteinium	Es	99	(254)	—	—	—	—
Erbium	Er	68	167.26	9.15	1522	2630	0.167
Europium	Eu	63	151.96	5.243	817	1490	0.163
Fermium	Fm	100	(237)	—	—	—	—
Fluorine	F	9	18.9984	1.696×10^{-3} (0°C)	−219.6	−188.2	0.753
Francium	Fr	87	(223)	—	(27)	—	—
Gadolinium	Gd	64	157.25	7.90	1312	2730	0.234
Gallium	Ga	31	69.72	5.907	29.75	2237	0.377
Germanium	Ge	32	72.59	5.323	937.25	2830	0.322
Gold	Au	79	196.967	19.32	1064.43	2970	0.131

Element	Symbol	Atomic Number Z	Molar Mass, g/mol	Density, g/cm³ at 20°C	Melting Point, °C	Boiling Point, °C	Specific Heat, J/(g·°C) at 25°C
Hafnium	Hf	72	178.49	13.31	2227	5400	0.144
Hassium	Hs	108	(265)	—	—	—	—
Helium	He	2	4.0026	0.1664×10^{-3}	−269.7	−268.9	5.23
Holmium	Ho	67	164.930	8.79	1470	2330	0.165
Hydrogen	H	1	1.00797	0.08375×10^{-3}	−259.19	−252.7	14.4
Indium	In	49	114.82	7.31	156.634	2000	0.233
Iodine	I	53	126.9044	4.93	113.7	183	0.218
Iridium	Ir	77	192.2	22.5	2447	(5300)	0.130
Iron	Fe	26	55.847	7.874	1536.5	3000	0.447
Krypton	Kr	36	83.80	3.488×10^{-3}	−157.37	−152	0.247
Lanthanum	La	57	138.91	6.189	920	3470	0.195
Lawrencium	Lr	103	(257)	—	—	—	—
Lead	Pb	82	207.19	11.35	327.45	1725	0.129
Lithium	Li	3	6.939	0.534	180.55	1300	3.58
Lutetium	Lu	71	174.97	9.849	1663	1930	0.155
Magnesium	Mg	12	24.312	1.738	650	1107	1.03
Manganese	Mn	25	54.9380	7.44	1244	2150	0.481
Meitnerium	Mt	109	(266)	—	—	—	—
Mendelevium	Md	101	(256)	—	—	—	—
Mercury	Hg	80	200.59	13.55	−38.87	357	0.138
Molybdenum	Mo	42	95.94	10.22	2617	5560	0.251
Neodymium	Nd	60	144.24	7.007	1016	3180	0.188
Neon	Ne	10	20.183	0.8387×10^{-3}	−248.597	−246.0	1.03
Neptunium	Np	93	(237)	20.25	637	—	1.26
Nickel	Ni	28	58.71	8.902	1453	2730	0.444
Niobium	Nb	41	92.906	8.57	2468	4927	0.264
Nitrogen	N	7	14.0067	1.1649×10^{-3}	−210	−195.8	1.03
Nobelium	No	102	(255)	—	—	—	—
Osmium	Os	76	190.2	22.59	3027	5500	0.130
Oxygen	O	8	15.9994	1.3318×10^{-3}	−218.80	−183.0	0.913
Palladium	Pd	46	106.4	12.02	1552	3980	0.243
Phosphorus	P	15	30.9738	1.83	44.25	280	0.741
Platinum	Pt	78	195.09	21.45	1769	4530	0.134
Plutonium	Pu	94	(244)	19.8	640	3235	0.130
Polonium	Po	84	(210)	9.32	254	—	—
Potassium	K	19	39.102	0.862	63.20	760	0.758
Praseodymium	Pr	59	140.907	6.773	931	3020	0.197
Promethium	Pm	61	(145)	7.22	(1027)	—	—
Protactinium	Pa	91	(231)	15.37 (estimated)	(1230)	—	—
Radium	Ra	88	(226)	5.0	700	—	—
Radon	Rn	86	(222)	9.96×10^{-3} (0°C)	(−71)	−61.8	0.092
Rhenium	Re	75	186.2	21.02	3180	5900	0.134
Rhodium	Rh	45	102.905	12.41	1963	4500	0.243
Rubidium	Rb	37	85.47	1.532	39.49	688	0.364
Ruthenium	Ru	44	101.107	12.37	2250	4900	0.239
Rutherfordium	Rf	104	261.11	—	—	—	—

Element	Symbol	Atomic Number Z	Molar Mass, g/mol	Density, g/cm³ at 20°C	Melting Point, °C	Boiling Point, °C	Specific Heat, J/(g · °C) at 25°C
Samarium	Sm	62	150.35	7.52	1072	1630	0.197
Scandium	Sc	21	44.956	2.99	1539	2730	0.569
Seaborgium	Sg	106	263.118	—	—	—	—
Selenium	Se	34	78.96	4.79	221	685	0.318
Silicon	Si	14	28.086	2.33	1412	2680	0.712
Silver	Ag	47	107.870	10.49	960.8	2210	0.234
Sodium	Na	11	22.9898	0.9712	97.85	892	1.23
Strontium	Sr	38	87.62	2.54	768	1380	0.737
Sulfur	S	16	32.064	2.07	119.0	444.6	0.707
Tantalum	Ta	73	180.948	16.6	3014	5425	0.138
Technetium	Tc	43	(99)	11.46	2200	—	0.209
Tellurium	Te	52	127.60	6.24	449.5	990	0.201
Terbium	Tb	65	158.924	8.229	1357	2530	0.180
Thallium	Tl	81	204.37	11.85	304	1457	0.130
Thorium	Th	90	(232)	11.72	1755	(3850)	0.117
Thulium	Tm	69	168.934	9.32	1545	1720	0.159
Tin	Sn	50	118.69	7.2984	231.868	2270	0.226
Titanium	Ti	22	47.90	4.54	1670	3260	0.523
Tungsten	W	74	183.85	19.3	3380	5930	0.134
Unnamed	Uuu	111	(272)	—	—	—	—
Unnamed	Uub	112	(285)	—	—	—	—
Unnamed	Uut	113	—	—	—	—	—
Unnamed	Unq	114	(289)	—	—	—	—
Unnamed	Uup	115	—	—	—	—	—
Unnamed	Uuh	116	—	—	—	—	—
Unnamed	Uus	117	—	—	—	—	—
Unnamed	Uuo	118	(293)	—	—	—	—
Uranium	U	92	(238)	18.95	1132	3818	0.117
Vanadium	V	23	50.942	6.11	1902	3400	0.490
Xenon	Xe	54	131.30	5.495×10^{-3}	−111.79	−108	0.159
Ytterbium	Yb	70	173.04	6.965	824	1530	0.155
Yttrium	Y	39	88.905	4.469	1526	3030	0.297
Zinc	Zn	30	65.37	7.133	419.58	906	0.389
Zirconium	Zr	40	91.22	6.506	1852	3580	0.276

The values in parentheses in the column of molar masses are the mass numbers of the longest-lived isotopes of those elements that are radioactive. Melting points and boiling points in parentheses are uncertain.

The data for gases are valid only when these are in their usual molecular state, such as H_2, He, O_2, Ne, etc. The specific heats of the gases are the values at constant pressure.

Source: Adapted from J. Emsley, *The Elements,* 3rd ed., 1998, Clarendon Press, Oxford. See also www.webelements.com for the latest values and newest elements.

Periodic Table of the Elements

Elements 111, 112, 114, and 116 have been discovered but, as of 2003, have not yet been named. Evidence for the discovery of elements 113 and 115 has been reported. See www.webelements.com for the latest information and newest elements.

Chapter 33

CP **1.** (a) (Use Fig. 33-5.) On right side of rectangle, $\vec{E}$ is in negative y direction; on left side, $\vec{E} + d\vec{E}$ is greater and in same direction; (b) $\vec{E}$ is downward. On right side, $\vec{B}$ is in negative z direction; on left side, $\vec{B} + d\vec{B}$ is greater and in same direction. **2.** positive direction of x **3.** (a) same; (b) decrease **4.** a, d, b, c (zero) **5.** a **Q** **1.** (a) positive direction of z; (b) x **3.** (a) and (b) $A = 1, n = 4, \theta = 30°$ **5.** (a) same; (b) increase; (c) decrease **7.** a, b, c **9.** none **11.** B **P** **1.** (a) 515 nm; (b) 610 nm; (c) 555 nm; (d) 5.41×10^{14} Hz; (e) 1.85×10^{-15} s **3.** 7.49 GHz **5.** 5.0×10^{-21} H **7.** 0.10 MJ **9.** (a) 6.7 nT; (b) y; (c) negative direction of y **11.** 1.2 MW/m² **13.** (a) 6.7 nT; (b) 5.3 mW/m²; (c) 6.7 W **15.** (a) 1.03 kV/m; (b) 3.43 μT **17.** (a) 87 mV/m; (b) 0.29 nT; (c) 6.3 kW **19.** 5.9×10^{-8} Pa **21.** 1.0×10^7 Pa **25.** (a) 1.0×10^8 Hz; (b) 6.3×10^8 rad/s; (c) 2.1 m⁻¹; (d) 1.0 μT; (e) z; (f) 1.2×10^2 W/m²; (g) 8.0×10^{-7} N; (h) 4.0×10^{-7} Pa **27.** 1.9 mm/s **29.** (a) 4.68×10^{11} W; (b) any chance disturbance could move sphere from directly above source—the two force vectors no longer along the same axis **31.** (a) 0.17 μm; (b) toward the Sun **33.** 4.4 W/m² **35.** 3.1% **37.** (a) 1.9 V/m; (b) 1.7×10^{-11} Pa **39.** 20° or 70° **41.** 0.67 **43.** (a) 2 sheets; (b) 5 sheets **45.** 1.48 **47.** 180° **49.** 1.26 **51.** 1.07 m **53.** (a) 56.9°; (b) 35.3° **57.** 182 cm **59.** (a) 26.8°; (b) yes **61.** (a) 1.39; (b) 28.1°; (c) no **63.** 23.2° **65.** (a) 48.9°; (b) 29.0° **67.** (a) $(1 + \sin^2 \theta)^{0.5}$; (b) $2^{0.5}$; (c) yes; (d) no **69.** 49.0° **71.** (c) 137.6°; (d) 139.4°; (e) 1.7° **73.** 1.22 **77.** (a) 0.50 ms; (b) 8.4 min; (c) 2.4 h; (d) 5446 B.C. **79.** (a) $(16.7 \text{ nT}) \sin[(1.00 \times 10^6 \text{ m}^{-1})z + (3.00 \times 10^{14} \text{ s}^{-1})t]$; (b) 6.28 μm; (c) 20.9 fs; (d) 33.2 mW/m²; (e) x; (f) infrared **81.** (a) 83 W/m²; (b) 1.7 MW **83.** (a) 3.5 μW/m²; (b) 0.78 μW; (c) 1.5×10^{-17} W/m²; (d) 1.1×10^{-7} V/m; (e) 0.25 fT **85.** 35° **87.** (a) 55.8°; (b) 55.5° **89.** (a) z axis; (b) 7.5×10^{14} Hz; (c) 1.9 kW/m² **91.** (a) white; (b) white dominated by red end; (c) no refracted light **93.** 1.5×10^{-9} m/s² **95.** 0.034 **97.** 9.43×10^{-10} T **99.** (a) $-y$; (b) z; (c) 1.91 kW/m²; (d) $E_z = (1.20 \text{ kV/m}) \sin[(6.67 \times 10^6 \text{ m}^{-1})y + (2.00 \times 10^{15} \text{ s}^{-1})t]$; (e) 942 nm; (f) infrared **101.** (a) 1.60; (b) 58.0°

Chapter 34

CP **1.** $0.2d, 1.8d, 2.2d$ **2.** (a) real; (b) inverted; (c) same **3.** (a) e; (b) virtual, same **4.** virtual, same as object, diverging **Q** **1.** (a) a and c; (b) three times; (c) you **3.** (a) a; (b) c **5.** d (infinite), tie of a and b, then c **7.** convex **9.** (a) all but variation 2; (b) 1, 3, 4: right, inverted; 5, 6: left, same **P** **1.** 40 cm **3.** 1.5 m **5.** 351 cm **7.** 10.5 cm **9.** (a) +36 cm; (b) −36 cm; (c) +3.0; (d) V; (e) NI; (f) opposite **11.** (a) +24 cm; (b) +36 cm; (c) −2.0; (d) R; (e) I; (f) same **13.** (a) −16 cm; (b) −4.4 cm; (c) +0.44; (d) V; (e) NI; (f) opposite **15.** (a) −20 cm; (b) −4.4 cm; (c) +0.56; (d) V; (e) NI; (f) opposite **17.** (b) 0.56 cm/s; (c) 11 m/s; (d) 6.7 cm/s **19.** (b) plus; (c) +40 cm; (e) −20 cm; (f) +2.0; (g) V; (h) NI; (i) opposite **21.** (a) convex; (b) −20 cm; (d) +20 cm; (f) +0.50; (g) V; (h) NI; (i) opposite **23.** (a) concave; (c) +40 cm; (e) +60 cm; (f) −2.0; (g) R; (h) I; (i) same **25.** (a) convex; (c) −60 cm; (d) +30 cm; (f) +0.50; (g) V; (h) NI; (i) opposite **27.** (b) −20 cm; (c) minus; (d) +5.0 cm; (e) minus; (f) +0.80; (g) V; (h) NI; (i) opposite **29.** (a) convex; (b) minus; (c) −60 cm; (d) +1.2 m; (e) −24 cm; (g) V; (h) NI; (i) opposite **31.** (a) concave; (b) +8.6 cm; (c) +17 cm; (e) +12 cm; (f) minus; (g) R; (i) same **33.** (a) 2.00; (b) none **35.** (c) −33 cm; (e) V; (f) same **37.** (d) −26 cm; (e) V; (f) same **39.** (c) +30 cm; (e) V; (f) same **41.** (a) 45 mm; (b) 90 mm **43.** 1.86 mm **45.** 5.0 mm **47.** 22 cm **49.** (a) +40 cm; (b) ∞ **51.** (a) −4.8 cm; (b) +0.60; (c) V; (d) NI; (e) same **53.** (a) −48 cm; (b) +4.0; (c) V; (d) NI; (e) same **55.** (a) +36 cm; (b) −0.80; (c) R; (d) I; (e) opposite **57.** (a) −8.6 cm; (b) +0.39; (c) V; (d) NI; (e) same **59.** (a) −30 cm; (b) +0.86; (c) V; (d) NI; (e) same **61.** (a) +55 cm; (b) −0.74; (c) R; (d) I; (e) opposite **63.** (a) +84 cm; (b) −1.4; (c) R; (d) I; (e) opposite **65.** (a) −18 cm; (b) +0.76; (c) V; (d) NI; (e) same **67.** (a) −7.5 cm; (b) +0.75; (c) V; (d) NI; (e) same **69.** (a) D; (b) −5.3 cm; (d) −4.0 cm; (f) V; (g) NI; (h) same **71.** (a) C; (b) plus; (d) −13 cm; (e) +1.7; (f) V; (g) NI; (h) same **73.** (a) C; (d) −10 cm; (e) +2.0; (f) V; (g) NI; (h) same **75.** (a) C; (b) +80 cm; (d) −20 cm; (f) V; (g) NI; (h) same **77.** (a) C; (b) +3.3 cm; (d) +5.0 cm; (f) R; (g) I; (h) opposite **79.** (a) D; (b) minus; (d) −3.3 cm; (e) +0.67; (f) V; (g) NI **81.** (a) +3.1 cm; (b) −0.31; (c) R; (d) I; (e) opposite **83.** (a) +24 cm; (b) +6.0; (c) R; (d) NI; (e) opposite **85.** (a) −5.5 cm; (b) +0.12; (c) V; (d) NI; (e) same **87.** (a) −4.6 cm; (b) +0.69; (c) V; (d) NI; (e) same **89.** (a) 13.0 cm; (b) 5.23 cm; (c) −3.25; (d) 3.13; (e) −10.2 **91.** (a) 3.5; (b) 2.5 **93.** (a) 2.35 cm; (b) decrease **95.** (a) 3.00 cm; (b) 2.33 cm **97.** (a) 6.0 mm; (b) 1.6 kW/m²; (c) 4.0 cm **99.** (a) 20 cm; (b) 15 cm **101.** 2.2 mm² **107.** (a) 40 cm; (b) 20 cm; (c) −40 cm; (d) 40 cm **109.** (a) +36 cm; (b) 1.2 cm; (c) real; (d) inverted **111.** (a) −50 cm; (b) 5.0; (c) virtual; (d) inverted **113.** (a) +8.6 cm; (b) +2.6; (c) R; (d) NI; (e) opposite **115.** (a) +7.5 cm; (b) −0.75; (c) R; (d) I; (e) opposite **117.** (a) +24 cm; (b) −0.58; (c) R; (d) I; (e) opposite **119.** (a) 0.60 m; (b) +0.20; (c) real; (d) left; (e) not inverted **125.** (a) 8.0 cm; (b) 16 cm; (c) 48 cm **127.** (a) $\alpha = 0.500$ rad: 7.799 cm; $\alpha = 0.100$ rad: 8.544 cm; $\alpha = 0.0100$ rad: 8.571 cm; mirror equation: 8.571 cm; (b) $\alpha = 0.500$ rad: −13.56 cm; $\alpha = 0.100$ rad: −12.05 cm; $\alpha = 0.0100$ rad: −12.00 cm; mirror equation: −12.00 cm **129.** 42 mm **131.** (b) P_n **133.** (a) $(0.5)(2 - n)r/(n - 1)$; (b) right **135.** 2.67 cm **137.** (a) 3.33 cm; (b) left; (c) virtual; (d) not inverted

Chapter 35

CP **1.** b (least n), c, a **2.** (a) top; (b) bright intermediate illumination (phase difference is 2.1 wavelengths) **3.** (a) 3λ, 3; (b) $2.5\lambda, 2.5$ **4.** a and d tie (amplitude of resultant wave is $4E_0$), then b and c tie (amplitude of resultant wave is $2E_0$) **5.** (a) 1 and 4; (b) 1 and 4 **Q** **1.** (a) $2d$; (b) (odd number)$\lambda/2$; (c) $\lambda/4$ **3.** (a) peak; (b) valley **5.** (a) decrease; (b) decrease; (c) decrease; (d) blue **7.** (a) intermediate closer to maximum, $m = 2$; (b) minimum, $m = 3$; (c) intermediate closer to maximum, $m = 2$; (d) maximum, $m = 1$

9. (a) maximum; (b) minimum; (c) alternates   **11.** c, d
P   1. 1.56   **3.** 4.55×10^7 m/s   **5.** (a) 3.60 μm; (b) intermediate closer to fully constructive   **7.** (a) 155 nm; (b) 310 nm
9. (a) 1.70; (b) 1.70; (c) 1.30; (d) all tie   **11.** (a) 1.55 μm; (b) 4.65 μm   **13.** (a) 0.833; (b) intermediate closer to fully constructive   **15.** 16   **17.** 648 nm   **19.** 2.25 mm   **21.** 0
23. 7.88 μm   **25.** 72 μm   **27.** 6.64 μm   **29.** 2.65
31. (17.1 μV/m) $\sin[(2.0 \times 10^{14} \text{ rad/s})t]$ **33.** $27 \sin(\omega t + 8.5°)$
35. 70.0 nm   **37.** (a) 0.117 μm; (b) 0.352 μm   **39.** 120 nm
41. 509 nm   **43.** 478 nm   **45.** 560 nm   **47.** 409 nm   **49.** 161 nm
51. 273 nm   **53.** (a) 552 nm; (b) 442 nm   **55.** 338 nm   **57.** 329 nm   **59.** 248 nm   **61.** 339 nm   **63.** 528 nm   **65.** 455 nm
67. 608 nm   **69.** 1.89 μm   **71.** 140   **73.** 0.012°
75. $[(m + \frac{1}{2})\lambda R]^{0.5}$, for $m = 0, 1, 2, \ldots$ **77.** 1.00 m   **79.** 588 nm
81. 1.00030   **83.** (a) 22°; (b) refraction reduces θ **85.** (a) 50.0 nm; (b) 36.2 nm   **87.** $x = (D/2a)(m + 0.5)\lambda$, for $m = 0, 1, 2, \ldots$
89. 0.23°   **91.** 450 nm   **93.** (a) 1.75 μm; (b) 4.8 mm
95. 33 μm   **97.** (a) 1500 nm; (b) 2250 nm; (c) 0.80   **99.** 600 nm
101. (a) 88%; (b) 94%   **103.** $I_m \cos^2(2\pi x/\lambda)$ **105.** 8.0 μm
107. (a) 1.80 μm; (b) 9   **109.** 51.6 ns   **111.** (a) 1.6 rad; (b) 0.79 rad   **113.** (a) dark; (b) dark; (c) 4   **115.** (a) 1; (b) 4.0 fs; (c) 7.5
117. (a) 48.6°; (b) away; (c) 1.49 m   **119.** (a) 42.0 ps; (b) 42.3 ps; (c) 43.2 ps; (d) 41.8 ps; (e) 4   **121.** $I_0[1 + 8 \cos^2(\phi/2)]$, with $\phi = (2\pi d/\lambda) \sin \theta$

Chapter 36

CP   1. (a) expand; (b) expand   **2.** (a) second side maximum; (b) 2.5   **3.** (a) red; (b) violet   **4.** diminish   **5.** (a) left; (b) less
Q   1. (a) A, B, C; (b) A, B, C **3.** (a) 1 and 3 tie, then 2 and 4 tie; (b) 1 and 2 tie, then 3 and 4 tie   **5.** (a) $m = 5$ minimum; (b) (approximately) maximum between the $m = 4$ and $m = 5$ minima   **7.** (a) decrease; (b) same; (c) remain in place
9. (a) larger; (b) red   **11.** (a) A; (b) left; (c) left; (d) right
P   1. (a) 700 nm; (b) 4; (c) 6   **3.** 60.4 μm   **5.** (a) 70 cm; (b) 1.0 mm   **7.** (a) 2.5 mm; (b) 2.2×10^{-4} rad   **9.** 1.77 mm
11. (a) 0.18°; (b) 0.46 rad; (c) 0.93   **13.** 160°   **15.** (b) 0; (c) -0.500; (d) 4.493 rad; (e) 0.930; (f) 7.725 rad; (g) 1.96
17. (d) 52.5°; (e) 10.1°; (f) 5.06°   **19.** 1.6×10^3 km   **21.** (a) 1.3×10^{-4} rad; (b) 10 km   **23.** 50 m   **25.** (a) 1.1×10^4 km; (b) 11 km   **27.** (a) 19 cm; (b) larger   **29.** (a) 0.346°; (b) 0.97°   **31.** (a) 17.1 m; (b) 1.37×10^{-10} **33.** (a) 8.8×10^{-7} rad; (b) 8.4×10^7 km; (c) 0.025 mm   **35.** 3   **37.** 5
39. (a) 9; (b) 0.255   **41.** (a) 5.0 μm; (b) 20 μm   **43.** (a) 7.43×10^{-3}; (b) between the $m = 6$ minimum (the seventh one) and the $m = 7$ maximum (the seventh side maximum); (c) between the $m = 3$ minimum (the third one) and the $m = 4$ minimum (the fourth one)   **45.** 3   **47.** (a) 62.1°; (b) 45.0°; (c) 32.0°   **49.** (a) 2.1°; (b) 21°; (c) 11

51. (a) 6.0 μm; (b) 1.5 μm; (c) 9; (d) 7; (e) 6   **53.** 1.09×10^3 rulings/mm   **55.** (a) 470 nm; (b) 560 nm   **57.** 0.15 nm
59. (a) 0.032°/nm; (b) 4.0×10^4; (c) 0.076°/nm; (d) 8.0×10^4; (e) 0.24°/nm; (f) 1.2×10^5 **61.** 3.65×10^3 **63.** (a) 10 μm; (b) 3.3 mm   **65.** (a) 25 pm; (b) 38 pm   **67.** 0.26 nm
69. (a) 0.17 nm; (b) 0.13 nm   **71.** (a) $0.7071a_0$; (b) $0.4472a_0$; (c) $0.3162a_0$; (d) $0.2774a_0$; (e) $0.2425a_0$ **73.** (a) 15.3°; (b) 30.6°; (c) 3.1°; (d) 37.8°   **75.** 4.7 cm   **77.** (a) 625 nm; (b) 500 nm; (c) 416 nm   **79.** 3.0 mm   **81.** (a) 13; (b) 6   **83.** 59.5 pm
85. 4.9 km   **87.** 1.36×10^4 **89.** 36 cm   **97.** 2   **99.** (a) fourth; (b) seventh   **103.** (a) 2.4 μm; (b) 0.80 μm; (c) 2   **107.** 9

Chapter 37

CP   1. (a) same (speed of light postulate); (b) no (the start and end of the flight are spatially separated); (c) no (because his measurement is not a proper time)   **2.** (a) Eq. 2; (b) $+0.90c$; (c) 25 ns; (d) -7.0 m   **3.** (a) right; (b) more
4. (a) equal; (b) less   **Q   1.** (a) C_1'; (b) C_1' **3.** (a) 4 s; (b) 3 s; (c) 5 s; (d) 4 s; (e) 10 s   **5.** c **7.** b **9.** (a) 3, tie of 1 and 2, then 4; (b) 4, tie of 1 and 2, then 3; (c) 1, 4, 2, 3   **11.** (a) a tie of 3, 4, and 6, then a tie of 1, 2, and 5; (b) 1, then a tie of 2 and 3, then 4, then a tie of 5 and 6; (c) 1, 2, 3, 4, 5, 6; (d) 2 and 4; (e) 1, 2, 5   **P   1.** 0.990 50   **3.** 0.446 ps   **5.** 2.68×10^3 y
7. (a) 0.999 999 50   **9.** 1.32 m   **11.** (a) 87.4 m; (b) 394 ns
13. (a) 0.999 999 15; (b) 30 ly   **15.** (a) 26.26 y; (b) 52.26 y; (c) 3.705 y   **17.** (a) 25.8 μs; (b) small flash   **19.** (a) 138 km; (b) -374 μs   **21.** (a) 1.25; (b) 0.800 μs
23. (a) 0.480; (b) negative; (c) big flash; (d) 4.39 μs
25. (a) $\gamma[1.00 \mu\text{s} - \beta(400 \text{ m})/(2.998 \times 10^8 \text{ m/s})]$; (d) 0.750; (e) $0 < \beta < 0.750$; (f) $0.750 < \beta < 1$; (g) no   **27.** (a) 0.35; (b) 0.62   **29.** $0.81c$ **31.** (a) 1.25 y; (b) 1.60 y; (c) 4.00 y
33. 1.2 μs   **35.** $0.13c$ **37.** 22.9 MHz   **39.** (a) 550 nm; (b) yellow   **41.** (a) 1.0 keV; (b) 1.1 MeV   **43.** (a) 196.695; (b) 0.999 987   **45.** $2.83mc$ **47.** (a) 0.222 cm; (b) 701 ps; (c) 7.40 ps   **49.** (a) 0.707; (b) 1.41; (c) 0.414   **51.** 18 smu/y
53. 1.01×10^7 km   **55.** (a) $\gamma(2\pi m/|q|B)$; (b) no; (c) 4.85 mm; (d) 15.9 mm; (e) 16.3 ps; (f) 0.334 ns   **57.** 110 km
59. (a) 2.08 MeV; (b) -1.21 MeV   **61.** (a) $vt \sin \theta$; (b) $t[1 - (v/c) \cos \theta]$; (c) $3.24c$ **63.** (d) 0.801   **65.** (a) 1.93 m; (b) 6.00 m; (c) 13.6 ns; (d) 13.6 ns; (e) 0.379 m; (f) 30.5 m; (g) -101 ns; (h) no; (i) 2; (k) no; (l) both   **69.** (b) $+0.44c$
71. 8.7×10^{-3} ly   **73.** 2.46 MeV/c   **75.** 189 MeV
77. (a) 5.4×10^4 km/h; (b) 6.3×10^{-10} **79.** $0.27c$
81. (a) 5.71 GeV; (b) 6.65 GeV; (c) 6.58 GeV/c; (d) 3.11 MeV; (e) 3.62 MeV; (f) 3.59 MeV/c   **83.** (a) 256 kV; (b) $0.745c$
85. 7   **87.** $0.95c$

Figures are noted by page numbers in *italics*, tables are indicated by t following the page number

Some Physical Properties

Air (dry, at 20°C and 1 atm)

Density	1.21 kg/m^3
Specific heat at constant pressure	1010 J/kg·K
Ratio of specific heats	1.40
Speed of sound	343 m/s
Electrical breakdown strength	3×10^6 V/m
Effective molar mass	0.0289 kg/mol

Water

Density	1000 kg/m^3
Speed of sound	1460 m/s
Specific heat at constant pressure	4190 J/kg·K
Heat of fusion (0°C)	333 kJ/kg
Heat of vaporization (100°C)	2260 kJ/kg
Index of refraction ($\lambda = 589$ nm)	1.33
Molar mass	0.0180 kg/mol

Earth

Mass	5.98×10^{24} kg
Mean radius	6.37×10^6 m
Free-fall acceleration at Earth's surface	9.8 m/s^2
Standard atmosphere	1.01×10^5 Pa
Period of satellite at 100 km altitude	86.3 min
Radius of the geosynchronous orbit	42 200 km
Escape speed	11.2 km/s
Magnetic dipole moment	8.0×10^{22} A·m^2
Mean electric field at surface	150 V/m, down

Distance to

Moon	3.82×10^8 m
Sun	1.50×10^{11} m
Star nearest the Sun	4.04×10^{16} m
Galactic center	2.2×10^{20} m
Andromeda galaxy	2.1×10^{22} m
Edge of the observable universe	$\sim 10^{26}$ m

The Greek Alphabet

Alpha	A	α	Iota	I	ι	Rho	P	ρ
Beta	B	β	Kappa	K	κ	Sigma	Σ	σ
Gamma	Γ	γ	Lambda	Λ	λ	Tau	T	τ
Delta	Δ	δ	Mu	M	μ	Upsilon	Y	υ
Epsilon	E	ϵ	Nu	N	ν	Phi	Φ	ϕ, φ
Zeta	Z	ζ	Xi	Ξ	ξ	Chi	X	χ
Eta	H	η	Omicron	O	o	Psi	Ψ	ψ
Theta	Θ	θ	Pi	Π	π	Omega	Ω	ω